Green Tribology

Emerging Materials and Technologies

Series Editor

Boris I. Kharissov

Biomaterials and Materials for Medicine: Innovations in Research, Devices, and Applications
Jingan Li

Advanced Materials and Technologies for Wastewater Treatment
Sreedevi Upadhyayula and Amita Chaudhary

Green Tribology: Emerging Technologies and Applications
T.V.V.L.N. Rao, Salmiah Binti Kasolang, Xie Guoxin, Jitendra Kumar Katiyar, and Ahmad Majdi Abdul Rani

Biotribology: Emerging Technologies and Applications
T.V.V.L.N. Rao, Salmiah Binti Kasolang, Xie Guoxin, Jitendra Kumar Katiyar, and Ahmad Majdi Abdul Rani

Bioengineering and Biomaterials in Ventricular Assist Devices
Eduardo Guy Perpétuo Bock

Semiconducting Black Phosphorus: From 2D Nanomaterial to Emerging 3D Architecture
Han Zhang, Nasir Mahmood Abbasi, and Bing Wang

Green Tribology

Emerging Technologies and Applications

Edited by

T.V.V.L.N. Rao

Salmiah Binti Kasolang

Guoxin Xie

Jitendra Kumar Katiyar

Ahmad Majdi Abdul Rani

CRC Press
Taylor & Francis Group
Boca Raton London New York

CRC Press is an imprint of the
Taylor & Francis Group, an **informa** business

First edition published 2022
by CRC Press
6000 Broken Sound Parkway NW, Suite 300, Boca Raton, FL 33487-2742

and by CRC Press
2 Park Square, Milton Park, Abingdon, Oxon, OX14 4RN

ISBN: 978-0-367-68860-8 (hbk)
ISBN: 978-0-367-68862-2 (pbk)
ISBN: 978-1-003-13938-6 (ebk)

DOI: 10.1201/9781003139386

Typeset in Times
by KnowledgeWorks Global Ltd.

Contents

Preface

Green tribology includes a focus on innovative lubricants, materials, surfaces and machining to reduce friction and wear for environmental conservation and sustainability. With the themes of "Emerging Technologies and Applications in Green Tribology," the book creates a platform for sharing knowledge emerging in the field of green tribology. The book chapters bring together the research expertise of the large international tribology community impacting the field of green tribology. The book focuses on the role of mathematics, chemistry, physics, materials and mechanical engineering in recent advancements and developments in green tribology. The scope of the book includes recent developments and the future outlook of the emerging technologies and applications in green tribology. An overview of the recent developments in green tribology in the areas of green lubricants, green composites, texture surfaces and green machining are presented. The chapters highlight the key findings on current trends and future developments of environmentally friendly (green) lubricants, tribological performance improvement with the advances in green/eco-friendly materials, superior tribological characteristics of textured/bio-inspired surfaces and minimum quantity lubrication. The ongoing trends and future prospects in green tribology are discussed. Wide themes in green tribology are addressed in this book in the interest of tribologists.

Editors

T.V.V.L.N. Rao received his PhD in Tribology of Fluid Film Bearings from the Indian Institute of Technology, Delhi, in 2000, and his MTech in Mechanical Manufacturing Technology from the National Institute of Technology (formerly known as Regional Engineering College) Calicut in 1994. Rao's current research interests are in bearings, lubrication and tribology. He has authored (and co-authored) over 120 publications to date. He has secured (as PI and Co-PI) several research grants from the Ministry of Higher Education, Malaysia, and a research grant from The Sumitomo Foundation, Japan. Rao is a member of the Editorial Board (2022) in Tribology and Lubrication Technology (TLT). He is a Guest Associate Editor in the *Journal of Engineering Tribology*, *Industrial Lubrication and Tribology*, *Tribology – Materials, Surfaces and Interfaces*, and *Arabian Journal for Science and Engineering*. Rao is an executive member (2015–2021) of the Malaysian Tribology Society. He is a member of the Society of Tribologists and Lubrication Engineers, Malaysian Tribology Society and Tribology Society of India. Rao is currently Professor in the Department of Mechanical Engineering at Madanapalle Institute of Technology and Science. Prior to joining MITS, he served as Research Associate Professor at SRMIST (2017–2020), Visiting Faculty at LNMIIT (2016–2017), Associate Professor at Universiti Teknologi PETRONAS (2010–2016), and Assistant Professor in BITS Pilani at Pilani (2000–2004, 2007–2010) and Dubai (2004–2007) campuses.

Salmiah Binti Kasolang is currently the Rector of Universiti Teknologi MARA (UiTM) Pulau Pinang Branch. Her research interest is in tribology specifically in hydrodynamic lubrication. She graduated from the University of Wisconsin–Madison in 1992 and later pursued her master's degree in Manufacturing System Engineering at Universiti Putra Malaysia UPM. She did her PhD at the University of Sheffield under the supervision of Professor Rob Dwyer-Joyce. She has administrative experience for 10 years as the Deputy Dean (7 years from 2009 to 2015) and Dean (3 years from 2015 to 2017). She is actively leading a tribology research group in UiTM with more than 100 indexed publications. Her engagement with MYTRIBOS has enabled her to link up with other tribologists in Malaysia. Currently, she is the President of the Malaysian Tribology Society (MYTRIBOS). She is also the President of the Society of Mechanical Engineering Liveliness SOMEL that promotes many aspects of engineering, including but not limited to education, research, industry-community engagement, engineering art, and engineering life style.

Guoxin Xie received his doctoral degree at Tsinghua University, China, in 2010, majoring in Mechanical Engineering. After that, he spent two years at State Key Laboratory of Tribology, Tsinghua University, China, for postdoctoral research. From 2012 to 2014, he worked at the Royal Institute of Technology, Sweden, for another two years of postdoctoral research. Since 2014, he has worked at Tsinghua University as a Tenured Associate Professor. His research interests include solid lubrication, electric contact lubrication, thin film lubrication, etc. He has published more than 80 referred papers in international journals. He won several important academic awards, such as Chinese Thousands of Young Talents, the Excellent Doctoral Dissertation Award of China, and Ragnar Holm Plaque from KTH, Sweden. He is currently Associate Editor of FRICTION, and he will be the Director of the Young Committee of Chinese Tribology Institution.

Jitendra Kumar Katiyar is presently working as a Research Assistant Professor in the Department of Mechanical Engineering, SRM Institute of Science and Technology Kattankulathur Chennai, India. His research interests include tribology of carbon materials, polymer composites, self-lubricating polymers, lubrication tribology, modern manufacturing techniques, and coatings for advanced technologies. He obtained his bachelor's degree from UPTU Lucknow with Honors in 2007. He obtained his master's degree from the Indian Institute of Technology Kanpur, India, in 2010 and his PhD from the same institution in 2017. He is a life member of the Tribology Society of India, Malaysian Society of Tribology, Institute of Engineers, India, and the Indian Society for Technical Education (ISTE), etc. He has authored/co-authored/published more than 25 articles in reputed journals, 30+ articles in international/national conferences, and 12+ book chapters, and he has two books published, *Automotive Tribology and Tribology in Materials and Application* by Springer and *Engineering Thermodynamics* for the undergraduate level by Khanna Publication. He has served as a guest editor for a special issue in *Tribology Materials, Surfaces and Interfaces*, *Journal of Engineering Tribology Part J*, *Arabian Journal for Science and Engineering*, and *Industrial Lubrication and Tribology*. He is also an active reviewer in various reputed journals related to materials and tribology. He has delivered more than 30+ invited talks on various research fields related to tribology, composite materials, surface engineering, and machining. He has received research grants from various government organizations such as MHRD and SERB. He has organized 5+ FDP/Short Term Courses in tribology and the International Tribology Research Symposium.

Ahmad Majdi Abdul Rani received his PhD in Mechanical and Manufacturing Engineering from Loughborough University, UK, his MSc in Industrial Engineering, and his BSc in Manufacturing from Northern Illinois University, USA. His research interests are in biomedical engineering, mechanical and manufacturing engineering, and tribology. He has supervised and is currently supervising more than 20 postgraduate (PhD and MSc) students. He recently received the Leadership in Innovation Fellowship – Newton Award, UK. He has secured 3 patents and won more than 15 gold awards in exhibitions such as ITEX, MaGRIs (MOSTI), MARS, PENCIPTA, IME, SPDEC, CoRIC, etc. He has authored over 150 publications to date. He has secured (as PI and Co-PI) several research grants from the Ministry of Higher Education, Malaysia (FRGS, PRGS, ERGS, MyBrain) and Universiti Teknologi PETRONAS (YUTP, I-GEN, STIRF). He has been associated with the Department of Mechanical Engineering at Universiti Teknologi PETRONAS since 1998. He was Head of the Department of Mechanical Engineering at Universiti Teknologi PETRONAS from 2009–2011.

Contributors

Adnan Ahmed
Mechanical Engineering Department
University of Engineering and Technology
Peshawar, Pakistan

Tauseef Ahmed
Mechanical Engineering Department
Universiti Teknologi PETRONAS
Seri Iskandar, Malaysia

Mohammad Azad Alam
Mechanical Engineering Department
Universiti Teknologi PETRONAS
Seri Iskandar, Malaysia

Arslan Ahmed
Department of Mechanical Engineering
COMSATS University Islamabad Sahiwal Campus
Islamabad, Pakistan

Mohammad Azeem
Mechanical Engineering Department
Universiti Teknologi PETRONAS
Seri Iskandar, Malaysia

Sayed Khadija Bari
Bio Fuels Division
CSIR – Indian Institute of Petroleum
Dehradun, India

Skylab P. Bhore
Rotor Dynamics and Vibration Diagnostics Lab
Department of Mechanical Engineering
Motilal Nehru National Institute of Technology
Allahabad, India

H. Chen
Department of Mechanical Engineering
Tsinghua University
Beijing, China

Sangeeta Das
Department of Mechanical Engineering
Girijananda Chowdhury Institute of Management and Technology
Guwahati, India

Shubhajit Das
Department of Mechanical Engineering
National Institute of Technology
Yupia, India

I.M. Rizwanul Fattah
School of Information, Systems and Modelling
Faculty of Engineering and Information Technology
University of Technology
Sydney, Australia

H.J. Gong
Department of Mechanical Engineering
Tsinghua University
Beijing, China

Rajeev Nayan Gupta
Department of Mechanical Engineering
National Institute of Technology
Silchar, India

A.P. Harsha
Department of Mechanical Engineering
Indian Institute of Technology (Banaras Hindu University)
Varanasi, India

Nilesh D. Hingawe
Rotor Dynamics and Vibration Diagnostics Lab
Department of Mechanical Engineering
Motilal Nehru National Institute of Technology
Allahabad, India

Mian Imran
Department of Mechanical Engineering
Institute of Space Technology
Islamabad, Pakistan

Z.J. Ji
Department of Mechanical Engineering
Qinghai University
Xining, China

M.A. Kalam
Department of Mechanical Engineering
Faculty of Engineering
University of Malaya
Kuala Lumpur, Malaysia

Salmiah Binti Kasolang
Universiti Teknologi MARA
Shah Alam, Malaysia

Jitendra Kumar Katiyar
Department of Mechanical Engineering
SRM Institute of Science and Technology
Kattankulathur, India

Hafiz Usman Khalid
Mechanical Engineering Department
Universiti Teknologi PETRONAS
Seri Iskandar, Malaysia

Mohammad Rehan Khan
College of Electrical and Mechanical Engineering
National University of Science and Technology
Islamabad, Pakistan

M.Y. Li
Department of Mechanical Engineering
Tsinghua University
Beijing, China

Y.B. Li
Department of Mechanical Engineering
Tsinghua University
Beijing, China

Abdul Munir Hidayat Syah Lubis
Department of Mechanical Engineering
Universiti Teknikal Malaysia
Melaka, Malaysia

Ponnekanti Nagendramma
Bio Fuels Division
CSIR – Indian Institute of Petroleum
Dehradun, India

Norhuda Hidayah Nordin
Department of Manufacturing and Materials Engineering
Kuliyyah of Engineering
International Islamic University Malaysia
Selangor, Malaysia

Tej Pratap
Department of Mechanical Engineering
Motilal Nehru National Institute of Technology
Allahabad, India

Ahmad Majdi Abdul Rani
Department of Mechanical Engineering
Universiti Teknologi PETRONAS
Seri Iskandar, Malaysia

T.V.V.L.N. Rao
Department of Mechanical Engineering
Madanapalle Institute of Technology and Science
Madanapalle, India

Anjan Ray
Analytical Sciences Division
CSIR – Indian Institute of Petroleum
Dehradun, India

Ali Raza
Department of Mechanical Engineering
COMSATS University Islamabad
Sahiwal Campus
Islamabad, Pakistan

Y.L. Ren
Department of Mechanical Engineering
Tsinghua University
Beijing, China

M.J.M. Ridzuan
School of Mechatronic Engineering
International Islamic University
Malaysia Perlis
Perlis, Malaysia

Neha Sharma
Academy of Scientific and Innovative
Research (AcSIR) and Tribology and
Combustion Division
CSIR – Indian Institute of Petroleum
Dehradun, India

L.N. Si
Department of Mechanical Engineering
North China University of Technology
Beijing, China

N.A. Sukindar
Department of Manufacturing and
Materials Engineering
International Islamic University
Malaysia
Selangor, Malaysia

Mohd Hafis Sulaiman
Department of Manufacturing and
Materials Engineering
Kuliyyah of Engineering
International Islamic University
Malaysia
Selangor, Malaysia

X.H. Sun
Department of Mechanical Engineering
Tsinghua University
Beijing, China

Gananath D. Thakre
Tribology and Combustion Division
CSIR – Indian Institute of Petroleum
Dehradun, India

Guoxin Xie
Department of Mechanical Engineering
Tsinghua University
Beijing, China

H.H. Ya
Mechanical Engineering Department
Universiti Teknologi PETRONAS
Seri Iskandar, Malaysia

L. Zhang
Department of Mechanical Engineering
Tsinghua University
Beijing, China

Leong Zhaoyuan
Department of Materials Science and
Engineering
Faculty of Engineering
The University of Sheffield
Sheffield, United Kingdom

1 Recent Developments in Green Tribology

T.V.V.L.N. Rao
Madanapalle Institute of Technology & Science
Madanapalle, India

Salmiah Binti Kasolang
Universiti Teknologi MARA
Shah Alam, Malaysia

Guoxin Xie
Tsinghua University
Beijing, China

Jitendra Kumar Katiyar
SRM Institute of Science and Technology
Kattankulathur, India

Ahmad Majdi Abdul Rani
Universiti Teknologi PETRONAS
Seri Iskandar, Malaysia

CONTENTS

DOI: 10.1201/9781003139386-1

1.1 INTRODUCTION

Green tribology encompasses friction, wear and lubrication principles of tribology of surfaces under relative motion impacting environment and biology. The recently highlighted perspectives and challenges in green tribology are in the areas of biomimetic surfaces, materials, and green lubricants [1]. Recently, the potential of green nano-tribology has been analysed and a path towards efficient, innovative and sustainable green nano-tribology has been established [2]. The basic concepts in green nano-tribology identified are (nano-) surfaces, (nano-) agents, and (nano-) processes. More recent progress in green tribology is directed towards eco-friendly materials, bio-based ceramics, water based fluids, bio-lubricants, and fibre-reinforced composites [3].

Growing interests around energy and environment have generated worldwide attention for efficient and cleaner systems towards sustainability [4]. Sustainable development thinking in green tribology design elucidates the significance of understanding tribology from an environmental viewpoint [5]. Recent advances in green tribology in lubricants, composites, surfaces and machining cause friction and wear reduction leading to sustainable development.

An overview of the recent developments in green tribology presented in Figure 1.1 is in the areas of green lubricants, green composites, textured surfaces and green machining.

1.2 GREEN LUBRICANTS

Bio-based lubricants have been acknowledged to play a significant role in overcoming environmental pollution and hazards. Bio-based lubricants are renewable and biodegradable and are a favourable alternative for numerous applications with superior lubricant properties [6]. Nevertheless, prior chemical alteration is essential to rise above the low temperature properties and oxidative stability. The appropriate base oil and additive formulations are required for development of bio-based lubricants to exceed the performance of conventional lubricants. Synthetic and vegetable oil-based esters extend the ideal choice in developing environment friendly lubricant products [7]. In recent times, intelligent materials and structures for lubrication with properties of bionic functions emulating the living systems have stimulated immense interest [8]. Functional lubricating materials with the feedback mechanism have the ability to control lubrication.

The desired lubricants in nano-technology applications require functioning at severe operating conditions. Ionic liquids have excellent thermal and electrical

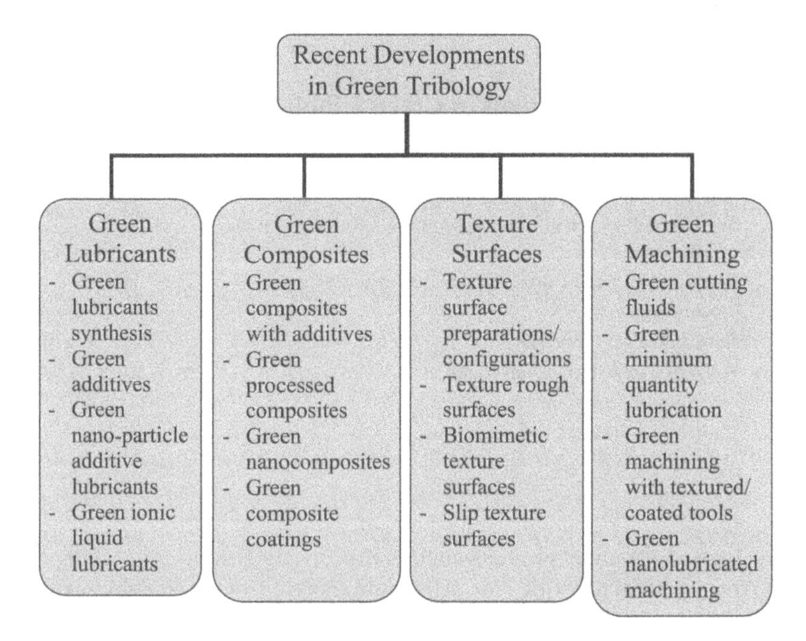

FIGURE 1.1 An overview of recent developments in green tribology

conductivity and have been explored as green lubricants as they do not emit volatile organic compounds [9]) Sustainable bio-based ionic liquid (IL) lubricants are environmentally friendly and have overcome the shortcomings related to conventional bio-lubricants. The outcome of adjusting cation–anion combinations is investigated in bio-based ionic liquids to understand their tribological performance and industrial viability [10]. Cation–anion combinations with the presence of large alkyl cation chain length and large aromatic anion ring size in bio-based ionic liquids can effectively provide friction and wear reduction. Ionic liquids as sustainable and environmentally friendly lubricants have an excellent potential to replace the conventional lubricants. Tailor-made advanced ionic liquid lubricants will play a vital role in augmenting tribological properties of sliding interfaces [11]. The sustainability of bio-based ionic liquid (IL) lubricants as an alternative to conventional bio-lubricants has also been confirmed in tribological tests. Phosphonium-based ionic liquids are further appropriate for tribological applications as these ionic liquids can be tailored rendering sustainability to the applications [12].

Recent developments in green tribology on lubricants are presented in Table 1.1.

1.2.1 Green Lubricants Synthesis

Bio-based lubricants blended from renewable and biodegradable resources have enormous possibility to replace conventional lubricants. Epoxidised palm stearin methyl ester bio-based lubricants demonstrated reduced friction coefficients and wear scar diameter [13]. The green concept demands lubricants to be from biodegradable resources and environmentally friendly. Synthesised biodegradable esters

TABLE 1.1
Recent Developments in Green Tribologyon Lubricants

Green Lubricants	Authors
Bio-lubricants potential for a range of applications	[6]
Biodegradable synthetic ester base stocks for new lubricants formulation	[7]
Novel micro-/nano-composites (containers) lubricating technologies	[8]
Ionic liquids as lubricants	[9]
Bio-based ionic liquids lubricants	[10]
Ionic liquids as sustainable lubricants	[11]
Ionic liquids as lubricants properties	[12]

were investigated for their physico-chemical properties, lubricity, biodegradability and toxicity characteristics [14]. The characterised biodegradable esters were found to have good potential for use in industrial applications. The shortcomings of bio-based lubricants restrict their tribological applications. The cold flow properties of the coconut oil have been improved with free movement of the fatty acid esters but the tribological properties are diminished compared to conventional coconut oil [15]. Oleate ester obtained from the transesterification process between oleic acid and alcohols results in minimum coefficient of friction and wear-scar diameter [16].

However, additives are used to improve the friction- and wear-reducing properties. Glycerol aqueous blends, with lower friction coefficient, have a great possibility as environmentally friendly lubricants in several applications [17]. The viscosity of glycerol in all lubrication regimes can be enriched by adding water. A canola oil and boric acid lubricant mixture demonstrated excellent potential in industrial sheet metal-forming applications [18]. The tribological properties are enhanced (reduced friction coefficient and wear) using commercialised palm oil mixed with zinc dioctyldithiophosphate (ZnDoDP) and zinc diamyldithiocarbamate (ZDDC) [19].

Recent developments in green tribology on bio-based lubricants synthesis are presented in Table 1.2.

1.2.2 Green Additives

Usually, the biodegradable castor oil-based formulations as lubricating greases offer the friction coefficient in a tribological contact subject to the class of thickener agents utilised. Biodegradable castor oil-based formulations containing cellulose or chitin derivatives as thickener agents generate higher values of the friction coefficient compared to those attained with the lithium greases in a tribological contact depending on the thickener agents [20]. Leaf surface waxes extracted from plants as biodegradable additives can effectually advance the friction and wear reduction

TABLE 1.2
Recent Developments in Green Tribology on Bio-Based Lubricants Synthesis

Green Lubricants Synthesis	Authors
Palm stearin methyl ester in an epoxidation reaction	[13]
Biodegradable esters	[14]
Fatty acids into esters by the process of alkali esterification	[15]
Oleate ester bio-based lubricant	[16]
Glycerol as a green lubricant	[17]
Canola oil and boric acid powder green lubricant	[18]
Commercialised palm oil mixed with zinc dioctyldithiophosphate (ZnDoDP) and zinc diamyldithiocarbamate (ZDDC)	[19]

accomplishments as leaf surface waxes form a protective film on the worn surface in the tribological contact [21].

Recent developments in green tribology on green additives are presented in Table 1.3.

1.2.3 Green Nano-particle Additive Lubricants

There is an increasing demand to use bio-lubricants in new base formulations as they are biodegradable, non-toxic and environmentally friendly. Biodegradable castor oil-based formulations containing calcium–copper–titanate and zinc dialkyldithiophosphate nano-particles have shown an improvement in the friction, wear and extreme-pressure reduction accomplishments in a tribological contact [22]. Renewable bio-oil with graphene oxide sheets as additives show a prospect of improvement in tribological iron/steel contacts [23]. The optimal concentrations of graphene oxide sheets with smaller sizes in bio-oil form an entire lubricant film on the tribological interfaces which results in lower friction coefficient and wear. However, higher concentrations of graphene oxide sheets in bio-oil form significant aggregation on the tribological interfaces which results in higher friction coefficient and wear. The optimal concentrations of MoS_2 nano-particles in castor oil reduce the asperity contact on the tribological interfaces which results in lower friction coefficient and adhesive wear [24]. However, higher concentrations of MoS_2 nano-particle in castor oil form agglomeration on the tribological interfaces which results in higher friction coefficient and wear due to abrasive wear. Water-based lubricant with an optimum concentration

TABLE 1.3
Recent Developments in Green Tribology on Additives

Green Additives	Authors
Biodegradable castor oil and biogenic thickeners formulations	[20]
Green additive of leaf-surface wax	[21]

of hexagonal boron nitride nano-additives shows improvement with friction and wear reduction of tribological contacts [25]. The recommended lubrication mechanisms of water-based lubricant with hexagonal boron nitride nano-additives in a mixed lubrication regime are mending, rolling and polishing effects. However, increasing concentrations of hexagonal boron nitride nano-additives in water-based lubricant leads to significant agglomeration on the tribological interfaces which results in higher friction. Rice bran oil with turmeric oil and halloysite nano-clay additives is recommended as a bio-lubricant due to the curcumin in turmeric oil, which has shown good anti-oxidant behaviour [26]. Coconut oil is a promising lubricant in sustainable manufacturing as the use of coconut oil as a lubricant reduces the friction coefficient and wear rate [27]. Furthermore, the tribological performance of coconut oil can be improved with the addition of hexagonal boron nitride (h-BN) powder nano-particles resulting in further reduction of friction coefficient and wear rate.

Recent developments in green tribology on nano-particle additive lubricants are presented in Table 1.4.

1.2.4 Green Ionic Liquid Lubricants

Ionic liquids usually display biodegradability and are opted for good potential to be used as lubricants or lubricant additives. The effect of external electric fields on the increase in the film thickness of ionic liquid films confined within a nano-space due to the shorter alkyl side chain is more noticeable [28]. It is believed that the charged anions and cationic head groups are arranged near electrified walls to form ordered layers, and short alkyl side chains at the interfaces are lined up alongside the direction of external electric fields owing to induced dipoles. Halide-free ionic liquids involving two types of anions and four types of cations as additives in glycerol yield highly biodegradable polar lubricants [29]. Friction and wear reduction is achieved by a sizable amount for methyl sulphates for all temperatures, which display strong anion domination. The ionic liquids ammonium- and pyrrolidinium-based cations combined with methylsulphate, methylsulphonate have been investigated as lubricant additives for their aquatic low-toxicity [30]. The choline amino acid ionic liquids exhibit good physicochemical and tribological characteristics with outstanding environmentally friendly characteristics of biodegradability and low toxicity against

TABLE 1.4
Recent Developments in Green Tribology on Nano-Particle Additive Lubricants

Green Nano-Particle Additive Lubricants	Authors
Castor oil with calcium–copper–titanate and zinc dialkyldithiophosphate nano-particles	[22]
Bio-oil with graphene oxide sheets	[23]
Castor oil with MoS_2 nano-particles	[24]
Water-based lubricant with hexagonal boron nitride nano-additives	[25]
Rice bran oil with turmeric oil and halloysite nano-clay additives	[26]
Coconut oil with hexagonal boron nitride (h-BN) powder nano-particles	[27]

aquatic organisms [31]. Synthesised polyol ester is blended with environmentally friendly ionic liquids derived from aspartic acid and glutamic acid to be used as lubricant additives [32]. The polyol ester ionic liquids form blends compatible as lubricant additives, resulting in friction- and wear-reducing performance characteristics between the contact interfaces. The fatty acid ionic liquids are designed and synthesised by quaternary ammonium cations with bio-based fatty acid anions which prevent metal–to-metal contact and further friction and wear reduction [33]. The lubrication mechanism of fatty acid ionic liquids reveals effectual physical adsorption and tribochemical reaction films formed on the sliding interfaces due to the polarity of the fatty acid anions and the aliphatic tails. Environmentally friendly ricinoleic acid-based ionic liquid is used as a multifunctional lubricant additive in glycerol solution due to its remarkable anticorrosion and lubricating properties [34]. The lubrication mechanism of the ricinoleic acid-based ionic liquid is accredited to the stable adsorbed layers and tribofilms on the contact surface avoiding wear and corrosion. Tribological characteristics of trimethylolpropane trioleate with phosphorus-type ionic liquid additives on the reduction in friction coefficient and wear have been evaluated [35]. In situ MoS_2 quantum dots (less than 10 nm) in green ionic liquids displayed lasting dispersion stabilities with enhanced lubricating properties owing to thin film formation [36]. Eco-friendly nano-particles have demonstrated an ideal performance in enhancing the friction and wear characteristics in sliding interfaces. The carbon dots enriched with the big organic cations from the ionic liquids have been demonstrated to be the ideal candidate as additives in synthesised nano-lubricants for friction coefficient and wear reduction [37]. The layered boric acid nano-particle is a promising green lubricant additive due to its biodegradability and improvement in scuffing load capacity [38]. An adsorption layer as well as boron tribofilm is formed on the contact surfaces to improve lubrication performance.

Recent developments in green tribology on ionic liquid lubricants are presented in Table 1.5.

TABLE 1.5
Recent Developments in Green Tribology on Ionic Liquid Lubricants

Green Ionic Liquid Lubricants	Authors
External electric fields effect on ionic liquid films confined within a nano-space	[28]
Halide-free ionic liquids involving two types of anions and four types of cations as additives in glycerol	[29]
Ammonium- and pyrrolidinium-based cations combined with methylsulphate, methylsulphonate	[30]
Choline amino acid ionic liquids from choline cations and amino acid anions	[31]
Ionic liquids as bio-lubricant additives for polyol ester	[32]
Fatty acid ionic liquids by using quaternary ammonium cations with bio-based fatty acid anions	[33]
Ricinoleic acid-based ionic liquid as a lubricant additive in glycerol solution	[34]
Trimethylolpropane trioleate with phosphorus-type ionic liquid	[35]
In situ MoS_2 quantum dots (less than 10 nm) in green ionic liquids	[36]
Metal-free and eco-friendly carbon-based ionic liquid nano-particles as additives in lubricants	[37]
Layered boric acid nano-particle as a promising green lubricant	[38]

1.3 GREEN COMPOSITES

Green tribology emphasises environmental adaptability with an improved tribological performance. The environmental compatibility, efficiency and durability of composites are the key requirements for utilisation in demanding tribological systems. Natural fibre-reinforced polymer composites have emerged in tribological applications due to their sustainable environmentally friendly aspects and cost-effective economic potential [39]. The enhanced tribological properties (friction and wear performance) are mainly influenced by the treatment and orientation fibres. Natural fibre composites are explored in tribology from friction materials to friction modifiers [40]. The fibre composite failures are due to fibre/matrix debonding, cracks in matrix, fibre fragmentation and debris formation. The hybridisation of the natural fibre composites with transfer film formation is vital to ensure enhanced tribological and mechanical properties. Core–shell (micro-/nano-) particles, with diverse composition and synergy, are widely used in tribology due to improved friction and wear performance [41]. The friction- and wear-enhancing characteristics of core–shell (micro-/nano-) particles are attained by optimizing the shell material coated on the abrasive surface. Two-dimensional (2D) nano-material-based composites are developed incorporating 2D nano-materials in a composite for potentially enhanced tribological and mechanical properties [42]. Advanced 2D nano-composites can act as excellent reinforcements as they possess ultralow friction, high elasticity modulus, and high strength. Thermoplastic polymer multiscale composite materials with high performance are used to address the tribological challenges [43]. High performance polymers such as polyetheretherketone (PEEK) and aromatic thermosetting polyester (ATSP) are used in bulk and as coatings for challenging tribological applications with superior performance [44]. The superior performance demonstrated by high performance polymers is due to the capability to form a transfer film on the surface. Advanced polymeric coatings are classified as tribological, superhydrophobic and self-healing [45]. The tribological coatings exhibit low friction with excellent wear resistance, the superhydrophobic coatings prevent wetting due to their surface structure with low surface energy and the self-healing coatings recover their functionality after damage. Polymer composite coatings with tailored properties (mechanical and tribological) reduce friction and wear. Functional fillers extend the service life and enhance the mechanical and tribological characteristics of the coatings [46]. Fillers provide low friction by promoting the formation of transfer films or liquid shear films. The polymer composite coating adhesion between the coating and substrate adhesion can be enhanced through treatment (mechanical, chemical and energy) of the substrate.

Recent developments in green tribology on composites are presented in Table 1.6.

1.3.1 Green Composites with Additives

The bio-based polymer composite has been developed using porous composites with self-lubricating properties [47]. The bio-based polymer composite using acrylic resin and short wood fibres led to a decrease in the coefficient of friction for applications in bearings. Sour-weed natural fibre reinforced with polyester matrix reinforced with

TABLE 1.6
Recent Developments in Green Tribology on Composites

Green Composites	Authors
Tribology of natural fibre-reinforced composites	[39]
Tribology of hybrid natural fibre composites	[40]
Synthesis methods and the tribology of core–shell (micro-/nano-) particles	[41]
Tribology of 2D nano-material reinforcements	[42]
Tribology of multiscale thermoplastic polymer composites	[43]
Tribology of high performance bulk polymers and coatings	[44]
Tribology of advanced polymeric coatings	[45]
Tribology of polymer matrices and composite coatings with functional fillers	[46]

optimum fibre's size resulted in enhanced tensile strength, impact strength, hardness and specific wear rate [48]. Hemp fibre reinforced with kevlar/carbon epoxy composites brought about noteworthy improvement of the mechanical and wear resistance properties [49].

Recent developments in green tribology on green composites with additives are presented in Table 1.7.

1.3.2 Green Processed Composites

The long fibre size of kenaf fibres and varying size proportions of an oil palm empty fruit bunch have significant influence on the specific wear rate of the epoxy composites [50]. Copper is an exceptional ingredient of friction materials but is a hazard for an aquatic environment [51]. Metal-free friction materials developed from hydrated calcium silicate as copper replacement are successful with a tribological properties characterisation. The phase transformation of the surface transfer layer is essential for controlling the mechanism of friction and wear properties of an activated carbon composite derived from palm kernel. The surface transfer layer is effective for producing low friction and wear only at limited applied loads [52]. The predominant wear mechanisms of kenaf/epoxy composite are identified as micro-cracking and fibre debonding [53].

TABLE 1.7
Recent Developments in Green Tribology on Composites with Additives

Green Composites with Additives	Authors
Bio-based composites of acrylic resin and short wood fibres	[47]
Sour-weed natural fibre reinforced with polyester matrix	[48]
Hemp fibre reinforced with kevlar/carbon epoxy composites	[49]

TABLE 1.8
Recent Developments in Green Tribology on Processed Composites

Green Processed Composites	Authors
Kenaf and oil palm empty fruit bunch epoxy composites	[50]
Metal-free brake pads for copper replacement	[51]
Activated carbon composite derived from palm kernel	[52]
Kenaf/epoxy composite friction material	[53]

Recent developments in green tribology on processed composites are presented in Table 1.8.

1.3.3 Green Nano-Composites

Bio-based composites are vital from energy and environment perspectives. Nano-modified epoxy polymers produced from adding nano-silica to the epoxy polymer resulted in improved tensile strength based on the degree of dispersion of the spherical nano-silica particles in the epoxy matrix [54]. Bio-based epoxy composites with cellulose nano-fibres showed improved tribological and mechanical properties compared to untreated conventional composites [55]. Cellulose nano-fibre bio-based epoxy composites form a uniform tribolayer which reduces the direct contact of contact surfaces and thus provides friction- and wear-reducing properties of the composites. The "metal-reservoir" nano-composite coating for lower viscosity oils exhibited lower friction based on the controlled the boundary films due to transition metal carbides stability [56]. The corrosion- and wear-resistant nano-composite coatings through the electrical discharge method effectively reduced the coefficient of friction [57]. Graphene synthesised from fruit cover plastic waste and oil palm fibre is an ideal source of solid carbon for graphene synthesis. Synthesised graphene substantially reduces the friction coefficient and wear rate on contact surfaces even at higher sliding speeds [58].

Recent developments in green tribology on nano-composites are presented in Table 1.9.

TABLE 1.9
Recent Developments in Green Tribology on Nano-Composites

Green Nano-Composites	Authors
Epoxy polymer with nano-silica	[54]
Bio-based epoxy composites with the plant-derived cellulose nano-fibre	[55]
Metal-reservoir nano-composite coatings	[56]
Nano-composite coatings through electrical discharge method	[57]
Graphene from fruit cover plastic waste (FCPW) and oil palm fibre (OPF)	[58]

TABLE 1.10
Recent Developments in Green Tribology on Composite Coatings

Green Composite Coatings	Authors
Hybrid of core-shell structure for epoxy composite coatings	[59]
Silk fibroin and titanates nano-composite coatings	[60]

1.3.4 Green Composite Coatings

The epoxy core-shell (CNF/MoS_2) structure is a promising reinforced composite coating which is attributed to the synergy of the core-shell (CNF/MoS_2) along with good interfacial adhesion of the reinforced structure (CNF/MoS_2) with the epoxy matrix [59]. The epoxy core-shell (CNF/MoS_2) composite coatings have superior friction- and wear-reducing properties compared with other reinforced composite coatings. The nano-mechanical characterisation of spin-coated biocompatible silk-based nano-composite coatings showed higher hardness and elastic modulus compared to previous silk coats [60]. The scratch resistance of a composite coating is improved with adding titanates nano-sheets.

Recent developments in green tribology for composite coatings are presented in Table 1.10.

1.4 TEXTURED SURFACES

Surface texturing has emerged as a potentially viable option to make contribution to surface engineering for the accomplishment of green tribology. Surface texturing has a potential to influence tribological contacts performance with significant improvement in load capacity and reduction in friction and wear. The benefits of surface texturing for improving the tribological contacts performance are being widely explored in recent times.

Laser Surface Texturing (LST) is the most advanced technique for surface texturing of micro-dimples at large numbers [61]. The micro-dimples can serve in cases of full or mixed lubrication (micro-bearing), starved lubrication (micro-reservoir), or in either lubricated or dry sliding (micro-debris trap). Laser surfaces texturing using the laser beam machining process is increasingly utilised due to its contribution to friction and wear reduction [62]. Texturing methods suited particularly for surfaces with contact areas larger than the texture width under either hydrodynamic or starved lubrication are maskless electrochemical texturing and etching [63].

Nano-scale surface texturing is a way to increase the tribological performance of micro-electro-mechanical systems (MEMs) and nano-scale technology significantly [64]. Nano-scale surface texturing will allow for precise control of the lubricant behaviour changes as the geometry of the texture is decreased toward the nano-scale. Both the load support and friction force decrease with the decreasing scale of the texture, and overall the effective coefficient of friction decreases. Surface texture functionality is ensured by correct optimisation of its geometrical parameters and therefore an appropriate choice of fabrication techniques is essential for each application [65]. The optimisation of surface texture geometrical parameters is essential for performance improvement of bearing sliders [66] and tribo contacts [67]. The geometrical parameters in surface texturing have been effectively used to improve tribological performance of sliding surfaces [68]. Optimisation of surface texture geometrical parameters is essential for each application to have improvements in the performance of sliding contacts. To accomplish the design of surface textures, a comprehensive study of texture design parameters (position, orientation, aspect ratio and texture density) is essential [69]. Surface texture design effects influence the mechanism of friction and wear reduction under dry contact and lubrication (boundary, mixed, elastohydrodynamic and hydrodynamic) conditions [70].

Surface texturing has gained a great deal of attention since precise texture design parameters will help to reduce friction and wear of sliding surfaces. The optimal design guidelines of surface textures in terms of texture design parameters will help to reduce friction and wear of sliding components [71]. Parallel bearing surfaces provide significant improvement in load capacity and reduction in the coefficient of friction with partial slip texturing patterns [72].

Recent developments in green tribology on texture surfaces are presented in Table 1.11.

1.4.1 Textured Surface Configurations

Surface textures and topography may significantly influence the tribological performance of contact surfaces. The need for further improvement of the performance of counterformal lubricating contact surfaces requires that surface topography and

TABLE 1.11
Recent Developments in Green Tribology on Textured Surfaces

Textured Surfaces	Authors
Potential of laser surface texturing in lubricated contacts	[61]
Laser surface texturing for reduction of friction and wear	[62]
Texturing methods for tribological applications with low cost and rapid texturing speed	[63]
Scale-dependent nano-scale surface textures lubrication theory	[64]
Texture manufacturing techniques classification	[65]
Surface texturing developments and optimisation in bearing sliders	[66]
Surface textures at the concentrated surface contacts	[67]
Surface texturing for piston cylinder assembly and mechanical seals	[68]
Texture modelling techniques classification	[69]
Potential of laser surface texturing in conformal and non-conformal lubrication	[70]
Surface texturing developments in machine elements	[71]
Partial slip texturing patterns in bearings	[72]

textures are optimised [73]. The virtual texturing is a tool to optimise the geometric surface texture pattern designs for the tribological performance improvement of contact surfaces. Appropriate dimple dimensions improve the friction characteristics of journal bearings particularly for low-viscous oils [74]. The lubrication produced in the dimpled space is the principal mechanism for step-up of performance in mixed lubrication regime. The dimple internal structure has an overwhelming impact on the load capacity [75]. Cylindrical dimples of the rectangular profile produce larger load capacity than those with triangular profile due to the pressure rise from the converging step profile.

An increasing interest is aimed at the analytical and numerical solutions of the textured lubricated contacts. Parametric analysis of the slider bearing in terms of texture dimensions reveals that the performance characteristics are independent of the number of cells [76]. Suitable preparation of textured surfaces with distinctive shapes and at unique locations is an efficient method to improve the performance of bearings [77]. The spherical textures or micro-groove surfaces of the journal bearing reduce the friction coefficient [78]. The textured journal bearing provides improved stability of the non-recessed hybrid journal bearing lubricated with non-Newtonian power-law fluids [79]. Laser surface texturing is a viable option for enhancing tribological characteristics. In the laser surface texturing of palm kernel-activated carbon epoxy composite, the friction coefficient decreased with increasing dimple diameter [80]. Partial textured micro-dimple geometries at proper location yield enhanced stability of journal bearings [81]. The dimples geometry and location are important due to their role in improving the tribological performance of sliding machine components. The turning operations are used for fabrication of dimples based on the process capabilities and cost under the green machining environment [82]. Optimal free-form and dimple textures in mechanical seals yield lower leakage and coefficient of friction [83]. The generalised elliptical dimple textures in high speed gas face seal yield improved sealing characteristics (lower leakage) compared with conventional elliptical dimple textures [84]. The textures have significant influence on the bearing characteristics. Partial textures yield maximum performance improvement in load capacity of the meso-scale air journal bearing [85]. Textured surface profile design helps to improve the tribological characteristics and increases the life span of the surfaces in relative motion. The textured surface with a triangle profile design yields the maximum load capacity and the minimum friction coefficient compared to other texture bottom profiles [86]. Surface velocities and the texture aspect ratio are the significant parameters based on load capacity improvement and friction coefficient reduction. Texture density has the least significant influence on the bearing performance characteristics.

Recent developments in green tribology on textured surface configurations are presented in Table 1.12.

1.4.2 Textured Rough Surfaces

Surface texturing of sliding surfaces is a way to transit to the hydrodynamic regime of lubrication. Spiral-groove surface textures of sliding surfaces enable

TABLE 1.12
Recent Developments in Green Tribology on Textured Surface Configurations

Textured Surface Configurations	Authors
Virtual dimple design exploration on the mixed lubrication characteristics for a counterformal contact	[73]
Stribeck curve of dimpled bearings using machining and chemical etching	[74]
Dimples internal profile structure on bearing lubrication effectiveness	[75]
Partially textured parallel bearing surfaces analysis for optimum parameters using analytical solution	[76]
Texture location effect in journal bearings using numerical solution	[77]
Texturing or grooving location effect in journal bearings	[78]
Textured surface analysis of non-recessed hybrid journal bearing	[79]
Laser surface texture effects on palm kernel activated carbon epoxy composite.	[80]
Partial micro-dimple geometries at different locations on the journal bearings	[81]
Dimples fabrication on a cylindrical surface using turning operation	[82]
Optimal free-form and dimple textures in mechanical seals	[83]
Ellipse dimple texture of gas face seal	[84]
Texture geometry and position on the performance of meso-scale air journal bearing	[85]
Square textured profiles on parallel slider bearing performance	[86]

the transition to the hydrodynamic regime of thick film lubrication [87]. The surface roughness of sliding surfaces produced the transition from the hydrodynamic regime of thick film lubrication to a mixed regime of lubrication at higher rotational speeds. Surface roughness effects may be ignored under the hydrodynamic regime of lubrication [88]. The surface roughness effects may also be ignored for the optimum texture parameters at minimum friction coefficient under the hydrodynamic regime of lubrication determined. Half-section dimples in the leading edge of the hybrid thrust pad bearing surface are beneficial in enhancing load capacity and reducing power losses [89]. Dimpled surfaces having transverse micro-roughness improve the dynamic characteristics of hybrid thrust pad bearings. The surface roughness pattern and standard deviation of asperity height have significant effects on the transient startup characteristics of the bearing [90]. The lubrication model of surface texture coupled with a non-Gaussian roughness distribution of cylinder liner surface has been analysed [91]. The surface texture coupled with a small negative skewness surface in cylinder liner improves the lubrication performance. There have been efforts recently to control friction and wear using surface texturing combined with surface topography modification. The mixed lubrication model of surface texture coupled with non-Gaussian roughness distribution considering the effects of skewness and kurtosis has been analysed [92]. The optimal surface texture coupled with skewness and kurtosis parameters influences the mixed lubrication performance.

Recent developments in green tribology on textured rough surfaces are presented in Table 1.13.

TABLE 1.13
Recent Developments in Green Tribology on Textured Rough Surfaces

Textured Rough Surfaces	Authors
Nano-texturing performance on hydrodynamic lubrication	[87]
Roughness and texture numerical investigations under asperity contact	[88]
Optimised micro-rough dimple hybrid thrust pad bearing	[89]
Mixed/hydrodynamic transient regime lubrication modelling in journal bearing considering surface roughness effects	[90]
Non-Gaussian roughness and surface textures distribution on cylinder liner	[91]
Mixed film lubrication modelling rough-textured interfaces	[92]

1.4.3 Biomimetic/Tailored Textured Surfaces

Biomimetic and/or tailored texturing of surfaces accomplishes the expectation of superior performance of tribological components. The biomimetic dragonfly wing micro-spike structure influences the friction reduction and whirling orbit stability in journal bearings [93]. Friction reduction and whirling orbit stability are due to the generation of micro-bubbles by the spikes which contract the oil film distribution and expand the gas-phase distribution. Biomimetics embraces ideas from nature and emulates it towards all-inclusive nano-structured design of structures and products for technological applications. Surface roughness has the greatest influence on oleophilicity behaviour in the case of pistia-motivated surfaces [94]. The tribological performance of sliding contacts with textures is improved by depositing an amorphous hydrogenated diamond-like carbon (DLC) coating [95]. The beneficial effects of amorphous hydrogenated diamond-like carbon (DLC)-coated textures are at an optimum texture diameter/depth as deep dimples weaken the benefits of coated textures in non-conformal contact under boundary/starved lubrication. Tailored texturing may lead to enhanced and impressive results of lubricated tribopair contacts. The design of tailored (anisotropic and non-uniform) texturing maximizing the load of the thrust pad largely affects the friction characteristics of the contact [96]. Tailored surface texturing shows friction characteristics are clearly sensitive to the relative sliding motion of the tribopair. The fabricated patterns (single and multi-scale) on shaft surface show greatly reduced coefficients of friction at smaller rotational speeds compared to the smooth surface [97]. Reduced coefficients of friction at smaller rotational speeds on the patterned shaft surface are due to the transition from a mixed to a hydrodynamic regime of lubrication. Multi-scale surface patterns offer excellent potential to control friction and guide the lubricant along the surface [98]. Multi-scale surface patterns demonstrated the strongest influences on the lubricant spreading along the surface. Multi-scale surface texturing is a promising technique to improve the tribological characteristics of meso-scale air bearing systems.

Recent developments in green tribology on textured rough surfaces are presented in Table 1.14.

TABLE 1.14
Recent Developments in Green Tribology on Biomimetic/Tailored Textured Surfaces

Biomimetic/tailored Textured Surfaces	Authors
Biomimetic dragonfly wing micro-spike structure on journal bearing	[93]
Pistia stratiotes-motivated surface roughness for oleophilicity	[94]
Amorphous hydrogenated diamond-like carbon (DLC)-coated surface textures	[95]
Surface micro-structural modifications on a thrust bearing	[96]
Multi-scale patterns by roller-coining and/or lasers on the journal bearings	[97]
Optimised multi-scale patterns on the journal bearings	[98]

1.4.4 Slip Textured Surfaces

Slip conditions in conjunction with texturing result in tribological performance enhancements in sliding contacts. Tribological performance enhancements in bearings are influenced by pressure and shear stress variation. The pressure distribution increases while the shear stress distribution decreases overall in the partial slip textured/grooved bearing compared to the conventional bearing [99]. A partial slip textured/grooved bearing with constant film thickness (parallel slider or concentric journal) results in significant enhancement in characteristics compared to the conventional bearing. The textured/slippage combined pattern has a favourable influence by increasing load and decreasing friction in slider bearings [100]. The textured/slippage combined pattern has a much lower coefficient of friction compared to a conventional (non-textured) surface. The textured/slippage combined pattern even with a negligible gap ratio (parallel gap) has a favourable influence on the lubrication behaviour by increasing load and decreasing friction in slider bearings [101]. The gap ratio in a textured/slippage combined pattern appears to promote the lower coefficient of friction compared to the slip pattern on the surface. Optimal slip parameters are calculated utilising the two-component slip model in a journal bearing with a range of eccentricity ratios [102]. The performance of the journal bearing evaluated for load and friction coefficient yields the location of the slip zone at the convergent region of the film. The slip zone in the circumferential direction should begin in the convergent region and end close to the minimum film thickness position, while the slip zone in the axial direction should be maximum, for improvement in bearing performance.

Recent developments in green tribology on slip textured surfaces are presented in Table 1.15.

1.5 GREEN MACHINING

Machining operations for various industrial applications have greatly increased with the increase in demand. With the rise of eco-awareness, sustainable machining for various industrial applications achieves environmental conservation. Sustainable machining involves enhancement in machining performance with green cutting

TABLE 1.15
Recent Developments in Green Tribology on Slip Textured Surfaces

Slip Textured Surfaces	Authors
Slider and journal partial slip textured/grooved bearing analysis	[99]
Slippage/textured combined pattern with optimum parameters in MEMS applications	[100]
Slip texture zones with optimum parameters in slider bearing	[101]
Slip zones with optimum partial extent in hydrodynamic journal bearing	[102]

fluids [103], green minimum quantity lubrication [104], green machining with textured/coated tools and green nano-lubricated machining [105].

Recent developments in green tribology on machining are presented in Table 1.16.

1.5.1 Green Cutting Fluids

Sustainable machining involves enhancement in machining performance with green cutting fluids, and therefore, the manufacturer tends to substitute mineral oil for environmentally friendly bio-based oil. The developments of green cutting fluids from environmentally friendly bio-based oil have brought about remarkable changes in the growing machining industry. Green cutting fluids give a positive impact on the environment and act as a cooling and lubrication agent in the machining process. The machining performance of modified jatropha oil (MJO) significantly improves lubricating effect compared to substitute synthetic ester (SE) as a machining lubricant [106]. Machining performance using green cutting fluids formulated with non-toxic emulsifiers and natural additives meets the specifications of commercial formulations [107]. Green cutting fluids synthesis from bio-based oils is a good basis for development of non-toxic eco-friendly products for sustainable machining.

Recent developments in green tribology on cutting fluids are presented in Table 1.17.

1.5.2 Green Minimum Quantity Lubrication

As the global demand for cutting fluids continues to rise in the manufacturing industry, the development of greener, more sustainable cutting fluids and approaches are highly desired. Recent trends in machining show an increasing of demand for the eco-friendly approaches in order to eliminate the health, energy and environmental issues caused by the application of conventional cutting fluids. Conventional

TABLE 1.16
Recent Developments in Green Tribology on Machining

Green Machining	Authors
Green metal-forming lubrication on the adhesion theory	[103]
Potential MQL advancements in terms of sustainability and machining performances	[104]
Sustainable machining using vegetable oil-based nano-fluids and the future prospects	[105]

TABLE 1.17
Recent Developments in Green Tribology on Cutting Fluids

Green Cutting Fluids	Authors
Bio-based modified jatropha oil (MJO) developed by transesterification of jatropha methyl ester (JME) to trimethylolpropane (TMP)	[106]
Sustainable and biodegradable cutting fluid prepared by combining non-toxic emulsifiers and natural additives	[107]

machining cannot fit the principles of green manufacturing in terms of health, energy and environmental issues and hence eco-friendly approaches such as minimum quantity lubrication (MQL) need to be implemented.

Minimum quantity lubrication (MQL) using nano-fluids significantly enhances the machinability, cooling and lubricating properties of base MQL [108]. Minimum quantity lubrication (MQL) using palm oil-based alumina nano-fluids resulted in the formation of an effective surface protecting film on the machined surface which influenced the tool wear under the mechanism of adhesion and abrasion [109]. Minimum quantity lubrication (MQL) using palm oil-based nano-fluids with graphene nanoplatelets (GNPs) could lessen the specific cutting energy compared to commercial lubricants because of reduction in the cutting force and improvement in the surface quality [110]. Conventional cutting fluids contaminated by bacteria, fungi and biocide contents are costly to dispose of and potentially harmful to the environment [111]. MQL machining with microbial-based cutting fluids containing microalgae species results in cutting forces, tool wear, surface finish and dimensional accuracy comparable to conventional cutting fluids. The MQL approach extends tool life with the maximum volume of material removal compared to machining with cryogenic CO_2 coolant [112]. The cryogenic CO_2 coolant lowers the tool wear rate due to effectiveness in lubrication compared to the cooling effect. The cryogenic CO_2 cooling approach is introduced for efficient lubricating and cooling performance.

Recent developments in green tribology on minimum quantity lubrication are presented in Table 1.18.

TABLE 1.18
Recent Developments in Green Tribology on Minimum Quantity Lubrication

Green Minimum Quantity Lubrication	Authors
Minimum quantity lubrication (MQL) using canola oil-based diamond nano-fluids	[108]
Minimum quantity lubrication (MQL) using alumina-enhanced palm oil medium	[109]
Minimum quantity lubrication (MQL) using palm oil-based nano-fluids with graphene nano-platelets (GNPs)	[110]
Small Quantity Lubrication (SQL) using microbial-based cutting fluids containing a microalgae species	[111]
Sustainable techniques using MQL on machinability of Inconel 718	[112]

1.5.3 Green Machining with Textured/Coated Tools

Green machining with textured/coated tools is desirable with high quality products in both metal cutting and forming industries. In metal cutting industries it is desired to obtain longer tool life and better surface quality, whereas in metal forming industries it is desired to obtain reduced die wear and improved surface finish.

The goals of green machining for high quality products with a cost effective process are achieved with a poly crystalline cubic boron nitride cutting tool under dry machining conditions [113]. Surface texturing is mainly used to improve cutting tool performance in green machining operations. Textures improve cutting performance by reducing the friction force, cutting forces and cutting temperature. Surface textures enhance lubricant availability at the contact point, reduce the contact area of tool-chip, reduce the crater/flank wear and trap the wear debris [114]. Micro-textured/dimpled surfaces lower the friction coefficient, reduce the thrust force and torque and minimise the chip clogging and chip evacuation force [115]. Micro-texturing of the drill tool improves cutting performance by minimizing frictional forces at the contact interfaces, and thereby the energy loss while machining Ti-6Al-4V. The surface coating is also used to improve cutting tool performance in green machining operations. The influence of tool life on the optimal coating properties is determined by the wear morphology of the crater [116].

Recent developments in green tribology on machining with textured/coated tools are presented in Table 1.19.

1.5.4 Green Nano-Lubricated Machining

The eco-friendly green nano-cutting fluids have the potential to improve machining performance. Green machining nano-fluids formulated using coconut oil dispersed with nano-graphene resulted in reduced machining forces, tool wear and temperatures, and improved surface finish [117]. The hot chrome steel rolling service life can be effectively improved using water-based nano-TiO_2 lubricants with significant reduction in coefficient of friction and wear [118]. Reduced coefficient of friction and wear of nano-TiO_2 water-based lubricants is due to the semisolid film and solid layer formation mechanisms. Green nano-particles are explored as lubricant additives due to improvement in tribological performance in metal forming applications. The improved lubrication provided by eco-friendly lubricant additives in a polymeric

TABLE 1.19
Recent Developments in Green Tribology on Machining with Textured/Coated Tools

Green Machining with Textured/Coated Tools	Authors
Green machining with poly crystalline cubic boron nitride cutting tool in dry machining	[113]
Surface micro-nano-texture effects on cutting tool performance	[114]
Sustainable machining with surface micro-texture drill tools for the machining of Ti-6Al-4V	[115]
Sustainable machining with coated tools for the machining of 316L stainless steel	[116]

TABLE 1.20
Recent Developments in Green Tribology on Nano-Lubricated Machining

Green Nano-Lubricated Machining	Authors
Green machining nano-fluids formulated by nano-graphene particles in coconut and canola oil	[117]
Environmentally friendly water-based nano-TiO_2 lubricants	[118]
Halloysite clay nanotubes (HNTs) as green lubricant additives	[119]

lubricant results in longer tool life and lower energy consumption for metal-forming applications [119]. Halloysite clay nano-tubes (HNTs) are attractive as natural green lubricant additives as they increased the scuffing load and surface finish and lowered wear loss and coefficient of friction due to the formation of a tribofilm.

Future outlooks in green tribology on nano-lubricated machining are presented in Table 1.20.

1.6 CONCLUSION

This chapter highlights the significance of green tribology in tribological systems for sustainable development. This chapter reviews the recent developments based on the literature on green lubricants, green composites, textured surfaces and green machining. The significant progresses in green tribology on noteworthy improvements in lubrication enrichment, friction and wear reduction are elaborated. Further outlooks render continual sustainable development with advances in green tribology.

REFERENCES

1. M. Nosonovsky and B. Bhushan. Green Tribology: Biomimetics, Energy Conservation and Sustainability, 2012, Springer, Germany.
2. I. C. Gebeshuber. Green nanotribology, Proc. Inst. Mech. Eng. C J. Mech. Eng. Sci., 226, 2, 2012, 347–358.
3. J. P. Davim. Progress in Green Tribology: Green and Conventional Techniques, 2017, De Gruyter, Germany.
4. K. Holmberg and A. Erdemir. The impact of tribology on energy use and CO_2 emission globally and in combustion engine and electric cars, Trib. Int., 135, 2019, 389–396.
5. M. Hadfield and C. Ciantar. Sustainable development thinking in tribology design, Tribol. Int. Eng. Ser., 48, 2005, 457–464.
6. A.Z. Syahir, N.W.M. Zulkifli, H.H. Masjuki, M.A. Kalam, Abdullah Alabdulkarem, M. Gulzar, L.S. Khuong and M.H. Harith. A review on bio-based lubricants and their applications, J. Cleaner Prod., 168, 2017, 997–1016.
7. P. Nagendramma and S. Kaul. Development of ecofriendly/biodegradable lubricants: An overview, Renew. Sustain. En. Rev., 16, 1, 2012, 764–774.
8. H. Gong, C. Yu, L. Zhang, G. Xie, D. Guo and J. Luo. Intelligent lubricating materials: A review, Comp. Part B: Eng., 202, 2020, 108450.
9. M. Palacio and B. Bhushan. A review of ionic liquids for green molecular lubrication in nanotechnology, Tribol. Lett., 40, 2010, 247–268.
10. C. J. Reeves, A. Siddaiah and P. L. Menezes. Friction and wear behavior of environmentally friendly ionic liquids for sustainability of biolubricants, J. Tribol., 141, 5, 2019, 051604.

11. S.A.S. Amiril, E.A. Rahim and S. Syahrullail. A review on ionic liquids as sustainable lubricants in manufacturing and engineering: Recent research, performance, and applications, J. Cleaner Prod., 168, 2017, 1571–1589.
12. T. Naveed, R. Zahid, R. A. Mufti, M. Waqas and M. T. Hanif. A review on tribological performance of ionic liquids as additives to bio lubricants, P. I. Mech. Eng. J.-J. Eng., 2020, https://doi.org/10.1177/1350650120973805.
13. A. N. Afifah, S. Syahrullail, N. I. W. Azlee and A. M. Rohah. Synthesis and tribological studies of epoxidized palm stearin methyl ester as a green lubricant, J. Cleaner Prod., 280, Part 1, 2021, 124320.
14. P. Nagendramma, B. M. Shukla and D. K. Adhikari. Synthesis, characterization and tribological evaluation of new generation materials for aluminum cold rolling oils, Lubricants, 4, 3, 2016, 23.
15. K. J. Babu, A. S. Kynadi, M. L. Joy and K. P. Nair. Enhancement of cold flow property of coconut oil by alkali esterification process and development of a bio-lubricant oil, P. I. Mech. Eng. J.-J. Eng., 232,3, 2018, 307–314.
16. Z. M. Zulfattah, N. W. M. Zulkifli, H. H. Masjuki, M. H. Harith, A. Z. Syahir, I. Norain, M. N. A. M. Yusoff, M. Jamshaid and A. Arslan. Friction and wear performance of oleate-based esters with two-, three-, and four-branched molecular structure in pure form and mixture, J. Tribol., 143, 1, 2021, 011901.
17. Y. Shi, I. Minami, M. Grahn, M. Björling and R. Larsson. Boundary and elastohydrodynamic lubrication studies of glycerol aqueous solutions as green lubricants, Tribol. Int., 69, 2014, 39–45.
18. M. A. Kabir, C. Fred Higgs, III and M. R. Lovell. A pin-on-disk experimental study on a green particulate-fluid lubricant, J. Tribol., 2008, 130, 4, 041801.
19. M.A. M. Alias, M. F. B. Abdollah, and H. Amiruddin. Lubricant and tribological properties of zinc compound in palm oil, Ind. Lub. Tribol., 71, 10, 2019, 1177–1185.
20. R. Sánchez, M. Fiedler, E. Kuhn and J. M. Franco. Tribological characterization of green lubricating greases formulated with castor oil and different biogenic thickener agents: a comparative experimental study, Ind. Lub. Tribol., 63, 6, 2011, 446–452.
21. Y. Xia, W. Zhang, Z. Cao and X. Feng. A comparative study on tribological properties of leaf-surface waxes extracted from coastal and inland plants, Ind. Lub. Tribol., 71, 4, 2019, 586–593.
22. R. N. Gupta and A. P. Harsha. Synthesis, characterization, and tribological studies of calcium–copper–titanate nanoparticles as a biolubricant additive, J. Tribol., 139, 2, 2017, 021801.
23. D. Wu, Y. Xu, L. Yao, T. You and X. Hu. Tribological behaviour of graphene oxide sheets as lubricating additives in bio-oil, Ind. Lubr. Tribol., 70, 8, 2018, 1396–1401.
24. R. Yu, J. Liu and Y. Zhou. Experimental study on tribological property of MoS_2 nanoparticle in castor oil, J. Tribol., 141, 10, 2019, 102001.
25. M. F. B. Abdollah, H. Amiruddin, M. Alif Azmi and N. A. M. Tahir. Lubrication mechanisms of hexagonal boron nitride nano-additives water-based lubricant for steel–steel contact, P. I. Mech. Eng. J.-J. Eng., 2020, https://doi.org/10.1177/1350650120940173.
26. E. Sneha, V. S. Sarath, S. Rani and K. B. Kumar. Effect of turmeric oil and halloysite nano clay as anti-oxidant and anti-wear additives in rice bran oil, P. I. Mech. Eng. J.-J. Eng., 2020, https://doi.org/10.1177/1350650120945074.
27. B. Sagbas. Tribological performance of peek with green lubricant enhanced by nano hexagonal boron nitride powder, Ind. Lub. Tribol., 72, 2, 2018, 203–210.
28. G. Xie, J. Luo, D. Guo and S. Liu. Nanoconfined ionic liquids under electric fields, Appl. Phys. Lett., 96, 2010, 043112.
29. M. Kronberger, V. Pejaković, C. Gabler and M. Kalin. How anion and cation species influence the tribology of a green lubricant based on ionic liquids, P. I. Mech. Eng. J.-J. Eng., 226, 11, 2012, 933–951.

30. S. Stolte, S. Steudte, O. Areitioaurtena, F. Pagano, J. Thöming, P. Stepnowski and A. Igartua. Ionic liquids as lubricants or lubrication additives: An ecotoxicity and biodegradability assessment, Chemosphere, 89, 9, 2012, 1135–1141.
31. S. Zhang, L. Ma, P. Wen, X. Ye, R. Dong, W. Sun, M. Fan, D. Yang, F. Zhou and W. Liu. The ecotoxicity and tribological properties of choline amino acid ionic liquid lubricants, Tribol. Int., 121, 2018, 435–441.
32. P. Nagendramma, P. K. Khatri, G. D. Thakre and S. L. Jain. Lubrication capabilities of amino acid based ionic liquids as green bio-lubricant additives, J. Mol. Liq., 244, 2017, 219–225.
33. M. Fan, L. Ma, C. Zhang, Z. Wang, J. Ruan, M. Han, Y. Ren, C. Zhang, D. Yang, F. Zhou and W. Liu. Biobased green lubricants: Physicochemical, tribological and toxicological properties of fatty acid ionic liquids, Tribol. Trans., 61, 2, 2018, 195–206.
34. D. Zheng, X. Wang, M. Zhang Z. Liu and C. Ju. Anticorrosion and lubricating properties of a fully green lubricant, Tribol. Int., 130, 2019, 324–333.
35. A. Z. Syahir, M. H. Harith, N. W. M. Zulkifli, H. H. Masjuki, M. A. Kalam, M. N. A. M. Yusoff, Z. M. Zulfattah and T. M. Ibrahim. Compatibility of ionic liquid with glycerol monooleate and molybdenum dithiocarbamate as additives in bio-based lubricant, J. Tribol., 142, 6, 2020, 061901.
36. C. Ju, D. Zheng, Q. Zhao, et al. Tribological properties of green ILs containing MoS_2 quantum dots with one-step preparation. Tribol. Lett. 68, 2020, 79.
37. C. Chimeno-Trinchet, M. E. Pacheco, A. Fernández-González, M. E. Díaz-García and R. Badía-Laíño. New metal-free nanolubricants based on carbon-dots with outstanding antiwear performance, J. Ind. Eng. Chem., 87, 2020, 152–161.
38. H. Wu, S. Yin, L. Wang, Y. Du, Y. Yang, J. Shi and H. Wang. Investigation on the robust adsorption mechanism of alkyl-functional boric acid nanoparticles as high performance green lubricant additives, Tribology International, 157, May 2021, 106909.
39. E. Omrani, P. L. Menezes and P. K. Rohatgi. State of the art on tribological behavior of polymer matrix composites reinforced with natural fibers in the green materials world, Eng. Sci. Tech. Int. J., 19, 2, 2016, 717–736.
40. S. Karthikeyan, N. Rajini, M. Jawaid, J. T. Winowlin Jappes, M. T. H. Thariq, S. Siengchin and J. Sukumaran. A review on tribological properties of natural fiber based sustainable hybrid composite, P. I. Mech. Eng. J.-J. Eng., 231, 12, 2017, 1616–1634.
41. H. Chen, L. Zhang, M. Li and G. Xie. Synthesis of core–shell micro/nanoparticles and their tribological application: A review, Materials, 13, 20, 2020, 4590.
42. Z. Ji, L. Zhang, G. Xie et al. Mechanical and tribological properties of nanocomposites incorporated with two-dimensional materials. Friction, 8, 2020, 813–846.
43. P. Saravanan and N. Emami. Chapter 13 – Sustainable tribology: Processing and characterization of multiscale thermoplastic composites within hydropower applications, Tribol. Polym. Compos. Charact. Prop. Appl., Elsevier Series on Tribology and Surface Engineering, 2021, 241–277.
44. E. E. Nunez, R. Gheisari and A. A. Polycarpou. Tribology review of blended bulk polymers and their coatings for high-load bearing applications, Tribol. Int., 129, 2019, 92–111.
45. P. Lan, E. E. Nunez and A. A. Polycarpou. Advanced polymeric coatings and their applications: Green tribology, Encycl. Renewable Sustainable Mater., 4, 2020, 345–358.
46. Y. Ren, L. Zhang, G. Xie et al. A review on tribology of polymer composite coatings. Friction, 9, 2021, 429–470.
47. E. I. Akpan, B. Wetzel and K. Friedrich, A fully biobased tribology material based on acrylic resin and short wood fibres, Tribol. Int., 120, 2018, 381–390.
48. V. K. Patel, S. Chauhan and J. K. Katiyar. Physico-mechanical and wear properties of novel sustainable sour-weed fiber reinforced polyester composites, Mater. Res. Express, 5, 4, 2018, 045310.

49. A. S. Negi, J. K. Katiyar, S. Kumar, N. Kumar and V. K. Patel. Physicomechanical and abrasive wear properties of hemp/kevlar/carbon reinforced hybrid epoxy composites, Mat. Res. Exp., 6, 11, 2019, 115304.
50. M. A. A. Bakar, A. H. Abdul, S. Kasolang and M. A. Ahmad. Specific wear rate of kenaf epoxy composite and oil palm empty fruit bunch (OPEFB) epoxy composite in dry sliding, J. Tekn., 58, 2012, 85–88.
51. V. Mahale, J. Bijwe and S. Sinha. Efforts towards green friction materials, Tribol. Int., 136, August 2019, 196–206.
52. D. N. F. Mahmud, M. F. B. Abdollah, N. A. B. Masripan, N. Tamaldin and H. Amiruddin. Frictional wear stability mechanisms of an activated carbon composite derived from palm kernel by phase transformation study, Ind. Lub. Tribol., 69, 6, 2017, 945–951.
53. A. Mustafa, M. F. B. Abdollah, H. Amiruddin, F. F. Shuhimi and N. Muhammad. Optimization of friction properties of kenaf polymer composite as an alternative friction material, Ind. Lub. Tribol., 69, 2, 2017, 259–266.
54. A. Jumahat, C. Soutis, S. A. Abdullah and S. Kasolang. Tensile properties of nanosilica/epoxy nanocomposites, Proc. Eng., 41, 2012, 1634–1640.
55. B. Barari, E. Omrani, A. D. Moghadam, P. L. Menezes, K. M. Pillai and P. K. Rohatgi. Mechanical, physical and tribological characterization of nano-cellulose fibers reinforced bio-epoxy composites: An attempt to fabricate and scale the 'Green' composite, Carb. Polym., 147, 2016, 282–293.
56. X. Wei, G. Zhang and L. Wang. "Metal-reservoir" carbon-based nanocomposite coating for green tribology. Tribol. Lett., 59, 2015, 34.
57. I. Taheridoustabad, M. Khosravi and Y. Yaghoubinezhad. Fabrication of GO/RGO/TiC/TiB2 nanocomposite coating on Ti–6Al–4V alloy using electrical discharge coating and exploring its tribological properties, Tribol. Int., 156, 2021, 106860.
58. N. A. M. Tahir, M. F. B. Abdollah, N. Tamaldin, M. R. B. M. Zin and H. Amiruddin. Effect of hydrogen on graphene growth from solid waste products by chemical vapour deposition: friction coefficient properties, Ind. Lub. Tribol., 72, 2, 2018, 181–188.
59. B. Chen, Y. Jia, M. Zhang, H. Liang, X. Li, J. Yang, F. Yan and C. Li. Tribological properties of epoxy lubricating composite coatings reinforced with core-shell structure of CNF/MoS_2 hybrid, Comp. Part A Appl. Sci. Manuf., 122, 2019, 85–95.
60. J. A. Arsecularatne, E. Colusso, E. D. Gaspera, A. Martucci and M. J. Hoffman. Nanomechanical and tribological characterization of silk and silk-titanate composite coatings, Tribol. Int., 146, 2020, 106195.
61. I. Etsion. State of the art in laser surface texturing, J. Tribol., 127, 1, 2005 248–253.
62. A. F. S. Baharin, M. J. Ghazali and J. A. Wahab. Laser surface texturing and its contribution to friction and wear reduction: a brief review, Ind. Lub. Tribol., 68, 1, 2016, 57–66.
63. H. L. Costa and I. M. Hutchings. Some innovative surface texturing techniques for tribological purposes. P. I. Mech. Eng. J.-J. Eng., 229, 4, 2015, 429–448.
64. R. L. Jackson. A scale dependent simulation of liquid lubricated textured surfaces, J. Tribol., 132, 2, 2010, 022001.
65. D. G. Coblas, A. Fatu, A. Maoui and M. Hajjam. Manufacturing textured surfaces: State of art and recent developments, P. I. Mech. Eng. J.-J. Eng., 229, 1, 2015, 3–29.
66. T. Ibatan, M. S. Uddin and M. A. K. Chowdhury. Recent development on surface texturing in enhancing tribological performance of bearing sliders. Surf. Coat Tech., 272, 2015, 102–120.
67. U. Sudeep, N. Tandon and R. K. Pandey. Performance of lubricated rolling/sliding concentrated contacts with surface textures: A review, J. Tribol., 137, 3, 2015, 031501.
68. A. Ahmed, H. H. Masjuki, M. Varman, M. A. Kalam, M. Habibullah and K. A. H. Al Mahmud. An overview of geometrical parameters of surface texturing for piston/cylinder assembly and mechanical seals, Meccanica, 51, 2016, 9–23.

69. D. Gropper, L. Wang and T. J. Harvey. Hydrodynamic lubrication of textured surfaces: A review of modeling techniques and key findings. Tribol. Int., 94, 2016, pp. 509–529.
70. C. Gachot, A. Rosenkranz, S. M. Hsu and H. L. Costa. A critical assessment of surface texturing for friction and wear improvement, Wear, 372, 2017, 21–41.
71. A. Rosenkranz, P. G. Grützmacher, C. Gachot and H. L. Costa. Surface texturing in machine elements – a critical discussion for rolling and sliding contacts, Adv. Eng. Mat., 21, 8, 2019, 1900194.
72. A. Senatore and T. V. V. L. N. Rao. Partial slip texture slider and journal bearing lubricated with Newtonian fluids: A review. J. Tribol., 140, 4, 2018, 040801-1-20.
73. Q. J. Wang and D. Zhu. Virtual texturing: Modeling the performance of lubricated contacts of engineered surfaces, J. Tribol., 127, 4, 2005, 722–728.
74. X. Lu, and M. M. Khonsari. An experimental investigation of dimple effect on the stribeck curve of journal bearings. Tribol. Lett., 27, 2007, 169.
75. C. Shen and M. M. Khonsari. Effect of dimple's internal structure on hydrodynamic lubrication. Tribol. Lett., 52, 2013, 415–430.
76. M. D. Pascovici, T. Cicone, M. Fillon and M. B. Dobrica. Analytical investigation of a partially textured parallel slider, P. I. Mech. Eng. J.-J. Eng., 223, 2, 2009, 151–158.
77. N. Tala-Ighil, M. Fillon and P. Maspeyrot. Effect of textured area on the performances of a hydrodynamic journal bearing, Tribol. Int., 44, 3, 2011, 211–219.
78. S. Kango, R. K. Sharma and R. K. Pandey. Comparative analysis of textured and grooved hydrodynamic journal bearing, P. I. Mech. Eng. J.-J. Eng., 228, 1, 2014, 82–95.
79. C. B. Khatri and S. C. Sharma. Influence of textured surface on the performance of non-recessed hybrid journal bearing operating with non-Newtonian lubricant, Tribol. Int., 95, 2016, 221–235.
80. M. Mohmad, M. F. B. Abdollah, N. Tamaldin and H. Amiruddin. The effect of dimple size on the tribological performances of a laser surface textured palm kernel activated carbon-epoxy composite, Ind. Lub. Tribol., 69, 5, 2017, 768–774.
81. S. Matele and K. N. Pandey. Effect of surface texturing on the dynamic characteristics of hydrodynamic journal bearing comprising concepts of green tribology, P. I. Mech. Eng. J.-J. Eng., 232, 11, 2018, 1365–1376.
82. J. A. Ghani, M. N. A. M. Dali, H. A. Rahman, C. H. C. Haron, W. M. F. Wan Mahmood, M. R. M. Rasani and M. Z. Nuawi. Analysis of dimple structure fabricated using turning process and subsequent reduction in friction, Wear, 426–427, Part B, 2019, 1280–1285.
83. X. Wang, M. Khonsari, S. Li, Q. Dai and X. Wang. Experimental verification of textured mechanical seal designed using multi-objective optimization, Ind. Lub. Tribol., 71, 6, 2019,766–771.
84. K. Guiyue, L. Xinghu, W. Yan, L. Mouyou, Z. Kanran and M. Mingfei. Parameter study and shape optimisation of a generalised ellipse dimple-textured face seal, Lub. Sci., 32, 1, 2020, 10–20.
85. N. D. Hingawe and S. P. Bhore. Tribological performance of a surface textured meso scale air bearing, Ind. Lub. Tribol., 72, 5, 2019, 599–609.
86. N. D. Hingawe and S. P. Bhore. Multi-objective optimization of the design parameters of texture bottom profiles in a parallel slider, Friction, 8, 4, 2020, 726–745.
87. T. Hirayama, H. Shiotani, K. Yamada, N. Yamashita, T. Matsuoka, H. Sawada and K. Kawahara. Hydrodynamic performance produced by nanotexturing in submicrometer clearance with surface roughness, J. Tribol., 137, 1, 2015, 011704.
88. C. Ma, Y. Duan, B. Yu, J. Sun and Q. Tu. The comprehensive effect of surface texture and roughness under hydrodynamic and mixed lubrication conditions. P. I. Mech. Eng. J.-J. Eng., 231, 10, 2017, 1307–1319.
89. V. Kumar and S. C. Sharma. Influence of dimple geometry and micro-roughness orientation on performance of textured hybrid thrust pad bearing. Meccanica, 53, 2018, 3579–3606.

90. S. Cui, L. Gu, M. Fillon, L. Wang and C. Zhang. The effects of surface roughness on the transient characteristics of hydrodynamic cylindrical bearings during startup, Tribol. Int., 128, 2018, 421–428.
91. B. Yin, H. Zhou, B. Xu and H. Jia. The influence of roughness distribution characteristic on the lubrication performance of textured cylinder liners, Ind. Lub. Tribol., 71, 3, 2019, 486–493.
92. C. Gu, X. Meng, S. Wang and X. Ding. Study on the mutual influence of surface roughness and texture features of rough-textured surfaces on the tribological properties. P. I. Mech. Eng. J.-J. Eng., 2020, https://doi.org/10.1177/1350650120940211.
93. K. Furukawa, M. Ochiai, H. Hashimoto and S. Kotani. Bearing characteristic of journal bearing applied biomimetics, Tribol. Int., 150, 2020, 106345.
94. N. A. Latif, S. Kasolang, M. A. Ahmad and M. A. A. Bakar. Effect of oleophilicity on Pistia inspired surface roughness, J. Mech. Eng., 8, 2019, 105–116.
95. A. Arslan, H. H. Masjuki, M. Varman, M. A. Kalam, M. M. Quazi, K. A. H. Al Mahmud, M. Gulzar and M. Habibullah. Effects of texture diameter and depth on the tribological performance of DLC coating under lubricated sliding condition, Appl. Surf. Sci., 356, 2015, 1135–1149.
96. A. Ancona, G. S. Joshi, A. Volpe, M. Scaraggi, P. M. Lugarà and G. Carbone. Non-uniform laser surface texturing of an un-tapered square pad for tribological applications. Lubricants, 5, 4, 2017, 41.
97. P. G. Grützmacher, A. Rosenkranz, A. Szurdak, F. König, G. Jacobs, G. Hirt and F. Mücklich. From lab to application – improved frictional performance of journal bearings induced by single-and multi-scale surface patterns, Tribol. Int., 127, 2018, 500–508.
98. P. G. Grützmacher, A. Rosenkranz, A. Szurdak, M. Grüber, C. Gachot, G. Hirt and F. Mücklich. Multi-scale surface patterning – an approach to control friction and lubricant migration in lubricated systems, Ind. Lub. Tribol., 71, 8, 2019, 1007–1016.
99. T. V. V. L. N. Rao. Analysis of single-grooved slider and journal bearing with partial slip surface, J. Tribol., 132, 1, 2010, 014501.
100. M. Tauviqirrahman, R. Ismail, J. Jamari and D. J. Schipper. Combined effect of texturing and boundary slippage in lubricated sliding contacts, Tribol. Int., 66, 2013, 274–281.
101. S. Susilowati, M. Tauviqirrahman, J. Jamari and A. P. Bayuseno. Numerical investigation of the combined effects of slip and texture on tribological performance of bearing, Tribol.– Mat. Surf. Inter., 10, 2, 2016, 86–89.
102. S. Cui, C. Zhang, M. Fillon and L. Gu. Optimization performance of plain journal bearings with partial wall slip, Tribol. Int., 145, 2020, 106137.
103. Z. G. Wang. Tribological approaches for green metal forming, J. Mat. Proc. Tech., 151, 1–3, 2004, 223–227.
104. N. N. Nor Hamran, J. A. Ghani, R. Ramli and C. H. Che Haron. A review on recent development of minimum quantity lubrication for sustainable machining, J. Clean. Prod., 268, 2020, 122165.
105. K. C. Wickramasinghe, H. Sasahara, E. A. Rahim and G. I. P. Perera. Green metalworking fluids for sustainable machining applications: A review, J. Clean. Prod., 257, 2020, 120552.
106. N. Talib and E.A. Rahim. Performance evaluation of chemically modified crude jatropha oil as a bio-based metalworking fluids for machining process, Proc. CIRP, 26, 2015, 346–350.
107. P. S. Suvin, P. Gupta, J.-H. Horng and S. V. Kailas. Evaluation of a comprehensive non-toxic, biodegradable and sustainable cutting fluid developed from coconut oil, P. I. Mech. Eng. J.-J. Eng., 2020, https://doi.org/10.1177/1350650120975518.
108. S. Yuan, X. Hou, L. Wang et al. Experimental investigation on the compatibility of nanoparticles with vegetable oils for nanofluid minimum quantity lubrication machining. Tribol. Lett. 66, 2018, 106.

109. B. Sen, M. K. Gupta, M. Mia, U. K. Mandal and S. P. Mondal. Wear behaviour of TiAlN coated solid carbide end-mill under alumina enriched minimum quantity palm oil-based lubricating condition, Tribol. Int., 148, 2020, 106310.
110. A. M. M. Ibrahim, W. Li, H. Xiao, Z. Zeng, Y. Ren and M. S. Alsoufi. Energy conservation and environmental sustainability during grinding operation of Ti–6Al–4V alloys via eco-friendly oil/graphene nano additive and minimum quantity lubrication, Tribol. Int., 150, 2020, 106387.
111. R. Teti, D. M. D'Addona and T. Segreto. Microbial-based cutting fluids as bio-integration manufacturing solution for green and sustainable machining, CIRP J. Manuf. Sci. Tech., 32, 2021, 16–25.
112. N. H. Abdul Halim, C. H. Che Haron and J. Abdul Ghani. Sustainable machining of hardened inconel 718: A comparative study, Int. J. Precis. Eng. Manuf., 21, 2020, 1375–1387.
113. S. Sivarajan and R. Padmanabhan. Green machining and forming by the use of surface coated tools, Proc. Eng., 97, 2014, 15–21.
114. A. Arslan, H. H. Masjuki, M. A. Kalam, M. Varman, R. A. Mufti, M. H. Mosarof, L. S. Khuong and M. M. Quazi. Surface texture manufacturing techniques and tribological effect of surface texturing on cutting tool performance: A review, Crit. Rev. Sol. State Mat. Sci., 41, 6, 2016, 447–481.
115. S. Niketh and G.L. Samuel. Surface texturing for tribology enhancement and its application on drill tool for the sustainable machining of titanium alloy, J. Clean. Prod., 167, 2017, 253–270.
116. Z. Lai, C. Wang, L. Zheng, W. Huang, J. Yang, G. Guo and W. Xiong. Adaptability of AlTiN-based coated tools with green cutting technologies in sustainable machining of 316L stainless steel, Tribol. Int., 148, 2020, 106300.
117. R. Padmini, P. V. Krishna, S. Mahith and S. Kumar. Influence of green nanocutting fluids on machining performance using minimum quantity lubrication technique, Mat. Today: Proc., 18, Part 3, 2019, 1435–1449.
118. H. Wu, F. Jia, J. Zhao, S. Huang, L. Wang, S. Jiao, H. Huang and Z. Jiang. Effect of water-based nanolubricant containing nano-TiO_2 on friction and wear behaviour of chrome steel at ambient and elevated temperatures, Wear, 426–427, Part A, 2019, 792–804.
119. L. Peña-Parás, D.o Maldonado-Cortés, P. García, M. Irigoyen, J. Taha-Tijerina and J. Guerra. Tribological performance of halloysite clay nanotubes as green lubricant additives, Wear, 376–377, Part A, 2017, 885–892.

2 Bio-Based Lubricant in the Presence of Additives

Classification to Tribological Behaviour

Ali Raza and Arslan Ahmed
COMSATS University Islamabad, Sahiwal Campus
Islamabad, Pakistan

M.A. Kalam
University of Malaya
Kuala Lumpur, Malaysia

I.M. Rizwanul Fattah
University of Technology Sydney
Ultimo, Australia

CONTENTS

DOI: 10.1201/9781003139386-2

2.1 BACKGROUND

The field of tribology has gained importance in the research community since 1966, and it was defined by researchers as "the practices and analysis of interactive rubbing surfaces when these surfaces are in motion." Tribology includes the behaviour of wear and friction of rubbing surfaces under lubrication [1]. With the use of tribology, it is easy now to save significant energy. In 1976, after establishing a survey it was proposed that 11% of U.S. energy expenditure could be reduced in industries, in the power production sector and in the transportation sector which are 80% of the total expenditure of energy of the United States [1].

Tribology gained further importance in the 1990s when wear characterization of various metals was conducted [2]. Almost 33.33% of total fuel energy is utilised in overcoming the tires, brakes and engine transmission friction in passenger locomotives [3]. For better understanding, frictional losses can be divided into four sections: 11.5% for overcoming the road tire rolling friction, 11.5% for overcoming engine system friction, 5% at the brakes and 5% at the power transmission system of the passenger locomotive engine as shown in Figure 2.1 [3].

Energy losses to overcome friction in an engine can be subdivided into four sections: (i) 10% for the functioning of the hydraulic process and pumping process, (ii) 15% in the utilization of the valve train, (iii) 30% in the proper functioning of seals and bearings and (iv) 45% in the piston assembly [3]. The most significant fuel energy consumption (45%) occurs in the piston assembly due to the presence of a ring pack mechanism. This part is most critical due to the reason that performance

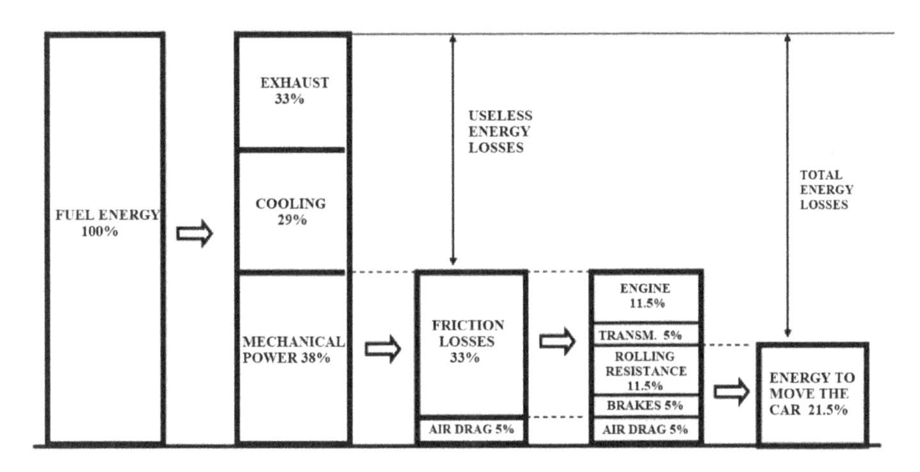

FIGURE 2.1 Flow sheet diagram of energy consumption in a car. (Adapted from Ref [3] with permission.)

of a piston is characterised by three functions of the ring pack which are as follows: first is the heat transfer between cylinder and piston, second is the transfer of oil to the combustion chamber from the crankcase and third is effective sealing between the crankcase and combustion chamber [4].

In Figure 2.2, two compression rings are presented at the top position and the oil ring is located at the bottom of the compression rings which are used to prevent the mixing of lubrication oil and fuel oil.

The diagram of the piston cylinder can be seen in Figure 2.3.

Chong investigated [6] friction in the piston ring pack mechanism during the presence of lateral motion of the piston and according to the author, the posture of the piston is very crucial for analysing the frictional forces. After some time, the author again investigated the frictional forces in the absence of lateral motion of the

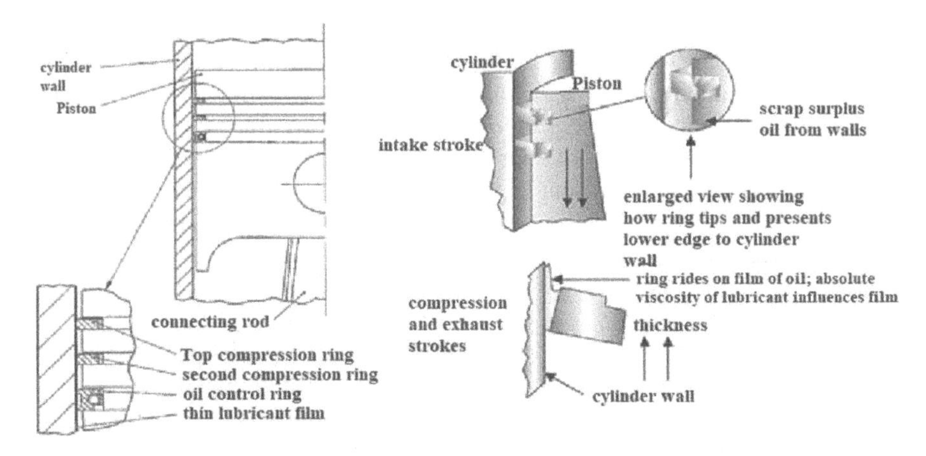

FIGURE 2.2 Function of piston ring and piston assembly. (Adapted from Ref [4] with permission.)

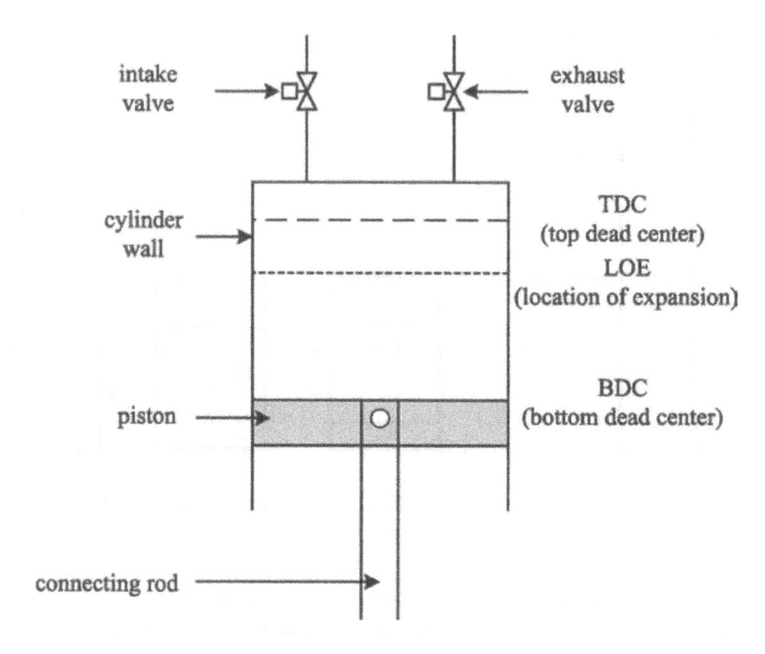

FIGURE 2.3 Diagram of piston cylinder. (Adapted from Ref [5] with permission)

piston [7] and proposed that lubrication affects the ring frictional forces during the absence of lateral motion of the piston.

Allmaier proposed that reduction of friction will occur by decreasing the surface area of the sliding crankshaft; however, this will cause an increase in knocking, seizure and wear [8]. Furthermore, the reduction in hydrodynamic losses was observed by decreasing viscosity. However, the harsh conditions of surface roughness were observed in the presence of thin oil film [9]. This phenomenon is shown in Figure 2.4.

As it is clear from previous research that the piston ring is most crucial in internal combustion engines because of variations of different conditions of lubrication, temperature, speed and load [4], this challenged researchers to investigate and improve lubrication with green atmosphere effects. Almost 208,000 million litres of oil were wasted for overcoming friction in cars worldwide in 2009 [3]. With such a huge amount of fuel wastage, a minor improvement in fuel efficiency can affect the fuel economy significantly [4]. The advantages of improvement of tribological properties in automobiles include increases in the output engine power, harmless and clean environment, longer life span of the automobiles, increases in reliability and durability with less maintenance cost and reduction in oil consumption. However still, 340 litres of fuel per annum are wasted for overcoming friction for 13,000 km distance for a passenger car as shown in Table 2.1.

Fuel consumption can additionally be decreased by using lightweight material such as aluminium. However after experimentation, it was proposed that aluminium alloys showed inferior results related to wear resistance and fatigue strength [10]. To solve problems raised by aluminium, some ceramics such as TiC [11], TiB_2 [12], and SiC [10] were used. However after some time, it was reported that these ceramics reduced wear resistance at elevated temperatures [13].

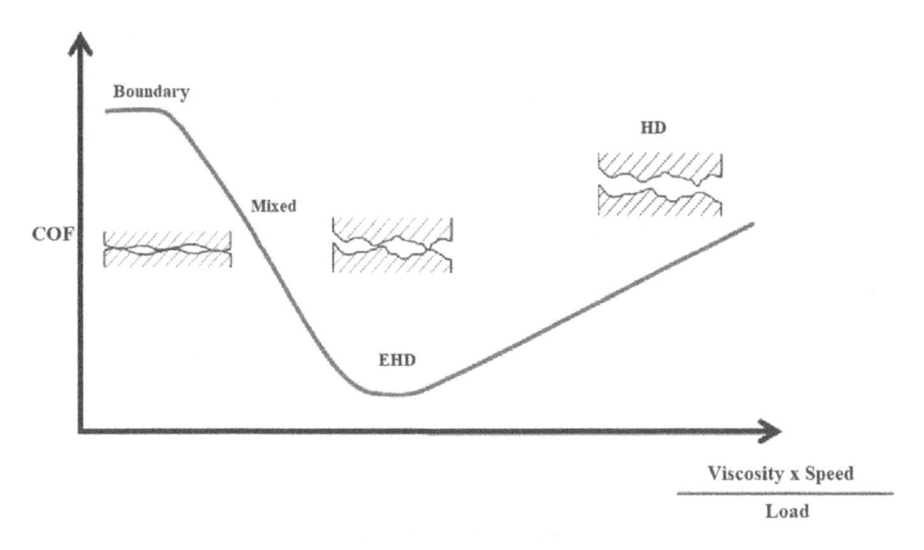

FIGURE 2.4 Different lubrication regimes. (Adapted from Ref [9] with permission.)

It is clear that the efficiency and performance of automobiles are directly affected by the friction which is produced by the sliding operations of the automobile. Wear is a prime disadvantage produced by the friction which additionally puts a limitation on the lifespan of automotive engine parts. Lubrication of moving parts is required for smooth sliding of parts. For decades, mineral oils have been in use for lubrication purposes in automobiles. However, because of their non-biodegradable nature, emission of toxic metals, such as iron nano-particles, zinc, etc., and depletion of reserves of crude oil are the main factors that demand alternatives to mineral oils [4, 14, 15].

TABLE 2.1
Friction Losses in One Global Average Passenger Car according to the Type of Tribocontact (Adapted from Ref [3] with permission)

Friction Losses Source type	Energy Consumed (MJ/car/a)	Fuel Used (l/car/a)	Percentage
Tyre road contact	1452	119	35
Hydrodynamic lubrication	1993	57	16.8
Mixed lubrication	899	26	7.6
EHD lubrication, sliding	747	21	6.3
EHD, sliding and rolling	979	28	8.2
EHD lubrication, rolling	356	10	3
Boundary lubrication	187	5	1.6
Viscous losses	771	22	6.5
Braking contact	1779	51	15
Total	11.863	340	100

Bio-based oils serve as an alternative to mineral oils due to some natural properties such as biodegradability, high lubricity, high flash point, high viscosity index (VI) and low evaporative losses [16, 17]. Bio-lubricants have chains of long fatty acids and they contain polar groups in their structure which impart the feasibility of hydrodynamic and boundary lubrication [18]. According to surveys, 350 oil-bearing crops are available worldwide; this may solve the problem of reserves of sources. Examples of bio-lubricants are jatropha [19], rice bran, neem, karanja, rapeseed, linseed, castor, mahua [20], palm [21], olive, canola, soybean, sunflower and coconut [22].

To improve the performance of bio-lubricants, some additives (nano-particles) of metals, metal oxides, chalcogenides, carbon-based additives, nitrides, ceramics and composites have been in use. The main advantage of nano-particles is that they do not react with any other particle of oil. Furthermore, high dispersion stability and high possibility of film formation on different types of surfaces are the advantages of nano-particles [23].

2.2 LUBRICANT TYPES AND CHARACTERISTICS

Any material used to reduce the wear and friction of rubbing surfaces in the field of tribology is known as a lubricant. There are different types of available lubricants, for example in oil and semi-solid form (grease). However, the most commonly used lubricants in the automotive industry are oil lubricants. Solid lubricants can be used in high-temperature applications where other lubricants spoil the system and pollute the environment. Examples of solid lubricants are TiO_2, MoS_2, graphite and WS_2 materials in nano- and micro-size [24–26]. Lubricating oils are prepared by producing special characteristics that are required under extreme pressure, load and extreme temperature and have an eco-friendly nature for the atmosphere.

With the reduction in wear and friction, lubricants additionally perform some other functions such as (i) sealing of the gap between the cylinder wall and compression ring, (ii) helping in the cleaning of the combustion engine and (iii) safe heat transfer [27]. Lubricants are prepared by a combination of different base oils and different additives. The function of additives is to impart properties to base oil which were not initially present in the base oil.

A lubricant will be considered good if it sustains conditions of elevated temperature and pressure [28–30]. Lubricant should not change its properties at different operating conditions because both elevated and normal operating conditions are present in an engine. Solidification of lubricants at low temperature, reduction of viscosity and increased oxidation of the lubricant at elevated temperature affect the wear and friction. So, a lubricant should be able to cope with different operating conditions. Additionally, a lubricant should be able to neutralise soot particles that are produced as a result of the combustion of fuel in the internal combustion engine [31].

Although there are different functions of a lubricant, the most necessary property of a lubricant is to reduce friction and wear between rubbing surfaces under different conditions. Degradation of lubricant over time is a problem related to lubricants. The quality of a lubricant is determined by analysing its degradation under operation with time. Degradation of lubricants occurs due to oxidation at elevated temperature and working of the engine in the open air. Different disadvantages of degradation of lubricants are sludge formation, thickening, corrosion and damaging of rubbing surfaces. Environmental effects and economic feasibility should additionally be considered while selecting a lubricant. The cost of lubricant is determined by measuring the volume of lubricant which has to be replaced in an engine [32].

2.2.1 Characteristics of Lubricating Oils

Different thermal, physical and chemical properties need to be considered for selecting lubricants [32]. A study by May et al. proposed fifteen important properties of lubricants in automotive engines which are: shear stability, cloud point, foaming characteristics, dynamic and kinematic viscosity, volatility, pour point, element content, flash point, density, ash, water tolerance, corrosiveness colour, homogeneity and elastomer compatibility [28].

2.2.2 Physical Properties of Lubricants

To determine the lifetime, the physical characteristics of lubricants are taken into consideration [28]. Specific gravity, density and viscosity are some physical properties that are used for the above purposes. Temperature and pressure both affect the viscosity of lubricants. Reduction in viscosity has been observed with an increase in temperature and an increase in viscosity has been observed by the increase in pressure. Variation of viscosity with change in temperature at constant atmospheric pressure is expressed by the viscosity index of a lubricant. So, the viscosity index is very important in an automotive engine [32] which depends upon shear rate [33–35].

2.2.2.1 Viscosity

Measurement of resistance during the flow of lubricant is known as the viscosity of a lubricant. The measure of opposition to shear of lubricants is known as the dynamic viscosity of a lubricant. Kinematic viscosity is determined when the flow is allowed under gravity through a capillary tube of known dimensions. The relation between kinematic viscosity and dynamic viscosity can be stated as follows [28]:

Kinematic viscosity,

$$v = \frac{\eta}{\rho} \tag{2.1}$$

Dynamic viscosity,

$$\eta = \frac{\tau}{\gamma} \tag{2.2}$$

where:

η is dynamic viscosity in the unit of Pa. s
ρ is lubricant density in the unit of kg/m^3
τ is lubricant shear stress in the unit of Pa
v is kinematic viscosity in the unit of m^2/s
γ is shear rate in the unit of s^{-1}

It is very important to note that base oils are considered to be Newtonian fluids; however, the addition of additives converts them to non-Newtonian fluids. Change in viscosity occurs even at a constant temperature because of the presence of shear rate [28].

2.2.2.1.1 Viscosity Index

The viscosity of a lubricant is significantly affected by change in temperature. To express this effect, the viscosity index is used. With the increase in temperature, decrease in viscosity occurs. Change in lubricant film thickness occurs because of a change in lubricant viscosity under mixed lubrication and hydrodynamic operating conditions [32]. With the help of the following equation, the lubricant viscosity index (VI) can be determined:

$$VI = \frac{(t-v)}{(t-s)} \times 100 \tag{2.3}$$

where:

v is the kinematic viscosity at 40°C in m^2/s
t is the value of reference oil having $VI = 0$ (from ASTM D2270 table)
s is the value of reference oil having $VI = 100$ (from ASTM D2270 table)

Lubricants of high viscosity index are better for lubrication of automotive engines because of the bearing capacity of the lubricant at elevated temperature [36].

2.2.3 Temperature Range of Lubricants

The temperature range and temperature-related characteristics are two different parameters that are considered while selecting a lubricant for a specific area of application. Decomposition of lubricant increases because of the oxidation of lubricant at high temperatures. Additionally, the solidification of lubricant occurs at low temperatures. Formation of emulsions formed by a combination of water and oil additionally creates a problem of deposits on rubbing surfaces. High temperature, which increases degradation, can be harmful to engines. Temperature-related characteristics of a lubricant include cloud point, fire point, pour point, flash point, oxidation stability, thermal stability and volatility [32].

2.2.4 Chemical Composition

Lubricant and its additives are composed of different chemical elements. Investigation of these elements is very necessary for analysing the quality of lubricating oil. Zinc, phosphorus, magnesium, barium and calcium are primary elements in lubricants. Molybdenum, potassium, sodium and boron are secondary elements present in lubricants. Some spectroscopic techniques such as EDX (Energy Dispersive X-ray analysis), XPS (X-ray Photoelectron Spectroscopy), etc. can be used to measure sulphur, boron and phosphorus. Accuracy of measurement is low in the case of phosphorus and sulphur as compared to other elements. The fitness of a lubricant for a specific purpose is determined by analysing the used elements of a lubricant. Lubricant life is determined by analysing the used concentrations of elements such as Cu, Sn, Fe and Pb.

2.2.5 Classification of Conventional Lubricant Base Oils

Mineral oils extracted from crude oil are used for base lubrication oils. Synthetic oils are those oils that are chemically prepared. Mostly paraffinic petroleum oils are used as mineral base oils because these oils have valid oxidation stability and high viscosity index. Some other required properties of base oils are volatility, carbon residue, cloud point, low temperature fluidity, oxidation stability and flashpoint. If there are several base oils which provide the required properties then the base oil with lower cost is selected for lubrication purposes. Viscosity index and viscosity are two important parameters in lubricant formulation. Hydrodynamic lubrication is determined by viscosity. Therefore, its value should be such that it is able to produce lubricant film; however, this should not be too high because a high value increases friction between sliding surfaces. The viscosity index is additionally taken into consideration because an engine operates at different temperatures [37].

In 1923, the Automotive Society of Engineers classified base oils according to their viscosity, such as heavy, medium and light. After some time, there were different methods that were used to improve the performance of base oils. Those methods are SO_2 treating, clay treating, acid treating, hydrotreating, solvent refining, hydrocracking, hydroisomerization and catalytic dewaxing. Base oil classification in Groups I–VI by the American Petroleum Institute was based on their chemistry [28, 38]. This classification is widely used in the industry of lubrication.

Groups I and II have the possession of mineral base oil. Group III base oils are known as synthetic base oils. After the refining of the solvent, we get a base oil that is free of aromatic compounds and is of better quality belonging to Group I base oils. In Group I base oils, the amount of saturation is less than 90%, the amount of sulphur is more than 0.03 wt% and there is an80–120 range viscosity index. Group II base oils are prepared by ahydroisomerization technique. The amount of saturation in these oils is greater than and equal to 90%, the amount of sulphur is up to 0.03 wt% and the range of the viscosity index is similar to Group I. Group III base oils additionally have the amount of saturation greater than and equal to 90% and the amount of sulphur is less than and equal to 0.03 wt%; however, the value of the viscosity index is greater than and equal to 120. So, these oils are known as very high viscosity index oils. Because of the presence of fully saturated hydrocarbon and less sulphur contents, Group III base oils are considered as dry and synthetic. Group IV includes polyalphaolefin (PAO) which is completely synthetic. Group VI base oils include polyinternal olefin (PIO) which is a completely synthetic base oil of its specific category. All remaining base oils which are not present in Groups I–IV and VI are referred to as Group V base oils. Bio-base oils are additionally included in Group V base oils.

2.2.6 Replacement of Mineral Oil by Bio-Lubricant

At present, in industrial and automotive sectors, mineral oil-based lubricants are being used extensively. They are deteriorating the environment in different manners such as CO_2 emissions that cause global warming and health hazards and disposal of waste mineral oil into soil and water that pollute them. So, to minimise these effects, there is a need for biodegradable lubricant oils. Because of the absence of sulphur compounds and bio-lubricants' harmless properties related to health and skin, bio-lubricants serve as a better alternative. The oxidation rate of fatty acids was predicted by Kodali in 2002 [39].

Biodegradability becomes an issue during the disposal of waste mineral lubricants. Almost 60% of the contents of mineral oil lubricants are non-biodegradable which creates the issue of environmental pollution that can be seen in Table 2.2.

However, in the case of bio-lubricants, issues related to pollution vanish. Bio lubricants additionally find their application where the loss of lubricant can occur in the surroundings.

Vegetable oils are considered as the most easily biodegradable oils which include animal contents. It is necessary to note that there is 100% biodegradability in vegetable oils because of the action of microorganisms which yield CH_4, CO_2, H_2O and mineral salts. In a biodegradability test referred to by the Organization for Economic Cooperation and Development, it has been observed that vegetable oils show 70–100% biodegradability whereas mineral oils show 15–35% biodegradability. Some recommendations related to biodegradability are [41]:

1. Primary biodegradability is the change in the chemical structure of oil by microorganisms which results in the form of change in some properties.
2. Ultimate biodegradability is the aerobic degradation that is obtained after total conversion of oil by microorganisms with the formation of H_2O, CH_4, CO_2, etc.
3. Readily biodegradable refers to almost 60% conversion of oil into CO_2 in 28 days. In conventional mineral lubricants, almost 5% of oil is readily biodegradable.
4. Inherently biodegradable refers to presence of any biodegradability in any given test of biodegradability.

TABLE 2.2
Contents of Mineral Oil Emissions (Adapted from Ref [40] with permission)

Oil Sample	Engine Oil	Coconut Oil	Palm Oil
CO_2	4.5	2.9	3.4
CO	0.92	0.67	0.73

2.2.7 Bio-Based Lubricants

Agriculture sources and biological sources are the origins of bio-lubricants. The main and foremost properties which make these base oils as an alternative to conventional mineral oil are lower toxicity to the environment, extraction from renewable resources and easy biodegradability. Bio-lubricant oils can be categorised into two different compositions. One class includes natural bio-based oil and the second class includes synthetic bio-based oil. Selection of bio-based oil is done from the following: synthetic or natural animal oils, synthetic or natural vegetable oil, vegetable oil in genetically modified (GM) form, synthetic or natural tree oil and combination of different bio-based oils [41]. Unsaturated and saturated ester base oils are classified as those bio-based oils which are easily biodegradable [42].

Some other bio-lubricants are vegetable oils in modified ester form, some triglycerides which include sunflower oil, rapeseed oil and ester oils in trans-esterified form and semi-saturated form. Some bio-oils such as palm oil, rapeseed oil and soybean oil show better antiwear properties, a high range of viscosity index, high value of flashpoints and lower friction. However, there is additionally a limit to the use of these bio-based oils because of inferior cold flow characteristics, higher oxidation and presence of hydrolytic stress. With the help of some modifications such as the addition of some additives, these limitations can be reduced [41]. Improvement in low temperature characteristics of sunflower, soybean oils, castor, and rapeseed oils have been observed by researchers [43].

It is important to note that mineral oil additives cannot fit as a bio-based oil additive. This is why bio-based oils require specially prepared additives for fruitful results. Bio-based oils were made the alternative because mineral oils harm the atmosphere, for example, use of soybean oil lubricant in the United States, use of rapeseed oil in Europe, and use of castor oil lubricant in lubrication of chain bar systems and two stroke engines [44]. To increase fuel economy, for extended life periods of the engine and better emissions, lubricants in automotive engines are subject to improvement. It has been observed that shifting towards biodegradability will reduce the lubricant characteristics related to sustainability as shown in Figure 2.5.

Bio-lubricant in an engine requires some specific properties such as the high value of viscosity index, low volatility and solubility of polar soluble contaminants for cleaning purposes [41]. Different modification techniques were used to improve oils that can be seen in Table 2.3.

2.2.8 Vegetable Oils as Bio-Lubricants

There are the following major properties that are required for vegetable oils to act as a bio-based lubricant [42]:

- There should be enough quantity for the production of bio-based lubricant.
- The extent of monosaturated fatty acid should be more than polysaturated fatty acids in bio-based lubricants.
- There should be a stable trading price for bio-based resources.

Different types of available biodegradable oils can be seen in Table 2.4.

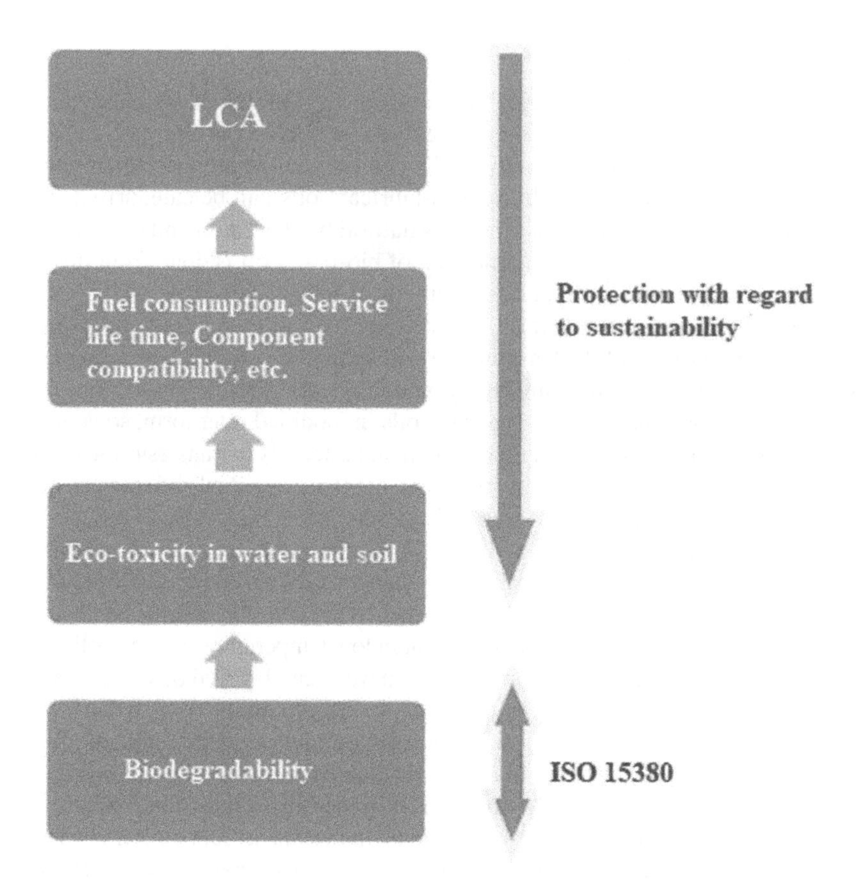

FIGURE 2.5 Factors of green tribology. (Adapted from Ref [40] with permission.)

2.2.9 Applications of Bio-Lubricant

Preference of bio-lubricants as an alternative to mineral lubricants finds its application in areas of loss lubrication, off road, military, mining, agriculture tractors, railroads, hydraulic systems, forestry systems and fishing equipment. In these areas of applications, problems of inferior cold flow behaviour and inferior thermal oxidation are solved by chemical modifications and genetic modifications. The trend of the use of bio-lubricants is encouraging because of the presence of high concentrations of oleic acid which improves wear and friction that can be seen in Figure 2.6 and Figure 2.7. This behaviour is because of the formation of dense monolayers on rubbing surfaces of metals especially for ferrous metals [46].

The area of application of a bio-lubricant is determined by the composition of the fatty acid as can be seen in Table 2.5. Rapeseed oil bio-lubricant is extensively used instead of petroleum products in the industrial field because of its physiochemical properties. Unmodified bio-lubricants cannot be used for a long time and are mostly utilised with fuel such as four stroke petrol engines. There is no wastage of lubricant in this case because of the consumption of lubricant oil with the fuel of

TABLE 2.3
Different Modification Techniques (Adapted from Ref [45] with permission)

Method	Description	Advantages	Disadvantages
Esterification/ trans-esterification	Transformation of ester into another ester through the interchange of alkoxy moiety	Improves oxidation stability and low-temperature characteristics	Requires feedstocks with high oleic acid content High reaction temperature
Estolide formation	Estolide formed either through the addition of a fatty acid to a hydroxyl-containing fat or by the condensation of fatty acid across the olefin functionality of a fat	Improves oxidative stability and low-temperature characteristics Low reaction temperature Can be used on a variety of vegetable oils	Expensive capping fatty acid often needed for the initial reaction
Selective hydrogenation	The reaction between molecular hydrogen atom with another compound or element. Transforms multiple unsaturated fatty acids into a single fatty acid	Reduces the degree of unsaturation Increases oxidative stability	Newly formed monoenoic acid can isomerise to form cis-acid and trans-acid that can alter the end product's properties significantly High reaction temperature
Epoxidation	The introduction of a new epoxide functional group (cyclic either) resulted from the double bond removal between two carbon atoms	Improves lubricity and oxidative stability Low reaction temperature	Increases pour point value Decreases viscosity index
Additional ring-opening and/or acetylation after epoxidation	Ring-opening: introduction of a hydroxyl group from the cleavage of the carbon oxygen bond	Production of final product with ideal lubricant properties	High production cost

TABLE 2.4
Different Types of Vegetable Oils (Adapted from Ref [46] with permission)

Natural Oil Type	Unsaturation Number (UN)	Room Temperature Viscosity (cP)	High Temperature Viscosity (cP)
Avocado	0.985	66.39	3.95
Canola (rapeseed)	1.287	61.79	4.12
Corn	1.381	53.13	3.77
Olive	0.948	67.09	4.26
Peanut	1.102	70.24	4.17
Safflower (high oleic)	1.010	66.27	4.08
Sesame	1.232	57.97	3.67
Vegetable (soybean)	1.451	53.77	3.85

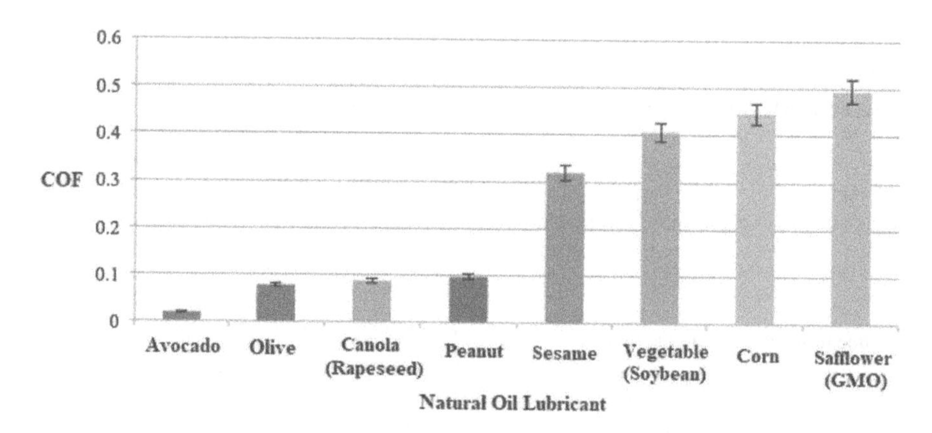

FIGURE 2.6 Frictional behaviour of different bio-lubricants at ambient conditions. (Adapted from Ref [46] with permission.)

the engine. The high affinity to metal surfaces of rapeseed oil because of the presence of the ester groups produces a protective film on tribosurfaces. For boundary lubrication, rapeseed oil as vegetable oil is very effective. This effectiveness is due to the presence of polar fatty acids which give a protective coating after reaction with metal surfaces. In the formulation of bio-based lubricants, these fatty acids are very important [41].

2.3 LUBRICANT ADDITIVES FOR AUTOMOTIVE ENGINES

The chemicals that are used to enhance some characteristics of engine lubricant are called additives. To increase the durability and performance of a lubricant, additives are added in a very minute amount as compared to the amount of base oil lubricant. The primary function of these lubricant additives is to enhance the tribological behaviour of rubbing surfaces. This is accompanied by the absorption of these additives into rubbing surfaces. Improvement in the extreme pressure properties of the engine is additionally observed with the addition of additives. Inhibition of

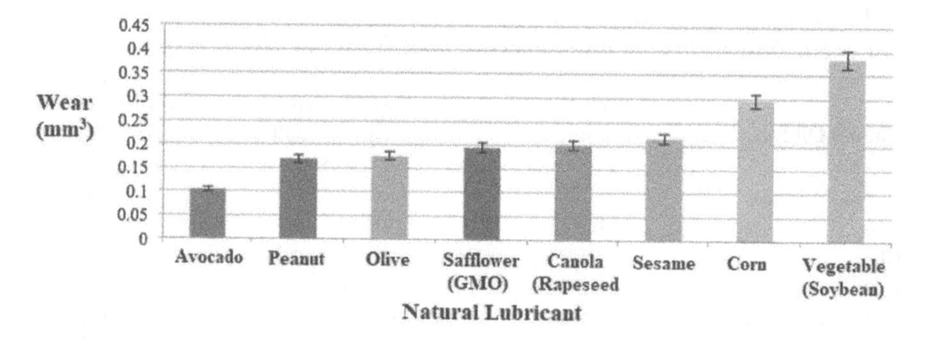

FIGURE 2.7 Wear behaviour of different bio-lubricants at ambient conditions. (Adapted from Ref [46] with permission.)

TABLE 2.5
Applications of Different Bio-Lubricants (Adapted from Ref [47] with permission)

Application	Properties	Tested Lubricants	Remarks
Applicable to all applications	High biodegradability Renewable	Vegetable oils/natural ester of various feedstocks	Environmentally benign Suitable for a variety of applications
	Good lubricity		Suitable for high-temperature applications
	Wide viscosity range High viscosity index Low volatility High thermal stability Good oxidation stability		• Suitable for a variety of applications • Only showed by derivatives of vegetable oils (chemical modification)
Engine oil	Low volatile organic compound emission	Castor oil, palm oil	Reduce engine emission, i.e. CO, HC
	Good lubricity	Karanja oil, rapeseed oil	Improves engine performance, i.e. BSEC, EGT, BTE and brake power
Hydraulic oil	Low compressibility	HOSO	Better pressure transmission
	Fast air release rate	Rapeseed oil	Less vibration and noise
Compressor oil	High thermal stability	Epoxidised soybean oil	Can tolerate the standard compressor discharge temperature (≈250°C)
Chainsaw oil	low volatility	Vegetable oil	Fewer harmful mist generation
MWF	Low volatility	Soybean oil emulsions	Fewer harmful mist generation
	Good anti-rust capability		Longer tool life
	Good falsifiability	Soybean, karanja, neem, rice bran	Stable emulsion at high temperature
	Good lubricity	Coconut oil, palm oil	• Reduce cutting temperature, surface roughness and tools flank wear • Longer tool life
Gear oil	Good lubricity Higher weld load	Complex esters PE tetraoleate ester	With the addition of suitable additives
Insulation fluid	High water solubility level	Vegetable oil	Decrease the effect of humidity on insulation strength
	High dielectrics constant	Corn, cottonseed, rapeseed, soybean	Better insulation properties

foaming, decrease in pour point, contamination prevention, corrosion prevention and reduction in oxidation were additionally observed by the addition of these additives. Improvement in the reduction of viscosity with increase in temperature is observed with some additives which are known as viscosity index improvers [32].

Categorization of lubricant additives is done into three different groups, that is, additives for surface protection, additives for oil protection and performance additives.

Additives for surface protection are most important in tribology. Extreme pressure additives, antiwear additives and friction modifiers are well-known surface protection additives. Antifoam additives, seal swell agents, pour point depressants and viscosity modifiers are some performance additives. Antioxidant additives are used for reduced base oil degradation, and these additives are known as oil protection additives. For proper lubricant formulation and for getting improved handling, the base oil is used for the addition of additive [28].

2.3.1 Types and Mechanisms of Additives

2.3.1.1 Surface Protection Additives

The additives which are used for improvement in friction and wear of contact surfaces are known as surface protection additives, and these additives are of extreme importance in tribology [32]. Corrosion inhibitors, rust inhibitors, detergents, dispersants, antiwear agents, friction modifiers and extreme pressure additives are additives that are included in this group. Their work is accomplished by a combination of a polar head group and a hydrocarbon tail. Some of the elements of S, P, O and N are present in polar head groups. A hydrocarbon tail is present in soluble form in the oil phase. Oxidised oil sludge, soot particles and some other deposits get charged and attract polar head groups. Polar head groups are additionally attracted by metal surfaces. Extreme pressure additives, friction modifiers and antiwear agents are formed by the different combinations of hydrocarbon tails with different polar head groups [48].

2.3.1.2 Friction Modifiers

The same lubricant is used for the lubrication of many engine components. Different tribo parts of the engine should be given satisfactory lubrication performance which operates in different lubrication regimes. There are two forces in a lubrication system; one is relative force between rubbing surfaces and the second is drag force in lubricant layers. Viscous drag forces can be reduced by using low viscosity lubricants but there is additionally a limit for reduction of viscosity of a lubricant. Reduction of viscosity beyond its limits causes problems in engines. Contacts that are under high load are piston skirt/cylinder liner, cam/follower and piston ring/cylinder liner which operate under boundary and mixed lubrication regimes. These are the places where friction modifiers are used to reduce friction [48].

Friction modifiers are classified into two groups according to their working mechanism. Their names are organo molybdenum friction modifiers and organic friction modifiers. Organic friction modifiers belong to the straight chain structure of hydrocarbons and are composed of a length of eighteen carbon atoms approximately. Polar head groups in them can be the following: Phosphonic acid, amide, amine, ester and carboxylic acid. Attachment of polar head groups with metal surfaces is common in all surface protection additives and this property is raised by the absorption mechanism [32]. A strong film is formed by adsorbed multilayers of polar head groups on rubbing surfaces. The reduction in the coefficient of friction is due to the presence of this tribo film. The organic modifiers do their functions immediately when attached to metal surfaces [48].

However, the working of the organo molybdenum friction modifier is different. In this mechanism, the frictional energy of tribo contacts converts Mo-containing additives to MoS_2. The formation of MoS_2 is feasible because it has a lamellar structure. These layers slide over each other, and friction is reduced. High temperature and high load are required in the case of Mo-type friction modifiers because of the requirement of a chemical reaction. Although these friction modifiers can be used for reduction in friction, there is a limit of extreme pressure beyond which their use becomes non-feasible for the reduction in friction [48, 49].

2.3.1.3 Antiwear Additives

Antiwear (AW) additives reduce friction between layers by rolling and sliding that can be seen in Figure 2.8.

The most frequently occurring types of wear are abrasive wear [51–53] and adhesive wear [54–56]. When the temperature exceeds the feasible limit of friction modifiers then antiwear additives are required for tribological performance. Antiwear additives are present in different types; however, the most commonly used antiwear additive for engine lubrication is zinc dialkyl dithiophosphate (ZDDP). Phosphate esters and tricresyl phosphate are other antiwear additives which are used for lubrication purpose in gas turbines [32]. Some other compounds such as dispersants and detergents having boron and borate esters additionally exhibit antiwear properties and show similarity to ZDDP in engine lubrication [48]. The concentration of antiwear additives is in the range of 1–3 wt.% in base oil [32]. Chemisorption reaction occurs at metal surfaces by antiwear additives at tribo contacts having high contact pressure. A polyphosphates protective layer forms on the surface because of the result of the chemical reaction. Destruction of this tribofilm is affected by the contact temperature which is responsible for initiation, traveling and destruction of tribofilm. Breakdown of tribofilm occurs when the contact temperature crosses the feasible limit of 250°C. This contact temperature is the sum of both frictional heating and lubricant heating [48].

The most commonly used AW engine lubricant is ZDDP. History, chemical structure, tribofilm formation and development are explained by Spikes [57]. Working on

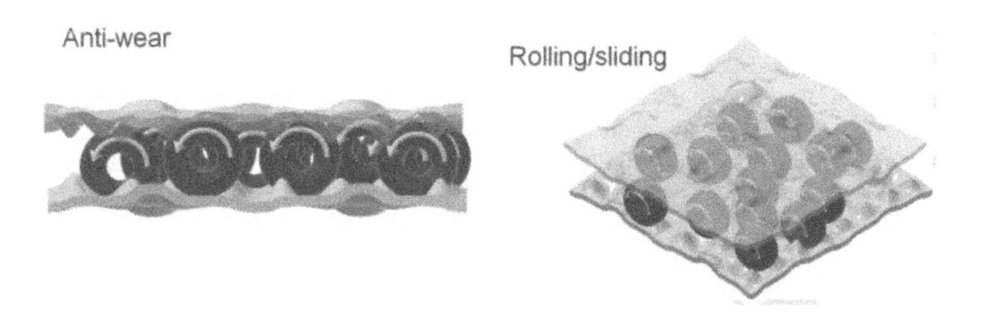

FIGURE 2.8 Antiwear mechanism. (Adapted from Ref [50] with permission.)

AW additives was proposed by Mackney [48]. Categorization of the whole process is done in four steps as explained by the following:

1. Cleaning of surface: removal of metallic oxides from contact surfaces at the initial stage of sliding and exposure of the metal surface to lubricant.
2. AW additives chemisorption into contact surfaces.
3. Formation of the protective layer by the chemical reaction between surface and additives in the lubricant.
4. Expansion of this protective layer to form a tribofilm.

ZDDP acts as an AW additive in automotive engine lubricants at most crucial parts such as valve train and the contact surface of piston ring cylinder liner [58]. ZDDP is very well known for the reduction of valve train wear problems and the degradation of oil in engines. However it additionally acts as a source of phosphorus and sulphur which create environmental pollution. So, to minimise this pollution effect, alternatives to ZDDP are required for antiwear lubricant additives [32].

AW additives affect the tribological behaviour of metal surfaces under lubrication. Mechanical mixing of antiwear additives to metal surfaces additionally affects transient lubricating action. ZDDP, which is a very well-known AW additive, has given two different forms of wear protection under different tribo conditions which were explained by Dienwiebel and Pohlmann [59]. Firstly, the formation of the AW film that caused a reduction in wear rate for a very short period in the presence of high stress; secondly, improvement in chemical and mechanical properties of rubbing surfaces when the stress is low or moderate. This happens because of the adsorption of lubricant on the surface and because of modification in shear stress, as a result of which friction, as well as wear, was reduced for a feasible long term of operation.

2.3.1.4 Extreme Pressure Additives

Extreme pressure additives are additionally known as surface active additives which act similarly as AW additives. The advantage of using extreme pressure additives is that they don't seize under extreme conditions of temperature and pressure or both. In the presence of shock loading or heavy load, extreme pressure additives give fruitful results as compared to AW additives along with extended lifetime in automotive engines [48].

Such extreme lubrication conditions are present in heavily loaded gears. For boundary lubrication in the presence of extreme pressure additives, one of the following elements must be present in the lubricant: Chlorine, phosphorus, iodine, sulphur and antimony. The most commonly used element is sulphur in extreme pressure additives lubrication. It is important to note that antiwear additives help in the prevention of wear and extreme pressure additives are used to control wear [48]. The mechanism of extreme pressure is depicted in Figure 2.9. Penetration of iron sulphides product at the asperities location was observed. Then because of relative motion, these sulphides are sheared easily and the result of this is the removal of iron salt from rubbing surfaces.

Because of the removal of these metallic sulphides, localised welding extent decreases which makes the surface smoother. Additive concentration additionally affects the behaviour of lubrication under extreme pressure additives. In the presence of a high concentration of additives, corrosive wear will occur. However, in a low concentration of additives, there will be no protection of contact surfaces [32, 48, 50].

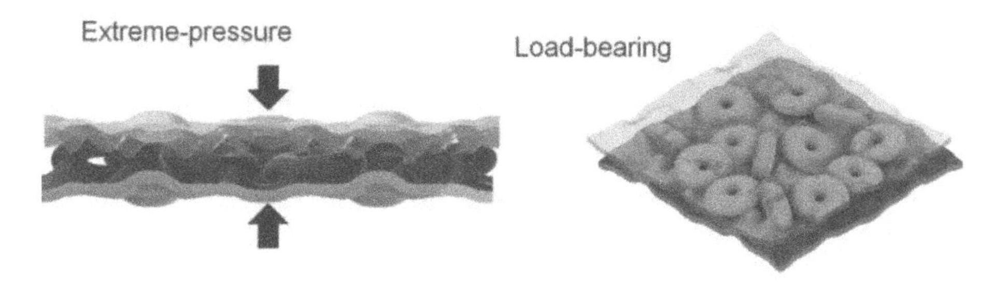

FIGURE 2.9 Extreme pressure additive mechanism. (Adapted from ref [50] with permission.)

2.3.2 Additives Summary

Different micro-/nano-additives were used by different researchers that can be shown in Table 2.6.

TABLE 2.6
Additives Summary

Nano-CuO and nano-graphite particles	[60]
CuO nano-particles	[61]
Micro WS_2, WS_2, WS_2/TiO_2 nano-particles	[62]
CuO and MoS_2 nano-particles	[63]
Nano- and micro-scale TiO_2	[64]
h-BN	[65]
Zeolite nano-crystals	[66]
SiO_2 nano-particles	[67]
CuO nano-particles	[68]
Ceria-Zirconia hybrid (Ce-Zr) nano-particle	[69]
MoS_2	[70]
TiO_2 + graphene	[71]
Nickel nano-particles	[72]
Nano-TiO_2/SiO_2 and TiO_2	[73]
Graphene	[74]
TiO_2, WS_2, and CuO	[75]
Nano-MoS_2 and WS_2	[76]
Graphene	[77]
Copper oxide and ZDDP nano-particles	[78]
Nano-TiO_2, WS_2 and CuO	[79]
hBN, WS_2, MoS_2 carbon nano-tubes and graphite	[80]
CuO, WS_2, and TiO_2	[81]
Graphene nano-platelets (GNPs)	[82]
Micron-sized hexagonal boron nitride	[83]
Trihexyltetradecylphosphonium bis(2-Ethylhexyl) phosphate	[84]
Graphene	[85]
Nano-tubes of WS_2	[86]
WS_2 and MoS_2 nano-particles	[87]
TiO_2 nano-particles	[88]

2.4 MICRO/NANO ADDITIVES

2.4.1 Copper Oxide

Researchers have used CuO nano-particles as additives in palm oil bio-lubricant and analysed extreme pressure and AW performances. A four-ball tribotester was used for this purpose. Wear analysis was conducted after experimentation with the help of a microscope of high resolution. After the analysis of results, it has been observed that these additives have given a 12% increase in performance. This experimentation showed that the CuO nano-particle is a lubricant additive with better extreme pressure and AW performance [60].

Another author proposed that because of the properties of low toxicity, renewability and high biodegradability of vegetable oils, vegetable oil can be considered as excellent bio-lubricating base oils. When coconut oil was used as a lubricant, it showed better results in the form of low coefficient of friction; however, it showed a high wear rate on contact surfaces as compared to mineral lubricant oils, while mustard oils gave good results in the form of lower wear. Therefore, Sajeeb and Rajendrakumar [61] used blends of these two oils with CuO nano-particles as additives to analyse tribological performance. After experimentation, improvement in wear and coefficient of friction was observed. An additive concentration of 0.2% was considered as an optimal concentration for better tribological performance [61].

Gulzar et al. [63] proposed that AW and extreme pressure properties were improved by using CuO as an additive in the following concentration %wt of nano-particles in (0.25%, 0.5%, 0.75%, 1%, 1.25% and 1.5%). Different blends were produced and to make certain of homogeneous dispersion of additives in palm oil, a magnetic stirrer or ultrasonic sonicator was used for two hours. The load was additionally varied during tribotesting and the load was kept in the range of 40–180 kg. For tribotesting, sonicator temperature condition was 30°C. Engine speed was held constant at 1.1 m/s and results were analysed in different concentrations of blends. A four-ball tribotester was used for experimentation. Steel of AISI 52100 grade was used as a material for the ball. Experimentation was conducted at the hardness in the range of 64 to 66 HRC. Toluene was used to clean the samples before any experimentation. After experimentation, scanning electron microscopy was used to analyse surface topography and X-ray diffraction was used to check and analyse the homogeneity of additives in palm oil [63].

In other research, it was proposed that the mechanical system is improved for the wear and friction aspect by the addition of nano-/micro-particles in the bio-lubricants. In this research bio-diesel and diesel fuel blends were used in the presence of copper oxide (CuO) nano-particles. For the stability of nano-particles, a surfactant that creates a cloak on the outer layer of the nano-particle was used. For uniform dispersion of nano-particles in lubricant and blends, an ultrasonic emulsifier of 400W power at 24 kHz frequency was utilised for 30 min under the 0.7s duty cycle. Bio-diesel is obtained from cooking oil with the help of a transesterification method under a methanol and potassium hydroxide catalyst. A four-ball tribometer is used for tribotesting. Nano-particle concentrations of 0, 25, 50 and 75 ppm size were investigated in the presence of three different blends at steady state conditions. Tribotesting was conducted at different rotational speeds of 600, 1200, and 1500 rev/min. Reduction of the friction coefficient up to 50 ppm concentration was observed because the wear cavity was filled by nano-particles. Moreover, a decrease in friction coefficient was because of the sphere shape of nano-particles which convert the case of sliding friction to rolling friction between rubbing surfaces. An increase in the quantity of nano-particles beyond this limit increased friction coefficient because of agglomeration which increases friction as a debris particle. It has been observed that with the addition of 75 ppm particles in fuel blends of B10 and B20, the friction coefficient was increased to a significant level. The main reason behind that is the presence of the oxygen content of copper oxide nano-particles which causes degradation and oxidation of the lubricant. However, the best results were observed with B50 fuel blends in the presence of 75 ppm particles as compared to all other situations. With the increase in bio-diesel concentration, a decrease in the friction coefficient was observed because of the presence of monoglycerides, diglycerides and free fatty acids which are essential components of biodiesel [68].

Another researcher analysed that the viscosity of lubricant oil increases because of the incorporation of additives such as WS_2, TiO_2, etc. in the form of micro- and nano-particles, and as a result of this coefficient of friction, wear reduces and load carrying capacity increases. With the help of the Krieger–Dougherty viscosity model, additive viscosities at different volume fractions are maintained. In this paper (Shinde and Soni [75]), the journal bearing served as the testing material for analysing the static performance of a synthetic lubricant with the presence of CuO additive in different volume fractions. Additionally, a comparison was made regarding the performance characteristics in the presence of additives and without the presence of the additive. The analysis was conducted to check CuO additive pressure distribution with 5% volume concentration in the presence of SAE30 base oil. It was observed that with the increment in the aggregate radius of nano-particles, pressure film increased which increased friction. It was observed that the plane pressure in the hydrodynamic journal bearing in the presence of 5% additives concentration was optimum and gave better results.

Nano-particles blended lubricants showed an increase in maximum pressure. It has been observed that in the presence of 1%, 2%, 3%, 4% and 5% CuO nanoparticles, maximum pressure does not vary significantly. The reason behind that is the aggregate radius of CuO nanoparticles is less than the effective radius [75].

In another study, SAE20W40 synthetic lubricant was chosen to make a comparison with modified rapeseed oil as lubricating oil. CuO was used for improving tribological properties. With the help of a journal bearing test rig, experimentation was conducted to check and analyse the effects of bio-lubricants and synthetic lubricants. Testing conditions were the following: 10kn load, speed of 3000 rpm. Testing was conducted by varying load in the range of 10 to 10,000 N. To apply the load at the radial direction to the system, a pneumatic servo motor was used. Experimentation was conducted in two phases. In the first phase, three different nano-lubricants were being investigated. The nano-lubricant CMRO with nano-CuO synthetic lubricant (SAE20W40) and CMRO were used in the second phase of experimentation and then the coefficient of friction and reduction in wear was investigated. It was observed that there was outstanding reduction in the coefficient of friction and wear scar diameter by the addition of CuO nano-particles. Other bio-lubricants and synthetic lubricants without CuO did not provide much better results as those lubricants with CuO particles. This statement can be proved with the experimental evidence that wear was reduced by about 47% and the coefficient of friction was reduced by about 27% as compared to commercial lubricant. A scanning electron microscope and atomic force microscope were used to analyse the surface morphology of bearing surfaces. After experimentation, it can be concluded that CMRO with nano-CuO served as the best alternator of commercial lubricating oils which can reduce the need for synthetic oils. Additionally, it is environmentally friendly [79].

It was proposed that RSM which is a response surface methodology was used by using D-optimal design to investigate tribological effects of bronze, copper and brass material journal bearings. RSM accuracy was found to be 95% and therefore RSM was used for a 95% confidence level. Tribotesting was conducted in the presence of rapeseed oil in modified form CMRO. CuO was the additive that was used for addition in bio-lubricants to improve their tribological performance. Variable factors for investigations of the coefficient of friction and wear scar during tribotesting were the following: load, sliding speed, nano-based different particles and bearing materials.

After experimentation, it was concluded that the lowest wear rate and coefficient of friction were in the case of CuO nano-particles in the presence of chemically modified rapeseed oil at 100N load and 2.0 m/s sliding speed. SEM showed that surface morphology was better in the case of bronze material in the presence of CuO blended chemically modified rapeseed oil as compared to brass/copper material for bearing steel [81]

2.4.2 Tungsten Disulphide

Lu et al. reduced the coefficient of friction by using WS_2 and WS_2/TiO_2 nano-particles in diisooctyl sebacate bio-base oil. WS_2/TiO_2 nano-particles gave better results as compared to WS_2 nano-particles. Eighty-one per cent reduction in wear volume was observed by 0.75% of WS_2/TiO_2 nano-particles [62].

In other research, it was proposed that the viscosity of lubricant oil increases because of the incorporation of the additive WS_2 in the form of micro- and nano-particles, and as a result of this coefficient of friction and wear reduces whereas load carrying capacity increases. With the help of the Krieger–Dougherty viscosity model, additive viscosities at different volume fractions are maintained. In the paper (Shinde and Soni [75]), the journal bearing served as testing material for analysing the static performance of a synthetic lubricant with the presence of WS_2 additive in different volume fractions into SAE30 base oil. It was observed that with the increase in the aggregate radius of nano-particles, pressure film increases. Results showed that in the presence of 1%, 2%, 3%, 4% and 5% WS_2 nano-particles volume concentration, results were better as compared to CuO and TiO_2 nano-particles [75].

In another study, the effect of different dispersion techniques and effects of different surfactants was analysed with wear and friction behaviour in the presence of WS_2 nano-sized additive in the presence of polyalphaolefin base oil as a lubricant. The material chosen for tribotesting was 8620 steel. This material finds applications in agricultural machinery and energy generation. Surfactants used in this study were polyvinylpyrrolidone (PVP) and oleic acid (OA). With the help of literature, it can be proved that polyvinylpyrrolidone (PVP) and oleic acid (OA) are frequently used in research for the stabilization of nano-particle additives. Previously, the author showed that OA was very important for the reduction in agglomeration in the presence of CuO and WC nano-particle additives but was, however, less effective in the case of WS_2 particles. From the study of the literature review, it was decided to make a 1% concentration of additives. Different samples of different additives were prepared. Samples without surfactants were subjected to sonication for 1-hour duration. For making samples in the presence of surfactants, samples were sonicated for 2 hours duration. To prepare samples of PVP, the first PVP was incorporated in distilled water; then the solution was subjected to sonication for a half hour duration and then this sample was dried in an oven at 60°C for 120-min duration, and then nano-particles were incorporated into oil and sonified for the 1 hour duration for uniform distribution. Different dispersion techniques were used for uniform dispersion as can be observed: (1) In the absence of a stabilizing agent, sonification for 1 hour; (2) in the presence of 1% weight OA surfactant, sonification for 1 hour. After the analysis of results, it was observed that the homogeneity of additives and agglomeration decrement did not improve results significantly. For the aspect of wear, it was observed that WS_2 gave a 45% reduction in wear depth in the presence of PVP surfactant as compared to the simple base oil. With the aspect of wear, results proved that reduction in agglomeration is an important factor that promotes high dispersion for better wear resistance. In addition to this, SEM additionally proved and gave a reason for the reduction mechanism in the form of a tribofilm layer in the case of a PVP surfactant for WS_2 nano-particles. Finally, we can say that wear resistance was improved for our experimental objective with present conditions under boundary lubrication conditions [76].

Basker et al. chose SAE20W40 synthetic lubricant to make a comparison with modified rapeseed oil as lubricant. A WS_2 nano-particle was used for improving tribological properties. With the help of a journal bearing test rig, experimentation was conducted to check and analyse the effects of bio-lubricants and synthetic lubricants [79].

Due to commercial usage of commercial lubricants, the atmosphere is deteriorating. For this reason, researchers are trying to find alternates to replace synthetic lubricants to save the environment. Vista et al. [80] used different types of additives (such as nitride, graphite, carbon nano-tubes, molybdenum disulphide and tungsten disulphide) of different shapes and sizes to improve the properties of naturally existing lubricants. For this purpose, an ionic liquid and a natural oil served as a base oil for lubrication. After experimentation, it was observed that different additives gave different types of responses in tribological performance. And tribological performance in the case of ionic liquids depends only on particle size and is independent of particle types. The colloidal mixture was prepared by using phosphonium-based ionic liquid and avocado oil as a base oil in the presence of the additive. Approximately 5% of weight was maintained by solid particles of different sizes, shapes and blends. Tribotesting was conducted by using the pin on a disc apparatus. 440C stainless steel grade was used for the pin and 2024 aluminium discs were used for plate material.

Operating conditions were the following: a normal load of 10N, the sliding velocity of 36 mm/s and testing of 13 hours.

After experimentation, the most effective parameter was particle type. However, in ionic liquids, the most effective parameter was particle size, and particle type had a negligible effect on performance. Smaller particles have better results. Therefore, nano-particles have a better effect as compared to micro-sized particles because larger sized particles increase wear scar, surface roughness and coefficient of friction [80].

It was proposed that RSM, which is a response surface methodology, was used by using D-optimal design to investigate tribological effects of bronze, copper and brass material journal bearings. RSM accuracy was found to be 95% and therefore RSM was used for a 95% confidence level. Tribotesting was conducted in the presence of rapeseed oil in modified form CMRO. WS_2 were the additives that were used for addition in bio-lubricants to improve their tribological performance [81].

2.4.3 Graphene

It was observed that previous research used nano-particles in lubricants to reduce the coefficient of friction and to reduce wear; however, now there is a scope of mixing of nano-particles in lubricant to further improve tribological effects. So, this research paper accompanied the preparation of nano-particles that are different and then by experimentation, analysis of their results. In this paper, tribological effects in the presence of nano-graphene were investigated. Tribological effects in nano-particle size of TiO_2 (A) + graphene and graphene were investigated in the presence of a pure base oil of group PBO-GII as a lubricant. A transmission electron microscopy (TEM) was used to analyse the morphology of the above mixed structure of nano-particles. Oleic acid (OA) was used as a surfactant to improve the uniform dispersion of mixed nano-particles into PBO-GII oil. Experimentation analysis was conducted on six different samples. Five out of six samples were without base oil. After experimentation, wear scar diameter was investigated by the image acquisition system. Wear morphology was analysed by scanning electron microscopy, Raman spectroscopy, elements mapping and X-ray spectroscopy were utilised to confirm the presence of a tribofilm of nano-particle mixture between mating parts. To visualise scar topography, a 3D optical surface analyser was used. Results showed that the combination of different nano-sized particles is a newly emerging field for research for analysis of lubricant durability and lubricating capacity. Results showed that the best combination in the above study is 0.4 wt% TiO_2 (A) + 0.2 wt% graphene which showed the best performance. Further, it has been observed that after an hour the average reductions in wear rate (WR), coefficient of friction (COF) and wear scar diameter (WSD) were 15.78%, 36.78% and 38.83% respectively for the above best combination [71].

In tribological applications, graphene is being used nowadays to reduce the coefficient of friction and wear rate. It is additionally being used in the form of a lubricant additive and as a filler in the coating field. After testing, it was observed to perform equally well for both humid and dry environments which makes graphene a unique solid additive because other solid additives do not possess this property. Because of less exposure to all the properties, there are many investigations that have to be done on graphene, which makes it very important in the research field. In this paper, graphene in micro-particles and nano-particles was being investigated for the reduction of wear and coefficient of friction. Graphene is in the form of an ultrathin multilayer. Because of low surface energy and smooth material, graphene can reduce the friction coefficient of many tribological systems. In mechanical, electronic and in hybrid systems, graphene is the most important attractive material which can reduce wear rate and coefficient of friction to a greater extent. Although graphene forms an ultimate thin layer, it is most suitable for high temperature, humid as well as dry conditions for running an engine [74].

1. In other research, SAE20W40 mineral oil was used as a base stock for modified jojoba oil as a lubricant for tribological analysis. In this process, jojoba oil was modified through two stages. Base esterification and acid esterification were the processes through which oil was modified. SAE20W40 mineral oil served as the baseline lubricating engine oil. Then blends of jojoba oil with mineral oil were prepared in 0, 10, 20, 30, 40 and 50% by volume. Friction and wear were objectives of interest for investigation. A four-ball tribometer was used for this purpose. Graphene in 0.075 wt% was added in the above lubricant as a nano-particle and results were again analysed by using a four-ball tribometer. The investigation was conducted to improve the coefficient of friction and wear scar in the three following different conditions: on the baseline lubricant,
2. bio-lubricant,
3. bio-lubricant in the presence of graphene nano-additives.

SEM scanning electron microscopy was used for surface analysis. After experimentation, it was observed that blends of 20% jojoba oil gave the best result in the form of low scar diameter as compared to all the above different compositions in the presence of extreme conditions. Lubricity was additionally improved after the addition of nano-graphene particles. Lesser wear scar diameter was observed in the presence of 10% and 20% volume blends of jojoba oil. However, 20% of jojoba oil was announced as the best composition to improve tribological properties as compared to other blends [77].

Another author worked with a blends formation of trimethylolpropane (TMP) with polyalphaolefin in the presence of nano-platelets of graphene (GNPs) as additives. Different lubrication samples were prepared by adding different concentrations of graphene nano-platelets in blends of 5% vol TMP ester and 95 vol% polyalphaolefin. A four-ball tribotester was used for the analysis of the effects of the above combination on sliding surfaces. X-ray energy dispersive technique and scanning electron microscopy were used to analyse sliding surfaces after tribotesting. AISI 52100 was the grade of steel balls having a 12.7 mm diameter with 64-66 HRC hardness. Friction torque data gave the values about the coefficient of friction.

After experimentation, it was observed that lower wear scar diameter and coefficient of friction were observed when we add 0.05 wt% graphene nanoplatelets in the above blended mixture. This concentration is assumed to be the most optimum concentration of GNP. Wear was reduced by 15% and friction was reduced by 5% respectively in the presence of 0.05wt% concentration of graphene nano-platelets [82].

2.4.4 Titanium Dioxide

Some researchers tried to reduce the coefficient of friction by using WS_2/TiO_2 nano-particles in bio-base oil. WS_2/TiO_2 nano-particles gave better results as compared to WS_2 nano-particles. Almost 81% reduction in wear volume was observed by 0.75% of WS_2/TiO_2 nano-particles [62].

Arumugam and Sriram tried to reduce the coefficient of friction and wear rate by using chemically modified rapeseed oil as a biodegradable oil in the presence and absence of TiO_2 particles in micro- and nano-size. A pin on a disc tribotester was used for experimentation. After the analysis of the result, it was proved that rapeseed oil with nano-particles gave a better tribological performance as compared to micro-sized particles or in the absence of additive particles [64].

Another research group explained tribological effects in the presence of nano-sized TiO_2. Tribological effects in nano-particle size of TiO_2 (A) + graphene, graphene and titanium dioxide anatase TiO_2 (A) were investigated in the presence of a pure base oil of group PBO-GII as a lubricant. Transmission electron microscopy (TEM) was used to analyse the morphology of the above mixed structure of nano-particles. Oleic acid (OA) was used as a surfactant to improve the uniform dispersion of mixed nano-particles into PBO-GII oil. Experimentation analysis was conducted on six different samples. Five out of six samples were without base oil. Results showed that the best combination in the above study is 0.4 wt% TiO_2 (A) + 0.2 wt% graphene. Furthermore, it has been observed that after an hour the average reductions in wear rate (WR), coefficient of friction (COF) and wear scar diameter (WSD) were 15.78%, 36.78% and 38.83% respectively for the above best combination [71].

Nano-TiO_2 and nano-TiO_2/SiO_2 additivated trimethylolpropane TMP ester have been tested by researchers. First of all, suspensions were analysed to check dispersion stability, and then nano-lubricants were analysed to check the absorbance level of nano-particles. Then nano-TiO_2 and nano-TiO_2/SiO_2 were blended into lubricant oil to prepare samples for tribotesting. Then a four-ball tribometer was used to analyse load carrying ability. Later, the above composite suspension was utilised in a piston ring–cylinder liner pair for tribotesting. A reciprocating test rig was utilised for this purpose. After tribotesting, characterization was performed by using EDX spectroscopy, atomic force microscopy and scanning electron microscopy. Extreme load capacity was determined by using a four-ball tribometer for tribotesting. The material of the ball was AISI 52100 steel. Approximately 12.7 mm diameter balls of 64-66 HRC hardness were used for tribotesting. Toluene was used for cleaning purposes of pots and balls to minimise the effect of error during experimentation. ASTM D2783 standards were used to select a procedure for experimentation. Initially, an approximately 40 kg load was implemented and then the magnitude of the applying load was increased until the welding of the balls. Each test was performed three times to assure the accuracy of experimentation. Improved tribological characteristics were observed with 0.75 wt% nano-TiO_2/SiO_2 in TMP oil against the ring cylinder liner rubbing mechanism. Improved surface topography was confirmed with the help of AFM, SEM and EDX [73].

In another study, it was observed that the journal bearing served as testing material for analysing the static performance of a synthetic lubricant with the presence of TiO_2 additive in different volume fractions. Additionally, a comparison was made about the performance characteristics in the presence of additives and without the presence of additives [75].

Basker et al. [81] used response surface methodology to investigate tribological effects of bronze, copper and brass material journal bearings in TiO_2 additivated modified rapeseed oil. RSM accuracy was found to be 95%, and therefore RSM was used for a 95% confidence level. Tribotesting was conducted in the presence of rapeseed oil in modified form as CMRO. TiO_2 was the additive that was used for addition in bio-lubricants to improve their tribological performance [81].

2.4.5 Molybdenum Disulfide

Gulzar et al. [63] proposed that AW and extreme pressure properties were improved by using MoS_2 nano-particles as an additive. Different blends were produced and to make certain the homogeneous dispersion of additives in palm oil, a magnetic stirrer or ultrasonic sonicator was used for two hours. The load was additionally varied during tribotesting and load was kept in the range of 40 to 180 kg. For tribotesting and the sonicator, the temperature condition was 30°C. Engine speed was held constant at 1.1 m/s, and results were analysed in different concentrations of blends. A four-ball tribotester was used for experimentation. The steel of AISI 52100 grade was used as a material for the ball. Experimentation was conducted at the hardness in the range of 64 to 66 HRC. Toluene was used to clean the samples before any experimentation. After experimentation scanning electron microscopy was used to analyse surface topography, and X-ray diffraction was used to check and analyse the homogeneity of additives in palm oil [63].

2.4.6 Hexagonal Boron Nitride

It was observed that the use of pure lubricant between mating parts is not efficient because of the degradation of viscosity after continuous usage of lubricant. Then the idea of introducing some additives in lubricant was introduced. Then different types of additives were used in different shapes, sizes and other properties. After experimentation, it was concluded that they reduced the coefficient of friction and wear. Easy sliding, layered structure, homogeneous thin film and good adhesion were observed between mating parts. Katiyar et al. [65] used multi-wall carbon nano-tubes in industrial-based mineral oil as an additive. His efforts resulted in the forming of a 20% increase in load-bearing capacity and a 70–76% reduction in wear as compared to pure lubricants. By experimentation, they concluded that WS_2 is less efficient as compared to MoS_2 in extreme conditions because MoS_2 gives better results in the form of thermal stability. Different blends of h-BN (0.1, 0.5, 1.0, 2.0, 2.5 and 3.0% wt) blended in pure coconut oil were used to check tribological effects. A probe sonicator was used for 30 minutes for assurance of homogeneity of additives in the lubricant. A four-ball tribometer was used for experimentation after blending. It was observed that flashpoint, fire point and viscosity were improved by 5%. SEM was performed to check and analyse wear scar on test samples. It was observed that surface roughness was reduced because of the presence of h-BN additives [65].

Vista et al. [80] also tested h-BN particles. For this purpose, an ionic liquid and a natural oil served as a base oil with h-BN particles for lubrication. After experimentation, it was observed that different additives gave different types of responses in tribological performance. Tribological performance in the case of ionic liquids depends only on particle size and is independent of particle types. The colloidal mixture was prepared by using phosphonium-based ionic liquid and avocado oil as a base oil in the presence of additives particles. Operating conditions were the following: the normal load of 10N, sliding velocity of 36 mm/s, testing for 13 hours. After experimentation, the most effective parameter was particle type in natural oils. However, in ionic liquids, the most effective parameter was particle size, and particle type had a negligible effect on performance. Nano-particles showed better effect as compared to micro-sized particles because larger sized particles increase wear scar, surface roughness, and coefficient of friction [80].

2.4.7 Zeolite Nano-Crystals

For better performance and to avoid absorbing acid which is formed during oxidation and to reduce depletion of additives, Zaarour et al. [66] used zeolite nano-crystals as a proactive agent. Tribotesting was conducted by preparing four lubricants with nano-sized particles. It has been observed that zeolite nano-crystals show strong affinity and protect ZDDP additive which was shown by FTIR (Fourier transform infrared) spectroscopy and 31P NMR (nuclear magnetic resonance) even after heating at 150°C for 24 hours. LTL (Linde Type L) zeolite nano-crystals of size 15–20 nm were prepared from inorganic materials. Preparation was conducted in the absence of structural directing agents such as proactive agents to increase the lifetime of lubricants for the use of the automotive engine. LTL (Linde Type L) nano-crystals showed: (i) depletion time increased and in this way the proactive domain was improved (ii) generation of a second layer of oxidation products which delayed the lubricant degradation [66].

2.4.8 Silicon Dioxide

SiO_2 nano-particles were used in sesame oil by Sabarinath et al. [67] in the presence of 1-hexyl-3-methylimidazolium tetrafluoroborate and 1-butyl-3-methylimidazolium tetrafluoroborate ionic liquids as hybrid antiwear additives. Experimentation was done on steel-steel contacts to improve tribological effects. A comparison was conducted between different wear scar diameters at different concentrations of the weight of nano-particles.

A four-ball tribometer was used for this purpose. Tribotesting was conducted with a 40 kg load and 1200 rev/min. It was observed that tribological properties of sesame oil were improved in the presence of ionic liquids. The results of tribotesting showed that SiO_2 nano-particles with a combination of 1-butyl-3-methylimidazolium tetrafluoroborate ionic liquid as a lubricant gave better results as compared to the combination with 1-hexyl-3-methylimidazolium tetrafluoroborate ionic liquid in sesame oil. It has been observed that wander walls' forces of attraction exist between glyceride molecules and ionic liquids which are responsible for the creation of tribofilm. Reduction in friction and wear is improved because of the formation of tribofilm produced by additive synergic action. It was also observed that corrosive and rheological properties were not affected significantly by the above combination of additives [67].

2.4.9 Ceria-Zirconia

Philip et al. [69] investigated a ceria-zirconia (Ce-Zr) hybrid in coconut oil. Tribotesting was conducted by using a pin on the disc tribometer mechanism. For the production of nano-particles of Ce-Zr, a chemically synthetic precipitation method was used. Response surface methodology under Box Behnken design was used for optimization and analysis of different independent factors of temperature, concentration, load and speed for optimised performance. Nano-hybrid lubricant samples were prepared at different concentrations of 0, 0.5 and 1% by weight in coconut oil for tribotesting to find the optimum concentration for best performance. A total of twenty-nine different combinations of parameters of speed, concentration, load and temperature were proposed by the RSM technique by using BBD. After analysis of the values obtained after experimentation, it was observed that the effect of speed on the coefficient of friction and wear scar diameter was insignificant. The concentration of nano-particles was proved to be the most significant factor. The optimum value for optimum performance was 0.62 wt.% of nano-particles [69].

2.4.10 Nickel

Nickel-based bio-lubricants were investigated for analysing antiwear properties in the presence of polyalphaolefin. Thermal decomposition was the process used for the synthesis of nickel nano-particles in PAO-6 from nickel formate. Surface capping agents were oleic acid and oleylamine. It was observed that the better antiwear behaviour was because of the low concentration of Ni. The combined response of nickel nano-particles and organic modifiers gave a lubrication mechanism. Nickel particles served as a source for the formation of a protective coating and filling of micro-roughness. So, low temperature grease of grade CATIM-201 as a lubricant was investigated in the presence of nickel particles for tribotesting. Tribotesting was conducted for analysis of a steel sliding pair. Tribotesting was conducted in the range of 60–90°C and the effect on antiwear properties was investigated. Tribotesting was conducted in the presence and absence of nano-nickel particles. Extreme pressure testing was conducted through a four-ball tribotester. The optimum particle size of nickel particles was in the 12–34.8 nm range. Antiwear and extreme pressure properties of low temperature grease were improved because of the presence of nickel particles. The best result was observed in 0.2%wt. nickel particles. The better performance can be associated with the micro-polishing effect produced by nickel particles [72].

2.4.11 Zinc Dialkyl Dithiophosphate

In this research soybean oil and palm olein RBD were used as bio-lubricants in the presence of additives to analyse their lubricating properties. Copper oxide and ZDDP nano-particles were used for tribotesting. Then ZDDP percentage of 1% was analysed with varying percentages in the range of 0.75–1% of copper oxide nano-particles. With the help of standard conditions of ASTM D4172, tribotesting was conducted. It was observed that from all above combinations, only ZDDP in 0.75% and CuO particles in 1% gave the best wear scar reduction which was 54.6% reduction, and the coefficient of friction was reduced to 29.2%. It was experienced that smooth surface roughness was in the above condition as compared to soybean oil and RDB palm olein. Mineral oil gave better results in the case of wear scar diameter; however, the coefficient of friction was reduced in the presence of bio-additive lubricating oil. From the above results, it can be said that nano-additives increase the tribological properties as compared to mineral oils as a lubricant [78].

2.5 TRIBOTESTING APPARATUS AND PROCEDURE

The tribotester is a machine that is used to investigate tribological characteristics such as friction coefficient and wear coefficient. There are the following types of tribotesters based on their design.

2.5.1 Pin on Disc Apparatus

Pin on disc tribotester is one in which load is applied on the pin while the base plate is subjected to rotate at some rpm under the loading conditions. A schematic diagram of a pin on disc apparatus can be seen in Figure 2.10.

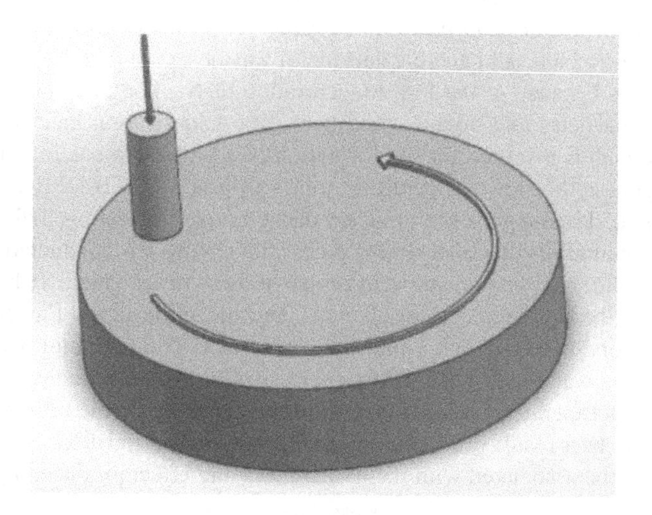

FIGURE 2.10 Schematic diagram of a pin on disc apparatus.

2.5.1.1 Tribotesting Conditions

An author used ASTM G-99 to check antiwear characteristics by rotating a disc against a stationary pin. The sliding distance of 1000 m with a variable speed of rotation from 1.4 to 5.6 m/s under the load of 100N was used [64].

Philip et al. [69] additionally used the ASTM G-99-05 standard for the analysis of thermophysical and rheological properties of a nano-lubricant such as fire point, viscosity and flashpoint. Zirconium oxychloride, distilled water and cerium (III) nitrate hexahydrate were in the precursors which were used for the generation of Ce-Zr hybrid nano-particles. To keep the solution basic and to increase the speed of reaction, ammonium hydroxide was used. Precursors were prepared in the standard beaker and at a temperature of 60°C under constant stirring. Then NaOH was added to maintain PH equal to 10. Nano-hybrid lubricant samples were prepared at different concentrations of 0, 0.5 and 1% by weight in coconut oil for tribotesting to find the optimum concentration for best performance. Agitation was conducted for an hour duration at 30°C. For uniform dispersion and stability, after 3 minutes, the process was stopped. The concentration of nano-particles was proved to be the most significant factor. The optimum value for optimum performance was 0.62 wt.% of nano-particles. Finally, test analysis on flashpoint, fire point and viscosity has shown the surfactant modifier as a credible modifier in the presence of coconut oil [69].

Another researcher used also the ASTM G-99 standard to analyse the physiochemical properties under variable rpm (1000–2000) speed of the disc, variable applied load (5–20 kg) and variable velocity of sliding (4.71–12.56 m/s) [87]. Some other authors used this apparatus and the experimental conditions of their research can be seen in Table 2.7.

2.5.2 Four-Ball Tribotester

In the four-ball tribotester, four balls are being used. One ball is held in the chuck, and the other balls are strongly clutched in the pot. The pot of balls is filled with the base oil during experimentation. The fixed ball in the chuck is pressed firmly against the other three stationary balls to wear out the surfaces. A high load is applied on the balls until all the balls weld together. The scars induced on the surfaces of each ball are measured to calculate the wear coefficient and friction coefficient. A schematic diagram of the four-ball tribotester is presented in Figure 2.11.

2.5.2.1 Tribotesting Conditions

Azman et al. [60] investigated CuO nano-particles as additives in palm oil biolubricant and analysed extreme pressure and AW performances. A four-ball tribotester under ASTM D4172 and ASTM D2783 were used for this purpose. After the analysis of results, it has been observed that those additives have given a 12% increment in performance. This experimentation showed that CuO nano-particles were a lubricant additive in respect to extreme pressure and AW performance [60].

Another researcher analysed the tribological characteristics by using the ASTM D5183-05 standard for friction and the ASTM D4172-94 standard for wear scar with the four-ball tribotester. When coconut oil was used as a lubricant, it showed a fruitful result in the form of a low coefficient of friction along with a problem of high wear rate on contact surfaces as compared to mineral lubricant oils, while mustard

TABLE 2.7
Pin on Disc Apparatus

Sr #	Lubricant	Additive	Parameters	ASTM Standards	Apparatus	Material Used For	Optimised Factors	References
1	Rapeseed oil	Nano- and micro-scale TiO_2	0.05% TiO_2 nano-particles with 0.05% TiO_2 micro-particles Speed of rotation = 1.4 to 5.6 m/s	G-99	Pin on disc apparatus	Real cylinder liner material **Disc** hardness of 87 HRB **pins** white cast iron 54 HRC	Antiwear	[64]
2	Coconut oil	Ceria zirconia hybrid (Ce-Zr) nano-particle	wt% 0, 0.5 and 1%	G-99-05	Pin on disc apparatus	**For Pin** aluminium alloy (Al-84%, Si-13%, and other elements-3%) **For Disc** steel (EN-31, 60 HRC)	COF SWR	[69]
3	Different bio-based oils	Graphene	Percentage variation in decimal and integer quantity	N/A	Pin on disc apparatus	**Disc** steel 52100 **Pin** steel 52100	Wear rate, COF	[74]
4	Avocado oil with Phosphonium-based ionic liquid	hBN, WS_2, MoS_2 carbon nano-tubes and graphite.	(hBN) in the size of 70nm, 0.5μm, 1.5μm and 5μm. WS_2 in 55nm and 0.6μm size. MoS_2 in 2μm size. and 5μm Carbon nano-tubes		pin on disc apparatus	**Disc** aluminium 2024 **Pin** 440C stainless steel	wear and friction characteristics	[80]

(Continued)

TABLE 2.7 (*Continued*)
Pin on Disc Apparatus

Sr #	Lubricant	Additive	Parameters	ASTM Standards	Apparatus	Material Used For	Optimised Factors	References
5	Chemically modified rapeseed oil (CMRO).	CuO, WS_2 and TiO_2	1. load 2. sliding speed 3. nano-based different particles 4. bearing materials	ASTM D4172	pin on disc apparatus	**Disc & Pin:** bronze, copper, and brass material journal bearings.	wear and COF	[81]
6	Canola oil	Micron-sized hexagonal boron nitride	Normal force (N)=10 sliding velocity (mm/s)=33 angular velocity (rpm)=21.5 distance travelled (m)= 100 duration (min) =50.5 environment ambient conditions lubricant amount (mL) Canola oil= 5 additives 5 wt% hBN additive particle sizes 5.0 μm, 1.5 μm, 0.5 μm &70 nm	N/A	Pin on disc apparatus	**Disc** copper and aluminium **Pin** steel and copper	COF and wear volume	[83]
7	SAE 20W40 grade motor oil	WS_2 and MoS_2 nano-particles	WS_2 and MoS_2 particles of 0.05 wt.% and 0.1 wt.%	ASTM G-99	Pin on disc apparatus		Physio-chemical properties	[87]

FIGURE 2.11 Schematic diagram of four-ball tribotester apparatus.

oil gave good results in the form of lower wear. So, blends of these two oils with CuO nano-particles as additives were used to analyse tribological performance. After experimentation, improvement in wear and coefficient of friction were observed. And 0.2% additive concentration was considered as an optimal concentration for better tribological performance [61].

Gulzar et al. [63] claimed that antiwear and extreme pressure properties were improved by using CuO as an additive in the following concentrations of nano-particles: 0.25%, 0.5%, 0.75%, 1%, 1.25% and 1.5%. Different blends were produced. The load was additionally varied during tribotesting and load was kept in the range of 40–180 kg. For tribotesting and sonicator temperature condition was 30°C. The engine speed was held constant at 1.1 m/s and we checked our results in different concentrations of blends. A four-ball tribotester with ASTM D2783 standard was used for experimentation [63].

Katiyar et al. [65] also used the ASTM-D4172 standard for tribotesting. Different blends of h-BN (0.1, 0.5, 1.0, 2.0, 2.5 and 3.0% wt) blended in pure coconut oil were used to check tribological effects. A probe sonicator was used for 30 minutes for assurance of homogeneity of additives in the lubricant. A four-ball tribotester was used for experimentation after blending. For the accuracy of the result, one concentration has been experimented with three times. After the tribotesting of each sample, culture bottles were used to collect samples. A redwood viscometer was used to determine the physiochemical properties such as fire point, the viscosity of the oil samples, and flashpoint. It was observed that flashpoint, fire point and viscosity were improved by 5% [65].

A four-ball tribotester was used by different researchers as can be seen in Table 2.8.

2.5.3 Ball on Disc Apparatus

The ball on disc tribotester is one in which load is applied on the spherical surface of the ball while the base plate is subjected to oscillate at a specified frequency under

TABLE 2.8
Four-Ball Tribotester

Sr #	Lubricant	Additive	Parameters	Apparatus	Material Used For	Optimised Factors	References
1	Additive palm oil	Nano-CuO and nano-graphite particles	0.1–0.6 wt%	Four-ball tribotester by using ASTM D4172 and ASTM D2783	N/A	(i) Extreme pressure properties (ii) Antiwear properties	[60]
2	Blends of mustard oil and coconut oil	CuO nano-particles	0.1, 0.2, 0.3 and 0.4 wt.%. blends composition is from 10 to 50%.	Four-ball tribotester ASTM 5183-05 for friction and ASTM D4172-94 for wear scar	**For ball**: GCr15 64 HRC	Friction and wear	[61]
3	Palm oil	CuO and MoS_2 nano-particles	%wt of nano-particles in (0.25%, 0.5%, 0.75%, 1%, 1.25%, 1.5%)	Four-ball tribotester ASTM D2783	**For ball:** AISI 52100 steel with a hardness of 64-66	AW and EP	[63]
4	Coconut oil	h-BN	(0.1, 0.5, 1.0, 2.0, 2.5 and 3.0 % wt)	Four-ball tribotester ASTM-D4172	N/A	AW/EP	[65]
5	Commercial lubricant	Zeolite nano-crystals	1 wt% of LTL zeolite nano-crystals stirring at 90° C for 35 days or at 150° C for 24 h		N/A	Lubricant oxidation	[66]
6	Sesame oil	SiO_2 nano-particles	wt% of 0.2, 0.4, 0.6, 0.8, 0.85	Four-ball tribotester	(AISI-52100), of 64–66 HRC	COF WSD	[67]

(Continued)

TABLE 2.8 (*Continued*)
Four-Ball Tribotester

Sr #	Lubricant	Additive	Parameters	Apparatus	Material Used For	Optimised Factors	References
7	Bio-diesel from waste cooking oil with diesel	CuO nano-particles	0, 25, 50 and 75 ppm nano-particle different rotational speed of 600, 1200 and 1500 rev/min	Four-ball tribotester	N/A	Coefficient of friction (COF)	[68]
8	Blending of bio- polyol ester-based lubricant with commercial lubricant	N/A	Blending is done in 10%, 15%, 20% and 25% by volume.	Four-ball tribotester	Hard steel balls	Wear and friction properties	[70]
9	Pure base oil group II PBO (PBO-GII)	TiO_2 + graphene	Sonicated for 2 hours at 80°C	Four-ball tribotester	Steel-52100 66 HRC	COF, WSD and SWR	[71]
10	Lubricant CATIM-201 (low temperature grease)	Nickel nano-particles	Load in the range from 0.01 to 500 mN. Ni particles 0.05, 0.1 and 0.2wt%.	Four-ball tribotester	Bearing steel	wear and friction	[72]
11	Trimethylolpropane TMP ester	Nano-TiO_2/SiO_2 and nano-TiO_2	TMP ester in concentrations of 0.25, 0.50, 0.75 and 1 wt%.	Four-ball tribotester	AISI 52100 steel of 64-66 HRC	Load-carrying capacity and antiwear properties	[73]
12	Modified jojoba oil	Graphene in 0.075 wt%.	Blending of jojoba oil with mineral oil in the ratio of 0, 10, 20, 30, 40 and 50% by volume	Four-ball tribotester	AISI 52100	Wear and friction characteristics	[77]
13	Soybean oil, palm olien RBD	Copper oxide and ZDDP nano-particles	1% ZDDP with different percentages of CuO particles of 0.75–1% range	Four-ball tribotester	Steel balls	Wear and friction characteristics	[78]
14	Palm oil (TMP) ester blended with PAO	Graphene nano-platelets (GNPs)	N/A	Four-ball tribotester	AISI 52100, 64-66 HRC	Wear and COF	[82]

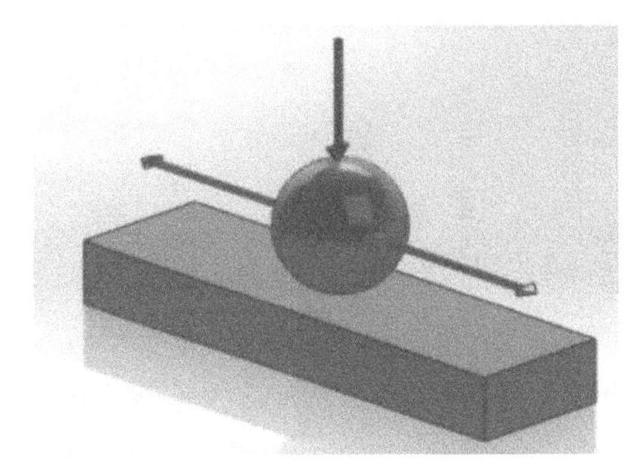

FIGURE 2.12 Schematic diagram of a ball on disc apparatus.

the loading conditions. A schematic representation of a ball on disc apparatus can be seen in Figure 2.12.

2.5.3.1 Tribotesting Procedure

A researcher tried to reduce the coefficient of friction by using WS_2 and WS_2/TiO_2 nano-particles in diisooctyl sebacate bio-base oil by using the ball on disc tribotester. WS_2/TiO_2 nano-particles gave better results as compared to WS_2 nano-particles. An almost 81% reduction in wear volume was observed by 0.75% of WS_2/TiO_2 nano-particles [62].

In another study, the effect of different dispersion techniques and effects of different surfactants were analysed with wear and friction behaviour in the presence of WS_2 nano-sized additive in the presence of polyalphaolefin base oil as a lubricant. From the study of the literature review, it was decided to make a 1% concentration of additives. Different samples of different additives were prepared. Samples without surfactants were subjected to sonication for 1-hour duration. For making samples in the presence of surfactants, samples were sonicated for 2 hours duration. To prepare samples of PVP, the first PVP was incorporated into distilled water and then the solution was subjected to sonication of half hour duration. Then this sample was dried in an oven at 60°C for 120-min duration and then nano-particles were incorporated into oil and sonified for an hour duration for uniform distribution. Different dispersion techniques were used for uniform dispersion as can be observed: (1) In the absence of a stabilizing agent, sonification for an hour; (2) in the presence of 1% weight OA surfactant, sonification for an hour. For the aspect of wear, it was observed that WS_2 gave a 45% reduction in wear depth in the presence of PVP surfactant as compared to the simple base oil [76].

Singh et al. [85] used the same apparatus for analysing the coefficient of friction. The experiments were performed at fixed ball speed 0.3 m/s and different loads 10N, 20N, 30N, 40N and 45N. Lithium grease was used in the presence of graphene as a lubricant on AISI steel 52100 [85].

Ball on disc apparatus was also used by another researcher with purified paraffin oil blended with TiO_2 nano-particles. Purified paraffin oil with 1 wt% TiO_2 particle, 1 wt% estisol 242 or 1 wt% oleic acid, 0.15 wt% oleylamine and 0.15 wt% pluronic RPE 2520 was used. Experimental conditions were: 4-h test runs with a normal force of $F_N = 2.5$ kN and a sliding velocity of 0.15 m/s in our ball disc contact. Different experimental conditions that were used by different researchers can be seen in Table 2.9.

TABLE 2.9
Ball on Disc Apparatus

Sr #	Lubricant	Additive	Parameters	Apparatus	Material Used For	Optimized Factors	References
1	Diisooctyl sebacate (DOS)	Micro-WS_2, WS_2, WS_2/TiO_2 Nano-particles	Ratio of TiO_2, WS_2/TiO_2 samples with mass ratios of 1:1, 1:2, and 2:1 of WS_2:TiO_2 %wt of nano-particles in (0.25%, 0.5%, 0.75%)	Ball on disc apparatus	**For ball:** GCr15 **For Disc:** GCr15 58–66 HRC in hardness	(i) Wear (ii) COF	[62]
2	Polyalphaolefin base oil, oleic acid (OA)	MoS_2 and WS_2 nano-sized	1% by weight concentration.	Ball on disc apparatus	8620 steel	Friction and wear	[76]
3	Polyalphaolefin (PAO) and SAE 5W-30 engine oil	Trihexyltetradecylphosphonium (2-ethylhexyl) phosphate	Two oil ionic liquid blends were prepared by adding 5 wt.% [P_{66614}] [DEHP] into the PAO base oil and the SAE 5W-30 engine oil, tests were conducted at room temperature (~23°C) under a normal load of 160 N and the oscillation frequency of 10 Hz with a 10 mm stroke for a sliding distance of 1000 m	Ball on disc apparatus	AISI 52100 steel	Wear and COF	[84]
4	Lithium grease	Graphene	The experiments were performed at fixed ball speed 0.3 m/s and at different loads 10N, 20N, 30N, 40N and 45N	Ball on disc apparatus	Steel disc (AISI 52100) and ball (AISI 52100)	COF	[85]
5	Purified paraffin oil	TiO_2 nano-particles	Purified paraffin oil with 1 wt% TiO_2 particle, 1 wt% estisol 242 or 1 wt% oleic acid, 0.15 wt% oleylamine, and 0.15 wt% pluronic RPE 2520. 4-h test runs with a normal force of $F_N = 2.5$ kN and a sliding velocity of 0.15 m/s in our ball disc contact.	Ball on disc apparatus	100Cr6 versus 100Cr6 disc ball contact	Antiwear and antifriction properties	[88]

2.6 CONCLUSION

Mineral oil and other lubricating oils are being used for lubrication purposes. They are produced from crude oil. Disposing of mineral oils leads to pollution in ecosystems. Furthermore, the use of conventional oils as a lubricant can emit traces of metals that are very harmful to living beings. Use of bio-waste oils and bio-lubricant oils in different proportions is a new thing which can give fruitful environmental and economical results. Bio-base oils serve as an alternative to mineral oil due to some natural properties such as being biodegradable and having high lubricity, high flash point, high viscosity index and low evaporative losses. Bio-lubricants have chains of long fatty acids, and they contain polar groups in their structure of vegetable oils which impart the feasibility of hydrodynamic and boundary lubrication. The performance of biodegradable oil in terms of lubricants cannot be compared with fully formulated synthetic lubricants. The reason is that fully formulated lubricants contain additives.

To make bio-oils comparable, researchers are working on micro- and nano-additives for biodegradable oils. The main advantage of these particles is that they don't react with any other particle of oil. Furthermore, high dispersion stability and high possibility of film formation are the advantages of particles on different types of surfaces.

The automobile sector is one of the major energy consuming sectors and the number of vehicles is increasing exponentially. It is imperative for sustainable growth that the consumption of mineral oil-based lubricants be reduced. Fuel efficient automotive engines and environmental pollution reduction will be the advantages of the adoption of biodegradable oils.

2.7 FUTURE WORK

- The use of different additive particles in bio-based lubricants such as castor oil, neem oil, coconut oil, mustard oil, etc. can be an area of future research for reducing the coefficient of friction and wear coefficient.
- Different additives can additionally be used in the blended form with other AW and extreme pressure additives in different compositions.
- The blended bio-lubricants can be used over rubbing surfaces of different metals to analyse tribological behaviour.

REFERENCES

1. Evans, C. (1978). Strategy for Energy Conservation through Tribology. American Society of Mechanical Engineers, New York, 1977. *Energy Policy*, *6*(2), 170–171.
2. Blau, P. J. (1997). Fifty Years of Research on the Wear of Metals. *Tribology International*, *30*(5), 321–331.
3. Holmberg, K., Andersson, P., & Erdemir, A. (2012). Global Energy Consumption due to Friction in Passenger Cars. *Tribology International*, *47*, 221–234.
4. Tung, S. C., & McMillan, M. L. (2004). Automotive Tribology Overview of Current Advances and Challenges for the Future. *Tribology International*, *37*(7), 517–536.
5. Baek, J. S., Groll, E. A., & Lawless, P. B. (2005). Piston-Cylinder Work Producing Expansion Device in a Transcritical Carbon Dioxide Cycle. Part I: Experimental Investigation. *International Journal of Refrigeration*, *28*(2), 141–151.

6. Tan, Y.-C., & Ripin, Z. M. (2011). Frictional Behavior of Piston Rings of Small Utility Two-Stroke Engine Under Secondary Motion of Piston. *Tribology International*, *44*(5), 592–602.
7. Tan, Y.-C., & Ripin, Z. M. (2014). Technique to Determine Instantaneous Piston Skirt Friction During Piston Slap. *Tribology International*, *74*, 145–153.
8. Nakada, M. (1994). Trends in Engine Technology and Tribology. *Tribology International*, *27*(1), 3–8.
9. Allmaier, H., Priestner, C., Reich, F. M., Priebsch, H. H., Forstner, C., & Novotny-Farkas, F. (2012). Predicting Friction Reliably and Accurately in Journal Bearings: The Importance of Extensive Oil-Models. *Tribology International*, *48*, 93–101.
10. Iwai, Y., Yoneda, H., & Honda, T. (1995). Sliding Wear Behavior of SiC Whisker-Reinforced Aluminum Composite. *Wear*, *181*, 594–602.
11. Gopalakrishnan, S., & Murugan, N. (2012). Production and Wear Characterisation of AA 6061 Matrix Titanium Carbide Particulate Reinforced Composite by Enhanced Stir Casting Method. *Composites Part B: Engineering*, *43*(2), 302–308.
12. Kumar, S., Chakraborty, M., Sarma, V. S., & Murty, B. S. (2008). Tensile and Wear Behaviour of in situ Al–7Si/TiB2 Particulate Composites. *Wear*, *265*(1–2), 134–142.
13. Rajan, H. B. M., Ramabalan, S., Dinaharan, I., & Vijay, S. J. (2014). Effect of TiB2 Content and Temperature on Sliding Wear Behavior of AA7075/TiB2 in situ Aluminum Cast Composites. *Archives of Civil and Mechanical Engineering*, *14*(1), 72–79.
14. Miller, A. L., Stipe, C. B., Habjan, M. C., & Ahlstrand, G. G. (2007). Role of Lubrication Oil in Particulate Emissions from a Hydrogen-Powered Internal Combustion Engine. *Environmental Science & Technology*, *41*(19), 6828–6835.
15. Ssempebwa, J. C., & Carpenter, D. O. (2009). The Generation, Use and Disposal of Waste Crankcase Oil in Developing Countries: A Case for Kampala District, Uganda. *Journal of Hazardous Materials*, *161*(2–3), 835–841.
16. Asadauskas, S., Perez, J. M., & Duda, J. L. (1996). Oxidative Stability and Antiwear Properties of High Oleic Vegetable Oils. *Lubrication Engineering*, *52*(12), 877–882.
17. Nagendramma, P., & Kaul, S. (2012). Development of Ecofriendly/Biodegradable Lubricants: An Overview. *Renewable and Sustainable Energy Reviews*, *16*(1), 764–774.
18. Quinchia, L. A., Delgado, M. A., Reddyhoff, T., Gallegos, C., & Spikes, H. A. (2014). Tribological Studies of Potential Vegetable Oil-Based Lubricants Containing Environmentally Friendly Viscosity Modifiers. *Tribology International*, *69*, 110–117.
19. Shahabuddin, M., Masjuki, H. H., Kalam, M. A., Bhuiya, M. M. K., & Mehat, H. (2013). Comparative Tribological Investigation of Bio-lubricant Formulated from a Non-edible Oil Source (Jatropha Oil). *Industrial Crops and Products*, *47*, 323–330.
20. Padhi, S. K., & Singh, R. K. (2011). Non-edible Oils as the Potential Source for the Production of Biodiesel in India: A Review. *Journal of Chemical and Pharmaceutical Research*, *3*(2), 39–49.
21. Ong, H. C., Mahlia, T. M. I., Masjuki, H. H., & Norhasyima, R. S. (2011). Comparison of Palm Oil, *Jatropha curcas* and *Calophyllum inophyllum* for Biodiesel: A Review. *Renewable and Sustainable Energy Reviews*, *15*(8), 3501–3515.
22. Sharma, R. V, & Dalai, A. K. (2013). Synthesis of Bio-lubricant from Epoxy Canola Oil Using Sulfated Ti-SBA-15 Catalyst. *Applied Catalysis B: Environmental*, *142*, 604–614.
23. Gulzar, M., Masjuki, H. H., Kalam, M. A., Varman, M., Zulkifli, N. W. M., Mufti, R. A., & Zahid, R. (2016). Tribological Performance of Nanoparticles as Lubricating Oil Additives. *Journal of Nanoparticle Research*, *18*(8), 223.
24. Watanabe, S., Noshiro, J., & Miyake, S. (2004). Tribological Characteristics of WS2/ MoS_2 Solid Lubricating Multilayer Films. *Surface and Coatings Technology*, *183*(2–3), 347–351. https://doi.org/10.1016/j.surfcoat.2003.09.063
25. Kim, H. J., Shin, D. G., & Kim, D. E. (2016). Frictional Behavior Between Silicon and Steel Coated with Graphene Oxide in Dry Sliding and Water Lubrication Conditions.

International Journal of Precision Engineering and Manufacturing – Green Technology, *3*(1), 91–97. https://doi.org/10.1007/s40684-016-0012-8
26. Donnet, C., & Erdemir, A. (2004). Historical Developments and New Trends in Tribological and Solid Lubricant Coatings. *Surface and Coatings Technology*, *180*, 76–84.
27. Barnes, A. M., Bartle, K. D., & Thibon, V. R. A. (2001). A Review of Zinc dialkyl dithiophosphates (ZDDPS): Characterisation and Role in the Lubricating Oil. *Tribology International*, *34*(6), 389–395.
28. Tung, S. C., & Totten, G. E. (2012). *Automotive lubricants and testing.* ASTM International.
29. Bartz, W. J. (1978). Tribology, Lubricants and Lubrication Engineering—A Review. *Wear*, *49*(1), 1–18.
30. Hutton, J. F. (1973). The Rheology of Petroleum-based Lubricating Oils and Greases: A Review. *The Rheology of Lubricants*, 16.
31. Gangopadhyay, A. (2017). A Review of Automotive Engine Friction Reduction Opportunities Through Technologies Related to tribology. *Transactions of the Indian Institute of Metals*, *70*(2), 527–535. https://doi.org/10.1007/s12666-016-1001-x
32. Stachowiak, G., & Batchelor, A. W. (2013). *Engineering tribology.* Butterworth-Heinemann.
33. Nabil, M. F., Azmi, W. H., Hamid, K. A., Mamat, R., & Hagos, F. Y. (2017). An Experimental Study on the Thermal Conductivity and Dynamic Viscosity of TiO_2–SiO_2 Nanofluids in Water: Ethylene Glycol Mixture. *International Communications in Heat and Mass Transfer*, *86*, 181–189.
34. Komatsuzaki, S., & Ito, R. (1975). Flow Properties of Lubricating Greases at High-Temperature. 1. Apparent Viscosity at High-Temperature. *Journal of Japan Society of Lubrication Engineers*, *20*(2), 97–105.
35. Bartz, W. J. (1976). About the influence of viscosity index improvers on the cold flow properties of engine oils. *TRIBOLOGY International*, *9*(1), 13–20. https://doi.org/10.1016/0301-679X(76)90064-5.
36. Bart, Jan C. J., Gucciardi, E., & Cavallaro, S. (2012). *Biolubricants: science and technology.* Elsevier.
37. Pillon, L. Z. (2016). *Surface activity of petroleum derived lubricants* (Vol. 127). CRC Press.
38. Bart, J C J, Gucciardi, E., & Cavallaro, S. (2013). *Lubricants: Properties and characteristics. Biolubricants.* Woodhead Publishing Series in Energy, *24*, 73.
39. Kodali, D. R. (2002). High Performance Ester Lubricants from Natural Oils. *Industrial Lubrication and Tribology.*
40. Anand, A., Haq, M. I. U., Vohra, K., Raina, A., & Wani, M. F. (2017). Role of Green Tribology in Sustainability of Mechanical Systems: A State of the Art Survey. *Materials Today: Proceedings*, *4*(2), 3659–3665.
41. Hamrock, B. J. (1994). *Fundamentals of fluid film lubrication.* McGraw-Hill. Inc., Hightstown, NJ, 8520.
42. Rudnick, L. R. (2013). *Synthetics, mineral oils, and bio-based lubricants: chemistry and technology.* CRC Press
43. Quinchia, L. A., Delgado, M. A., Franco, J. M., Spikes, H. A., & Gallegos, C. (2012). Low-Temperature Flow Behaviour of Vegetable Oil-based Lubricants. *Industrial Crops and Products*, *37*(1), 383–388.
44. Crawford, J., Psaila, A., & Orszulik, S. T. (1997). Miscellaneous additives and vegetable oils. In *Chemistry and technology of lubricants* (pp. 181–202). Springer.
45. McNutt, J. (2016). Development of Biolubricants from Vegetable Oils via Chemical Modification. *Journal of Industrial and Engineering Chemistry*, *36*, 1–12.
46. Reeves, C. J., Menezes, P. L., Jen, T.-C., & Lovell, M. R. (2015). The Influence of Fatty Acids on Tribological and Thermal Properties of Natural Oils as Sustainable Biolubricants. *Tribology International*, *90*, 123–134.

47. Syahir, A. Z., Zulkifli, N. W. M., Masjuki, H. H., Kalam, M. A., Alabdulkarem, A., Gulzar, M., Khuong, L. S., & Harith, M. H. (2017). A Review on Bio-based Lubricants and their Applications. *Journal of Cleaner Production*, *168*, 997–1016.
48. Salimon, J., Salih, N., & Yousif, E. (2012). Industrial Development and Applications of Plant Oils and their Biobased Oleochemicals. *Arabian Journal of Chemistry*, *5*(2), 135–145.
49. Bovington, C. H. (2010). Friction, wear and the role of additives in controlling them. In *Chemistry and technology of lubricants* (pp. 77–105). Springer.
50. Peña-Parás, L., Maldonado-Cortés, D., Kharissova, O. V, Saldívar, K. I., Contreras, L., Arquieta, P., & Castaños, B. (2019). Novel Carbon Nanotori Additives for Lubricants with Superior Anti-wear and Extreme Pressure Properties. *Tribology International*, *131*, 488–495.
51. Khrushchev, M. M., & Babichev, M. A. (1953). *Resistance to abrasive wear and the hardness of metals* (Vol. 15). US Atomic Energy Commission, Technical Information Service.
52. Rabinowicz, E., & Mutis, A. (1965). Effect of Abrasive Particle Size on Wear. *Wear*, *8*(5), 381–390.
53. Zhang, Z., Zhang, L., & Mai, Y.-W. (1997). Modeling Steady Wear of Steel/A12O3 Al Particle Reinforced Composite System. *Wear*, *211*(2), 147–150.
54. Archard, J. (1953). Contact and Rubbing of Flat Surfaces. *Journal of Applied Physics*, *24*(8), 981–988.
55. Suh, N. P. (1973). The Delamination Theory of Wear. *Wear*, *25*(1), 111–124. https://doi.org/10.1016/0043-1648(73)90125-7
56. Sin, H., Saka, N., & Suh, N. P. (1979). Abrasive Wear Mechanisms and the Grit Size Effect. *Wear*, *55*(1), 163–190. https://doi.org/10.1016/0043-1648(79)90188-1
57. Spikes, H. (2004). The History and Mechanisms of ZDDP. *Tribology Letters*, *17*(3), 469–489.
58. Mortier, R. M., Orszulik, S. T., & Fox, M. F. (2010). *Chemistry and technology of lubricants* (Vol. 107115). Springer.
59. Dienwiebel, M., & Pöhlmann, K. (2007). Nanoscale Evolution of Sliding Metal Surfaces During Running-in. *Tribology Letters*, *27*(3), 255–260.
60. Azman, N. F., Samion, S., Moen, M. A. A., Hamid, M. K. A., & Musa, M. N. (2019). The Anti-wear and Extreme Pressure Performance of CuO and Graphite Nanoparticles as an Additive in Palm Oil. *International Journal of Structural Integrity*. *10*(5), 714–725.
61. Sajeeb, A., & Rajendrakumar, P. K. (2019). Experimental Studies on Viscosity and Tribological Characteristics of Blends of Vegetable Oils with CuO Nanoparticles as Additive. *Micro & Nano Letters*, *14*(11), 1121–1125.
62. Lu, Z., Cao, Z., Hu, E., Hu, K., & Hu, X. (2019). Preparation and Tribological Properties of WS2 and WS2/TiO_2 Nanoparticles. *Tribology International*, *130*, 308–316.
63. Gulzar, M., Masjuki, H. H., Varman, M., Kalam, M. A., Mufti, R. A., Zulkifli, N. W. M., Yunus, R., & Zahid, R. (2015). Improving the AW/EP Ability of Chemically Modified Palm Oil by Adding CuO and MoS_2 Nanoparticles. *Tribology International*, *88*, 271–279.
64. Arumugam, S., & Sriram, G. (2013). Preliminary Study of Nano- and Microscale TiO_2 Additives on Tribological Behavior of Chemically Modified Rapeseed Oil. *Tribology Transactions*, *56*(5), 797–805.
65. Katiyar, J. K., Bhaumik, S., Ashok, A., & Sharma, A. K. (2018). Physiochemical Properties of Hexagonal Boron Nitride Blended Coconut Oil. *Proceedings of TRIBOINDIA-2018 – An International Conference on Tribology*, Available at SSRN: https://ssrn.com/abstract=3321035 or http://dx.doi.org/10.2139/ssrn.3321035
66. Zaarour, M., El Siblani, H., Arnault, N., Boullay, P., & Mintova, S. (2019). Zeolite Nanocrystals Protect the Performance of Organic Additives and Adsorb Acid Compounds During Lubricants Oxidation. *Materials*, *12*(17), 2830.

67. Sabarinath, S., Rajendrakumar, P. K., & Prabhakaran Nair, K. (2019). Evaluation of Tribological Properties of Sesame Oil as Biolubricant with SiO2 Nanoparticles and Imidazolium-Based Ionic Liquid as Hybrid Additives. *Proceedings of the Institution of Mechanical Engineers, Part J: Journal of Engineering Tribology*, *233*(9), 1306–1317.
68. Khorshidnia, H., & Shirneshan, A. (2019). Investigating the Tribological Behavior of Diesel-Biodiesel Blends with Nanoparticle Additives Under Short-term Tests. *ADMT Journal*, *12*(3), 19–24.
69. Philip, J. T., Koshy, C. P., Mathew, M. D., & Kuriachen, B. (2019). Tribological Characteristic Evaluation of Coconut Oil Dispersed with Surfactant Modified Ceria-Zirconia Hybrid Nanoparticles. *Tribology-Materials, Surfaces & Interfaces*, *13*(4), 197–214.
70. Kotturu, C. M. V. V, Srinivas, V., Vandana, V., Chebattina, K. R. R., & Seetha Rama Rao, Y. (2019). Investigation of Tribological Properties and Engine Performance of Polyol Ester–based Bio-lubricant: Commercial Motorbike Engine Oil Blends. *Proceedings of the Institution of Mechanical Engineers, Part D: Journal of Automobile Engineering*, *234*(5), 1304–1317.
71. Alghani, W., Ab Karim, M. S., Bagheri, S., Amran, N. A. M., & Gulzar, M. (2019). Enhancing the Tribological Behavior of Lubricating Oil by Adding TiO_2, Graphene, and TiO_2/Graphene Nanoparticles. *Tribology Transactions*, *62*(3), 452–463.
72. Zadoshenko, E. G., Burlakova, V. E., & Novikova, A. A. (2020). Effect of Nickel Nanopowder on Lubrication Behaviour of Low-Temperature Grease in Steel-Steel Tribosystem. *Tribology-Materials, Surfaces & Interfaces*, *14*(1), 51–58.
73. Gulzar, M., Masjuki, H. H., Kalam, M. A., Varman, M., Zulkifli, N. W. M., Mufti, R. A., Zahid, R., & Yunus, R. (2017). Dispersion Stability and Tribological Characteristics of TiO_2/SiO_2 Nanocomposite-Enriched Biobased Lubricant. *Tribology Transactions*, *60*(4), 670–680.
74. Srivyas, P. D., & Charoo, M. S. (2019). Graphene: An effective lubricant for tribological applications. *Lecture Notes in Mechanical Engineering*. https://doi.org/10.1007/978-981-13-6469-3_22
75. Shinde, V., & Soni, S. (2018). Effect of Lubricants with Nanoparticles on Performance of Hydrodynamic Journal Bearing. *Proceedings of TRIBOINDIA-2018, An International Conference on Tribology*, Available at SSRN: https://ssrn.com/abstract=3320357 or http://dx.doi.org/10.2139/ssrn.3320357
76. Jazaa, Y., Lan, T., Padalkar, S., & Louis, S. (2018). The Effect of Agglomeration Reduction on the Tribological Behavior of WS 2 and MoS 2 Nanoparticle. *Lubricants*, 6(4), 106. https://doi.org/10.3390/lubricants6040106
77. Kannan, K. T., & Rameshbabu, S. (2018). Tribological Properties of Modified Jojoba Oil as Probable Base Stoke of Engine Lubricant. *Journal of Mechanical Science and Technology*, *32*(4), 1739–1747. https://doi.org/10.1007/s12206-018-0330-6
78. Pillay, D. S., Azwadi, N., & Sidik, C. (2017). Tribological Properties of Biodegradable Nano-Lubricant. *Journal of Advanced Research in Fluid Mechanics and Thermal Sciences*, *33*(1), 1–13.
79. Baskar, S., Sriram, G., & Arumugam, S. (2016b). Tribological Analysis of a Hydrodynamic Journal Bearing Under the Influence of Synthetic and Biolubricants. *Tribology Transactions*, *60*(3), 428-436. https://doi.org/10.1080/10402004.2016.1176285
80. Vista, L. B., Reeves, C. J., Menezes, P. L., Lovell, M. R., & Jen, T. (2014). The Effect of Particulate Additives on the Tribological Performance of Bio-based and Ionic Liquid-based Lubricants for Energy Conservation and Sustainability Track. *Proceedings of the 2014 STLE Annual Meeting & Exhibition Disney's Contemporary Resort. M.* 2–4.
81. Baskar, S., Sriram, G., & Arumugam, S. (2016a). The Use of D-optimal Design for Modeling and Analyzing the Tribological Characteristics of Journal Bearing Materials Lubricated by Nano-based Biolubricants. *Tribology Transactions*, *59*(1), 44–54. https://doi.org/10.1080/10402004.2015.1063179

82. Sa, S., Azman, N., Wahidah, N., Zulki, M., Masjuki, H., & Gulzar, M. (2016). Study of tribological properties of lubricating oil blend added with graphene nanoplatelets. *Journal of Materials Research*, *31*(13)
83. Reeves, C. J., Menezes, P. L., Lovell, M. R., & Jen, T.-C. (2015). The Influence of Surface Roughness and Particulate Size on the Tribological Performance of Bio-Based Multi-Functional Hybrid Lubricants. *Tribology International*, *88*, 40–55.
84. Qu, J., Bansal, D. G., Yu, B., Howe, J. Y., Luo, H., Dai, S., Li, H., Blau, P. J., Bunting, B. G., & Mordukhovich, G. (2012). Antiwear Performance and Mechanism of an Oil-Miscible Ionic Liquid as a Lubricant Additive. *ACS Applied Materials & Interfaces*, *4*(2), 997–1002.
85. Singh, J., Anand, G., Kumar, D., & Tandon, N. (2016). Graphene Based Composite Grease for Elastohydrodynamic Lubricated Point Contact. *IOP Conference Series: Materials Science and Engineering*, *149*(1), 12195.
86. Kenig, S., Wagner, H., Zak, A., Moshkovith, A., Rapoport, L., & Tenne, R. (2011). The Mechanical and Tribological Properties of Epoxy Nanocomposites with WS2 Nanotubes. *Sensors & Transducers*, *12*, 53–65.
87. Srinivas, V., Manikanta, P. V, Satish, V., & Valluripally, D. (2014). Physicochemical Properties of Motor Oil Dispersed with WS2 and MoS_2 Nano Particles. *International Mechanical Engineering Congress.*
88. Bogunovic, L., Zuenkeler, S., Toensing, K., & Anselmetti, D. (2015). An Oil-Based Lubrication System Based on Nanoparticular TiO_2 with Superior Friction and Wear Properties. *Tribology Letters*, *59*(2), 29.

3 Tribological Investigations of Sustainable Bio-Based Lubricants for Industrial Applications

Neha Sharma, Sayed Khadija Bari, Ponnekanti Nagendramma, Gananath D. Thakre, and Anjan Ray
CSIR—Indian Institute of Petroleum
Uttarakhand, India

CONTENTS

3.1 INTRODUCTION

There is an emergent quest for energy-efficient technologies with reduced power consumption [1]. This technological requirement is in line with economic and ecological imperatives; less energy input per unit work output reduces the production costs and conserves the resources. As per available statistical data, ~30% of the

DOI: 10.1201/9781003139386-3

world's principal energy resources are devoted to overcoming friction, and the useful life of ~80% of the mechanical components is limited due to wear [2, 3]. Different studies have revealed that frictional losses account for ~2%–7% of Gross Domestic Product (GDP) per year, translating to ~1.7–6.0 trillion US dollars (USD) worldwide in 2018. For countries such as China, Germany, and Japan with a more extensive manufacturing capabilities base, the friction losses are higher as a percentage of GDP. It is estimated that ~90 billion USD can be saved, if friction is reduced by 18%, with an accompanying reduction in CO_2 emissions by 290 million tons a year.

The energy lost to overcome friction has been of significant concern in systems that involve moving parts. However, this can be efficiently mitigated with the use of appropriate lubricants. A lubricant facilitates the relative motion between interacting surfaces by minimizing friction and wear between them. The lubricants used are often non-aqueous liquids. They can be hydrocarbon-based, sometimes referred to as mineral oil (MO). Simultaneously, those not produced from petroleum are termed as synthetic (for instance, polyol polyesters and others derived from tree-borne oils, TBOs). MOs are extremely complex mixtures of long-chain hydrocarbons. They contain a range of species that may be alicyclic (naphthenic), linear alkanes (waxes and paraffin), branched alkanes (isoparaffins), olefinic, and aromatic moieties. These petroleum-derived lubricants are predominantly used in automotive and industrial environments due to their extended drain intervals, longer shelf life, and ease of formulation. Apart from these well-established advantages, MOs have higher boundary friction coefficients, lower viscosity indices, and higher volatility than synthetic and TBO-derived lubricants.

The post-consumption studies of MO-based lubricants have shown that their extended use can adversely affect the environment due to poor biodegradability and the ability to release toxic materials [4]. Increasingly stringent environment norms have recently encouraged industry to use more eco-friendly and "green" lubricants. Such eco-friendly lubricants mainly use TBO or synthetic polyol base-stock instead of hydrocarbons. These lubricants are typically biodegradable and non-carcinogenic and do not accumulate in effluents or solid wastes as MO-based lubricants do.

TBOs are triglycerides of esters and are classified as edible and non-edible oils. Due to high demand pressure on edible oils (mustard, groundnut, soybean, sunflower, etc.) in economies like India and China, non-edible oils such as jatropha (jatropha curcas) and Karanja (pongamia pinnata) are studied for lubrication purposes as well as diesel extenders. Non-edible oils are preferred over edible ones as they do not affect food security [5–7]. On the other hand, unlike edible oils, supply chains of TBO production are not yet cost-effective due to economies of scale. However, high Viscosity Index (VI) properties, high flash point, low aquatic toxicity, and high biodegradability suggest TBOs to be plausible options as energy-efficient lubricants [8]. TBOs have poor oxidation stability, which inhibits their direct use as lubricants. Hence, they are chemically modified by transesterification to obtain the necessary enhanced physical and chemical properties that enable their use as bio-lubricants [9–11]. The long-chain fatty acids and polar groups render TBOs suitable for boundary and hydrodynamic lubrication.

Arumugam and Sriram [12] investigated the lubrication behavior of rapeseed oil-based bio-lubricant (blended with castor and palm oil methyl ester) on a pin-on-disk arrangement. They reported that the bio-lubricant reduced friction and wear significantly. When used in diesel engines, the bio-lubricants reduced frictional losses, emission of trace metals, and brake-specific energy consumption. The type of lubricant used is one of the significant factors determining the lubrication efficiency of contacting surfaces. The lubricant's chemical structure and molecule polarity determine the molecules' attraction and migration at the active sites on the surfaces [13]. The superior tribological behavior of TBO-based lubricants [14, 15] is ascribed to the presence of polar ester structures of saturated and unsaturated fatty acids above Critical Micellar Concentration (CMC).

Another category of lubricants, polyol esters, are prepared from fairly pure and simple starting materials to produce molecular structures intended specifically for high-performance lubrication. Synthetic ester base stocks are both thermally and oxidatively stable compounds in comparison to neat TBOs; they have high viscosity indices and lack the undesirable and unstable impurities found in conventional petroleum-based oils. Polyol esters are much more adaptable in their application than MOs because of their much more comprehensive viscosity range, better low-temperature flow characteristics and a great extent of biodegradability with low aquatic toxicity. Polyol esters are broadly price-competitive with MOs, enabling them to be blended with MOs to boost their performance. Moreover, the additives used in MO formulations are equally compatible with polyol esters. The additive compatibility of polyol esters helps to trade-off the cost of a lubricant with its performance.

Fatty acid esters of polyglycerol (PG) are used as multi-functional additives (viscosity depressants, crystallization controllers, etc.) [16]. Over the past decade, the rapid increase in world bio-diesel production has raised concerns about a potential oversupply of glycerol in global markets and, thus, raised interest in exploring new value-added opportunities for glycerol. PGs themselves are quite stable thermally and oxidatively and generally have a shelf life of at least two years. They are used in their native state or re-transformed into derivatives as additives for many industrial applications. PG-based esters (PGEs) have been cited only occasionally in lubricant applications [17].

PGEs are synthesized from the reaction of different PG-oligomers with various fatty acids. They offer increased hydrophilicity and biodegradability relative to conventional polyol esters. PGEs are extremely safe, and therefore 80% or more of their industrial output is consumed in the food and cosmetic industry, which is a testimony to their benign nature [18]. Their primary function is to emulsify hydrophobic food ingredients, such as those found in chocolate or dairy products, and they are widely permitted by law in several countries [19]. The functional properties of PGEs are, as expected, dependent on their structure and composition, such as average PG chain length, polydispersity, and fatty acid composition and distribution. By inter-esterification with a triglyceride fat or fatty acid in varying proportions to the starting PG selected, the PG structure is altered to varying extent [20]. For a given PG, increased acid chain length in esterification lessens the hydrophilicity [21]. Within the range of

TABLE 3.1
Commercial Producers of PGEs

Country	Producer
Japan	Solvay Chemicals
	Lonza
	Spiga NORD
	Evonik
	Sakamoto
Europe	Hangzhou J & H Chemical
	Silver-Unchemical
China	Stokely Van camp
USA	Parry Enterprises
India	Fine Organics

PG-oligomers used commonly as starting material for PGEs, lower hydrophilicity is attributed to the PG chain's ether oxygen linkages relative to the terminal hydroxyl groups; hydrophobicity, therefore, increases with increasing PG chain length. In practice, industrial PG-oligomers are generally limited to an upper limit of PG-5 or PG-6 due to intractable viscosities as one goes to higher chain lengths. PG fatty acid esters exhibit higher heat stability than the equivalent glycerides [22]. According to the published and patented information, much work has been done on PGEs in food, personal care, pharmaceuticals, bio-technology, and agriculture. Several corporations are involved in the commercial production of PGEs (Table 3.1).

PGEs thus appear to meet the key criteria of functional performance, low environmental impact, and desired stability for substitution of conventional lubes and merit further study as a possible improvement over conventional MO-based lubricant. Figure 3.1 shows the flowchart for the synthesis of PGEs.

Table 3.2 outlines the strengths and weaknesses of major base oils used in lubricant formulations. Every property of the lubricant is justified by considering a weight factor depending on the application's operational requirement. The choice of TBO and synthetic esters for a given application is usually dependent on the lubricant's characteristic functionality desired for that particular application. For example, at ultra-high temperatures in power plants, lubricants with fire resistance characteristics

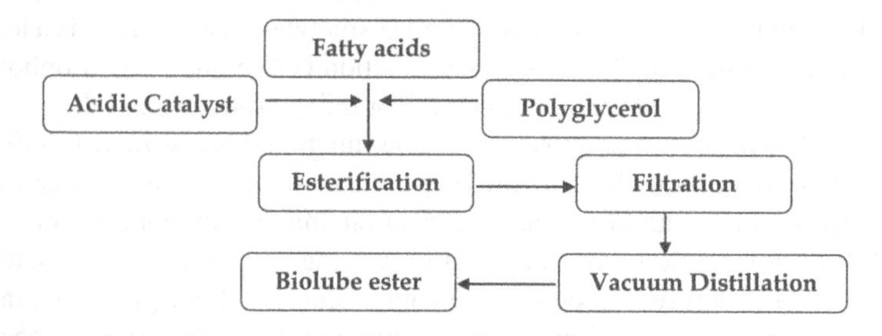

FIGURE 3.1 Flowchart for the synthesis of PGEs.

TABLE 3.2
Different Types of Base Fluids with Their Strengths and Weaknesses

S. No	Characteristics	MO	TBO	Neo Pentyl Ester (NPE)	Poly Glycerol Ester (PGE)
1	Viscosity Index	Moderate	Good	Very good	Moderate
2	Pour point	Poor	Good	Very good	Good
3	Liquid range	Moderate	Excellent	Very good	Very good
4	Ageing	Moderate	Excellent	Very good	Moderate
5	Thermal stability	Moderate	Excellent	Very good	Moderate
6	Volatility	Moderate	Excellent	Moderate	Moderate
7	Flashpoint	Poor	Excellent	Excellent	Moderate
8	Hydrolytic stability	Excellent	Excellent	Excellent	Good
9	Load-bearing capacity	Good	Excellent	Very good	Moderate
10	Non-Toxicity	Poor	Excellent	Good	Excellent
11	Biodegradability	Poor	Excellent	Very good	Excellent
12	Corrosion protection	Excellent	Excellent	Very good	Moderate
13	Solubility of additives	Excellent	Good	Very good	Good
14	Compatibility	Good	Excellent	Moderate	Good
15	Price relation against MO	Low	Medium	High	Moderate

are desired while automotives demand lubricants that assist in cold starting at low temperatures. In these cases, synthetic esters can be the best choice.

On the other hand, vegetable oil-based lubricants are ideal for lubrication of sawmill blade or chain drives. In these applications, the lubricant is used on a "once-through" basis and hence cost becomes a key consideration; low toxicity is also highly desired. They are also well suited for low to medium pressure hydraulic systems or lightly loaded gear drives operating at temperatures below 100°C, and there is little chance of water ingress or high contamination [23, 24].

It may be concluded that each type of TBO and synthetic ester is just a balancing act between desirable and undesirable characteristics. For example, the interaction between lubricant molecules and surfaces strongly influences the lubricant's behavior under the mixed lubrication regime. In such cases, the fluid's chemical and physical reactivity with the surfaces characterizes the contact friction. Many synthetic fluids exhibit excellent chemical and physical characteristics, including good oxidation and thermal stability. On the other hand, the same fluids may show moderate or poor lubricant performance under the mixed lubrication regime.

The esters derived from TBOs, and synthetic PGs and their derivatives, can be used in industrial applications where there is an increasing need for safer and more biologically-derived lubricants and additives. Climate, health, and safety concerns are currently bringing about a fresh look at various alternative lubricants. The market penetration of bio-based lubricants into the domain of petroleum-derived lubes has been slow, primarily because of cost and supply chain considerations. However, with significant improvement in biodegradable byproduct utilization from bio-fuels and other bio-based industries, bio-lubricants appear to be a promising alternative to conventional lubricants in selected applications in the future. The method used to produce bio-lubricant by utilizing PG may offer an efficient, economical, and greener route [25–28].

The studies discussed in this chapter include the tribological performance behavior of a lubricant derived as the oleyl ester of Karanja oil, hereafter referred to as Karanja Ester (KE), and a synthetic PGE, polyglycerol-4 polycaprylate (PGPC), as environmentally benign and biodegradable lubricants for industrial applications. The enhanced tribo-performance behavior of KE represented by its anti-friction and anti-wear characteristics is justified using molecular dynamic (MD) simulations and experimentation on tribo-testers. Superior anti-friction and anti-wear behavior of KE relative to MO is revealed from MD simulations, showing higher binding energy over the iron surface than MO. Quantum chemical calculations have also been deployed to evaluate the mean square displacement and adsorption of lubricant molecules over the iron surface. The studies show higher stability of KE lubricating film than MO over the iron surface. PGs, a family of inter-molecular glycerol ethers obtained by self-polycondensation of refined glycerine with the elimination of water molecules, are non-toxic, biodegradable, bio-compatible, and versatile building blocks due to their multiple hydroxyl groups. Such studies will help lubricant designers and tribologists design the required lubricant molecules and understand their lubrication characteristics.

3.2 EXPERIMENTAL STUDIES OF KE AND PGPC

Esterification of Karanja oil was performed by following a two-step method consisting of Step I: Saponification, and Step II: Esterification. The saponification of oil was carried out by refluxing alkaline IPA solution and Karanja oil mixture for 12 hours. The obtained saponified Karanja oil fatty acid was subjected to acid-catalyzed esterification in the presence of p-toluene sulfonic acid with oleyl alcohol. Toluene was used as a solvent in the reaction mixture. The reaction mixture was then refluxed for 7–8 hours, and the final product was obtained after concentration under reduced pressure. The esterification of PG was conducted in a round bottom flask fitted with a thermometer, a condenser for refluxing and a Dean and Stark receiver. A chemical reaction occurs between PG-4 and caprylic acid to produce an ester, eliminating water in a condensation reaction, referred to as Fischer esterification, as shown in Figure 3.2. All the reaction steps are reversible. Hence, according to Le Chatelier's principle, the forward reaction is favoured by taking an excess of PG or by eliminating water [29]. A mixture of one mole of PG-4, 0.9 moles of caprylic acid, and 1.5% (w/w %) of a catalyst by weight of the reaction mixture was refluxed with solvent toluene at 120±1°C. The refluxing was continued until the theoretical quantity of water was collected in the receiver. The reaction was completed in 9–10 hours. The acid value of the product was noted every 60 minutes. Reduction in Total Acid Number (TAN) values (Figure 3.3) indicated the reaction's progress and was taken to be concomitant with ester linkages formed. The reaction was checked for completion using a TLC plate, and when the theoretical quantity of water was obtained. The product was brought to room temperature, filtered, and then washed with water until pH became neutral. The solvent, toluene, was collected through vacuum distillation.

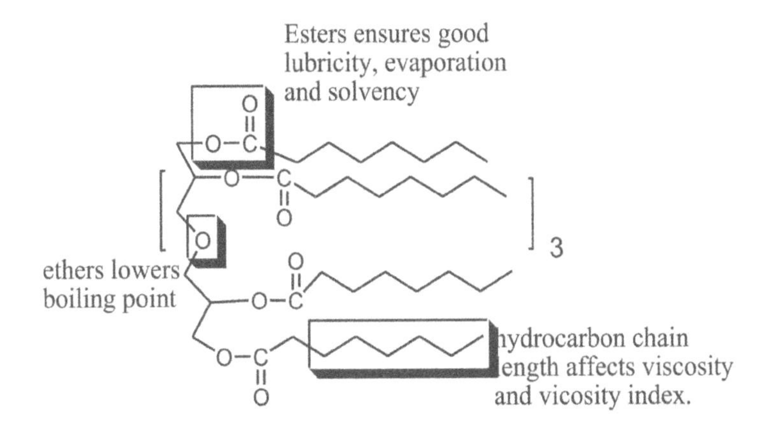

FIGURE 3.2 Synthesis of PGPC ester.

The product was percolated over basic alumina to remove the traces of unreacted acid and improve its acid value and colour.

3.2.1 Spectroscopic Analysis of KE and PGPC

The synthesized KE and PGPC esters were characterized by IR spectroscopy on a Perkin-Elmer IR spectrophotometer using KBr pellets. ^{1}H and ^{13}C NMR spectra were recorded using a Bruker Av III 500 MHz spectrometer.

3.2.1.1 IR and NMR Studies

The synthesized KE and PGPC ester were characterized by FTIR and FT-NMR spectroscopy. The peaks obtained from FTIR and FT-NMR spectroscopy of reactants and synthesized products are summarized in Tables 3.3 and 3.4 and clearly

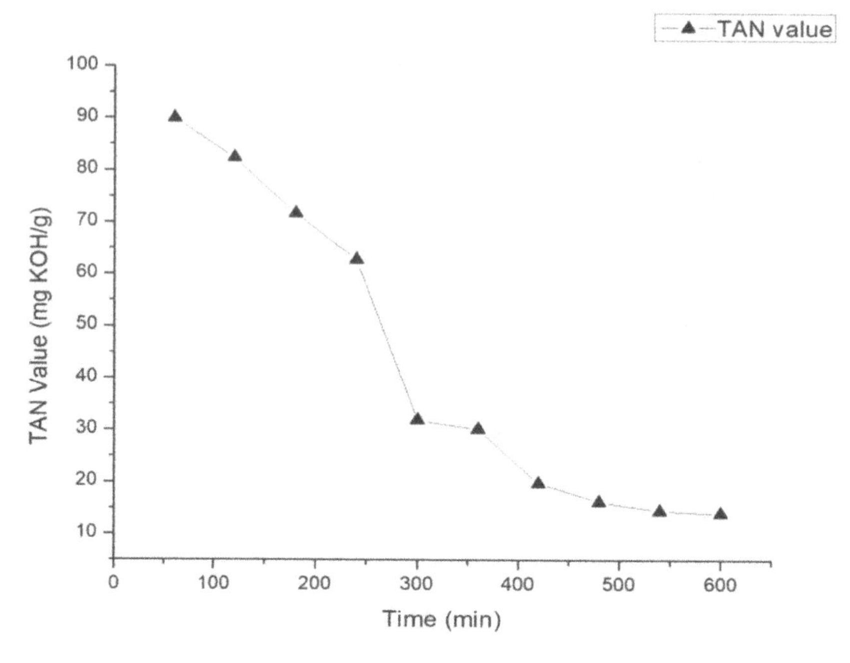

FIGURE 3.3 TAN of PGPC ester as a function of reaction time.

TABLE 3.3
IR and NMR Data of KE

Analysis	Functional Group	Peaks (cm^{-1})	Oleyl Alcohol (Reactant 1)	Karanja Oil Fatty Acid (Reactant 2)	KE (Product)
FTIR	sp^3C-H stretch	2850–3000	Present	Present	Present
	sp^2C-H stretch	3010–3100	Present	Present	Present
	C=O fatty acid stretch	1755–1760	Absent	Present	Absent
	C=O acyl ester	1735–1750	Absent	Absent	Present
	O-H bend	1100–1450	Present	Present	Absent
	-O-H stretch	3600–3500	Present	Present	Absent
^{1}H NMR (ppm)	t-CH_3	0.5–2	Present	Present	Present
	Methylene H	1.2–1.3	Present	Present	Present
	H adjacent C-O	4.11–4.32	Present	Present	Present
	-O-H-	3.5–3.75	Present	Absent	Absent
	-COOH	10–12	Absent	Present	Absent
	-COO-CH	3.3–5	Absent	Absent	Present
	=C-H	5.3	Present	Present	Present
^{13}C NMR (ppm)	-COOH-	160–173	Absent	Present	Absent
	-COOR-	135–150	Absent	Absent	Present
	-C-OH	45–75	Present	Absent	Absent
	sp^3C-H	16–34	Absent	Absent	Absent
	sp^3-CH_2	15–55	Present	Present	Present

TABLE 3.4
IR and NMR Data of PGPC

Analysis	Functional Group	Peaks (cm^{-1})	Poly Glycerol (Reactant 1)	Octanoic Acid (Reactant 2)	PGPC (Product)
FTIR	sp^3C-H bend	2850–3000	Present	Present	Present
	sp^2C-H stretch	3000–3100	Present	Present	Present
	C=O acid stretch (broad)	2400–3400	Absent	Present	Absent
	C=O acyl ester	1735–1750	Absent	Absent	Present
	C=O fatty acid	1755–1760	Absent	Present	Absent
	O-H bend	1100–1450	Present	Absent	Absent
	-O-H stretch	3600–3500	Present	Present	Absent
^{1}H NMR (ppm)	t-CH_3	0.5–2	Present	Present	Present
	Methylene H	1.2–1.3	Present	Present	Present
	H adjacent C-O	4.11–4.32	Present	Present	Present
	-O-H-	3.5–3.75	Present	Absent	Absent
	-COOH	10–12	Absent	Present	Absent
	-C-OO-CH	3.3–5	Absent	Absent	Present
^{13}C NMR (ppm)	-COOH-	160–173	Absent	Present	Absent
	-C-OH	45–75	Present	Absent	Absent
	-COOR-	135–150	Absent	Absent	Absent
	sp^3C-H	16–34	Absent	Absent	Absent
	sp^3-CH_2	15–55	Present	Present	Present

show that the intended reactions have occurred. The FTIR spectrum of KE in Figure 3.4(a) shows strong stretching bands with transmittance between 0 and 35% at wavenumber 1187 cm^{-1}, 1248 cm^{-1}, 1655 cm^{-1}, 1739 cm^{-1}, 2925 cm^{-1}, 3006 cm^{-1} corresponding to C-O (ester moiety stretch), C-C (alkane stretch), C=C (olefin stretch), C=O (carbonyl stretch), -CH (alkane stretch), =CH (olefin stretch), and bendings of alkane between 1350 and 1465 cm^{-1}, thus, confirming the formation of oleyl ester of Karanja oil.

FTIR spectrum of the PGPC, as shown in Figure 3.4(b) is consistent with the expected structure. The broad vibrational peaks of methylene and methyl units of the alkyl chain of PGPC are in the range of 2800–3000 cm^{-1}. The IR peaks appearing at around 1741 cm^{-1} and 1153 cm^{-1} were due to carbonyl and carbon-oxygen bond stretching of the ester group. The lower frequency strong band at1460 cm^{-1} is due to aliphatic esters. A peak at around 727 cm^{-1} is due to long alkyl chains present in the synthesized lube.

Proton shifts further confirmed the chemical structure of PGPC ester in ^{1}H and ^{13}C NMR spectra (Figures 3.5 and 3.6). Functional groups and electronegative heteroatoms attract the electrons, whereby the protons of the neighbouring carbon atoms experience downfield shifts. ^{1}H NMR spectra of PGPC and fatty acid esters have a set of CH-, CH_2-, and –OH PG protons. The –OCH_2 bond protons are confirmed from the signals of 4.11–4.32 ppm in both PGPC and KE cases. Protons located at the double bonds (vinylic protons) were assigned to the peaks at 5.17–5.3 ppm.

(a)

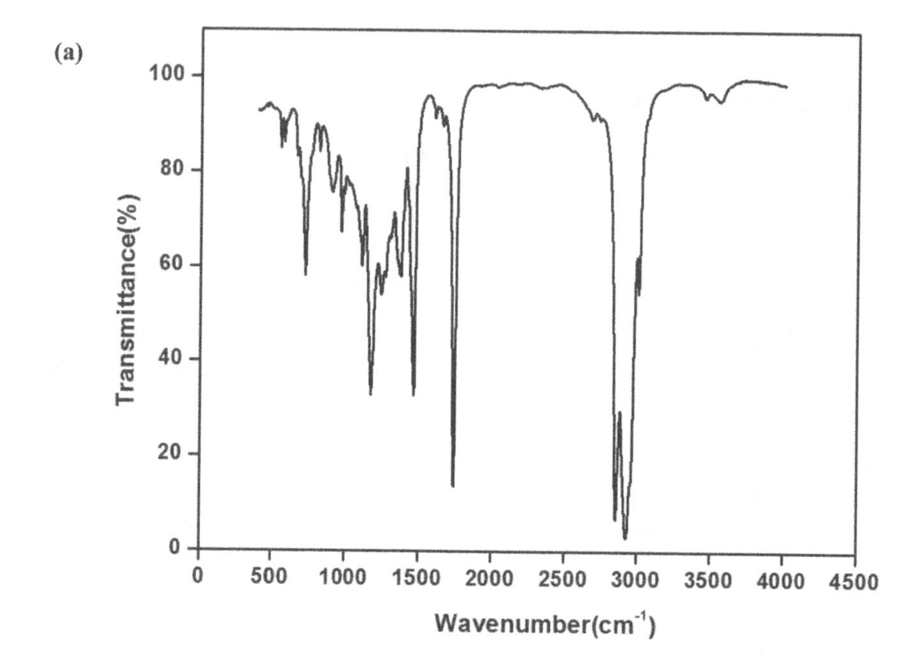

(b)

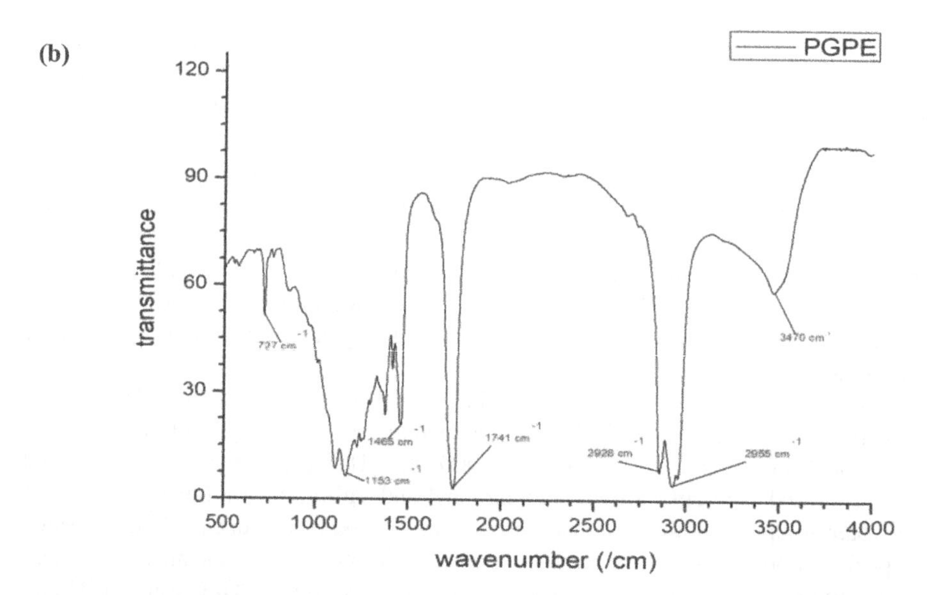

FIGURE 3.4 FTIR spectra observed of (a) KE and (b) PGPC ester using Fourier transform spectrophotometer of 400–4000 cm^{-1} range.

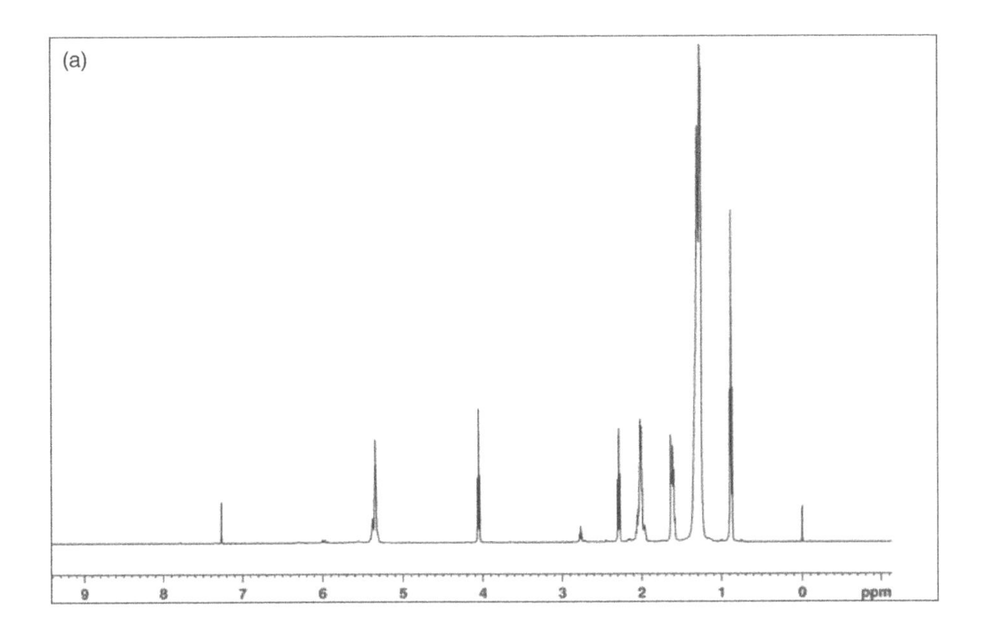

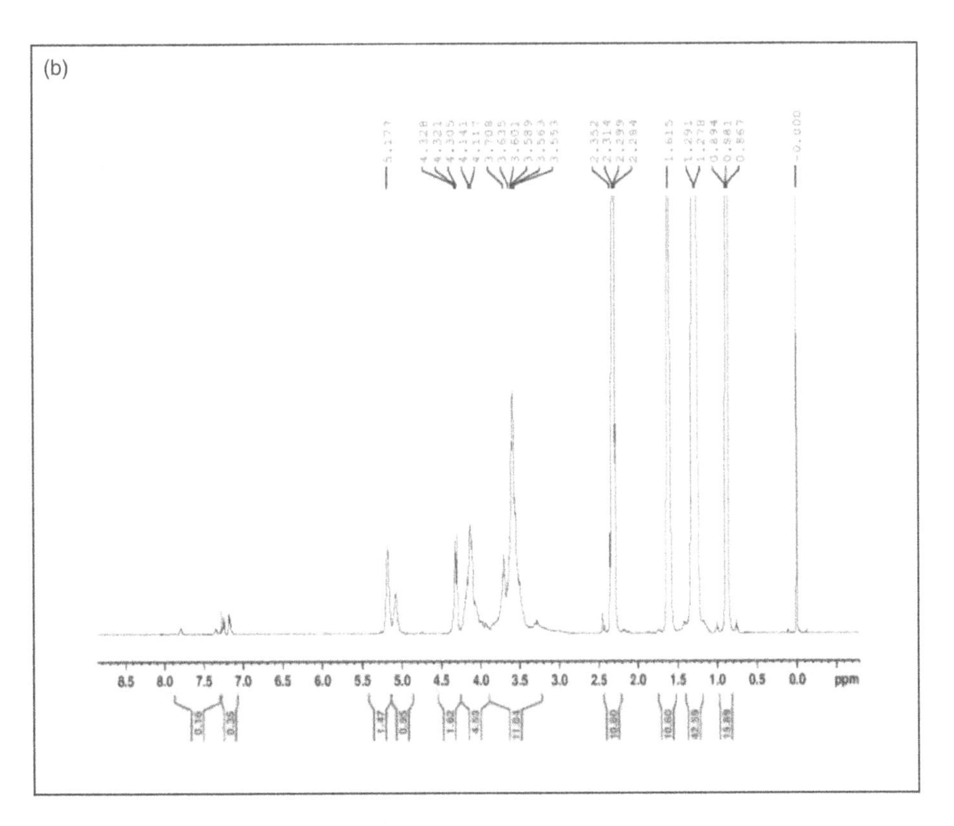

FIGURE 3.5 ^{1}H NMR of (a) KE and (b) PGPC ester.

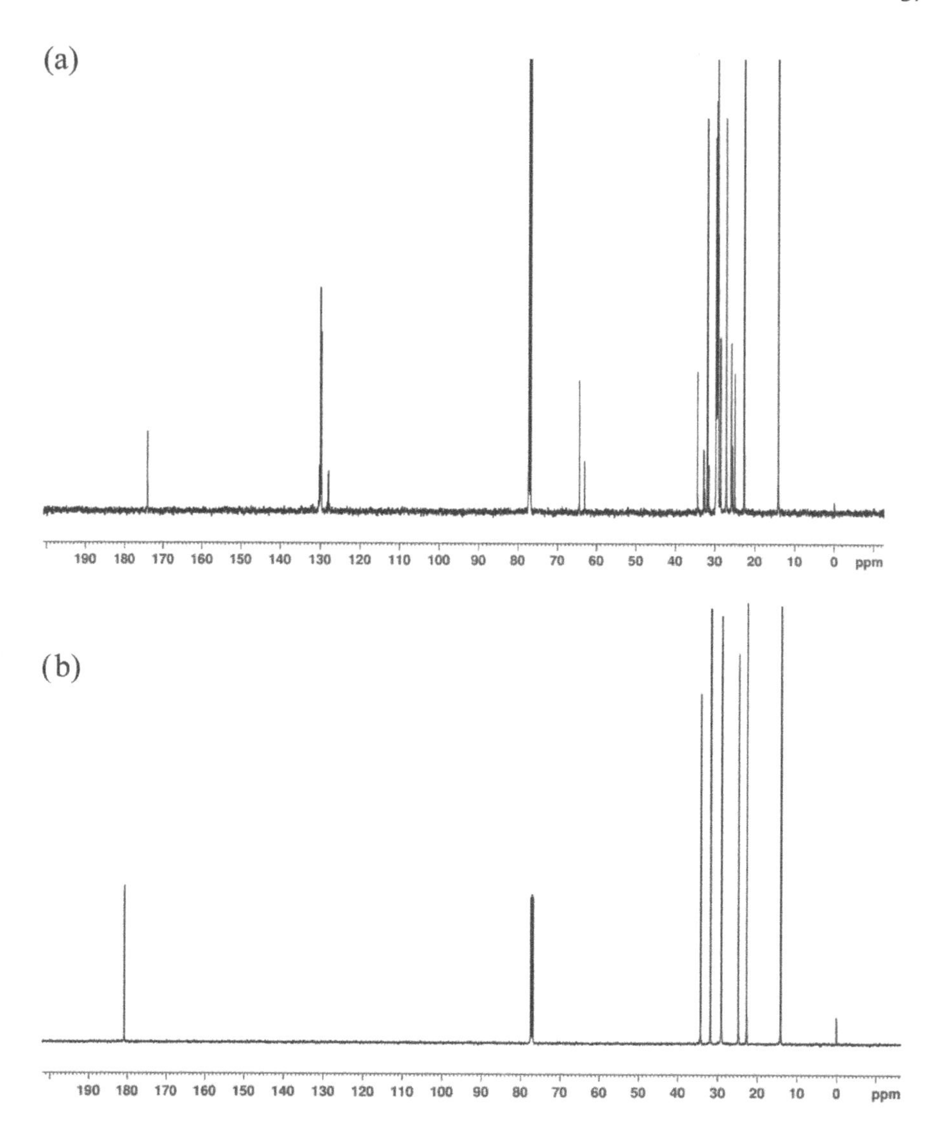

FIGURE 3.6 ^{13}C NMR of (a) KE (b) PGPC ester.

3.2.2 Physico-Chemical Characterization of KE and PGPC

The physico-chemical characteristics viz. TAN, kinematic viscosity, VI, and density of synthesized esters were determined per American Society of Testing Materials (ASTM) procedures D-974, D-445, D-2270, and D-4052, respectively. The dynamic/kinematic viscosity and density of PGPC ester were determined using the SVM3000 Anton Paar Stabinger viscometer at temperatures ranging from 30°-100°C. The molecular-level interactions of lubricants and additives provide suitable structure regulating correlations to their characteristic properties. The length of fatty acid or alcohol chains, branching, and the branched positions on the carbon

TABLE 3.5
Physico-Chemical Characterization of Lube Base Oils

S. No.	Characteristic Property	Mineral Oil	KE	PGPC
1	Density (g/cm^3)			
	at 15°C	0.845	0.899–0.920	0.979
	at 30°C	0.832	0.857–0.878	0.935
2	Kinematic viscosity (mm^2/s)			
	at 40°C	13.28	14.24–15.90	50.0
	at 100°C	3.07	3.84–4.41	7.27
3	Viscosity Index	80.98	174–208	124.9
4	TAN (mg KOH/g oil)	2	0.86	1.04

chains significantly influence the viscosity, VI, and compatibility of the lubricants. Similarly, the number of ester groups affects the evaporation and lubricity, while structural hindering influences the lubricant's thermal and hydrolytic stability [30]. Viscosity is resistance to flow due to cohesive forces between layers of lubricants. The viscosity values help us comprehend various potential applications of the developed esters [31]. Temperature, pressure, the extent of branching, functionality, and ease of rotation of the bonds that make up the molecule are factors that invariably affect the lubricant's viscosity. Other factors like density and concentration of additives in the base stock may also play a role [32, 33].

The viscometric properties of esters are dependent on their molecular weight and acid branching. The MO and prepared KE and PGPC have viscosities of 13.28, 14.24–15.90, 50 cSt at 40°C and 3.07, 3.84–4.41 and 7.27 cSt at 100°C, respectively, as shown in Table 3.5. The molecular weights of the PG and its esters were 387 g/mol and 1112 g/mol (Table 3.6). The higher the molecular weight, the higher is the viscosity [34]. The VI of the MO and synthesized KE and PGPC are calculated to be 80, 174–208, and 124. Acidity is practically negligible, which, apart from indicating completion of the reaction, is also directionally preferred as being less likely to induce corrosion by the lubricant on metal surfaces. Densities smoothly decrease with increase in temperature and are also affected by the length of the alkyl chain [35]. The TAN value of the synthesized esters is very low. However, certain acid components tend to increase the TAN value, which may change kinematic viscosity and degree of oxidation of the compound, affecting its stability [36].

TABLE 3.6
GPC Average Molecular Weight of PG and PGPC

		Molecular Weight (g/mol)	
S. No	Sample	Calculated	Theoretical
1	PG-4	387	300
2	PGPC	1112	1072

The molecular weight distribution studies of PG and PGPC were also carried out with a gel permeation chromatography system (Model: 1260 Infinity, Agilent Technologies. Ltd) equipped with an evaporative light scattering detector. An Agilent PL gel 5μm Mini Mix-D column with 250 mm length, 4.6 mm internal diameter, along with an Agilent PL gel 5 μm Mini MIX-D guard column of 50 mm length, 4.6 mm inner diameter was used. Tetrahydrofuran was used as a mobile phase at a flow rate of one ml per minute. The temperature was maintained at 32°C.

As a general rule, esters prepared by linear acids or alcohols have better lubricity than those made from branched acids and alcohol [37]. The presence of multiple hydroxyl groups results in higher molecular weight. The increase in molecular weight, number of functional groups in the esters, fatty acid chain length, and the degree of branching lead to improved viscosity as compared with other MO and linear ester lube base stocks. There should be an increase in branching, reduction in acyl chain length, and less symmetry in the molecule to reduce the pour point. Polyol esters offer these advantages over MOs [38]. It is also observed that as the chain length of esterified PG increases, its viscosity increases without drastically reducing polarity [39]. Ester and carbonyl groups may significantly impact viscosity in polyfunctional compounds containing hydrogen bonding, as the oxygen atoms can act as hydrogen bond acceptors. Thus, ether linkages help improve the viscosity of PGPCs compared to MO and neat TBOs [40].

3.3 TRIBOLOGICAL BEHAVIOR OF KE AND PGPC

Tribological behavior represented by anti-friction and anti-wear properties of the synthesized esters (KE, PGPC) and MO was established based on wear scar diameter (WSD) and friction coefficient (COF) using a four-ball tribotester (Figure 3.7). The MO used for the study was a Group II MO base-stock and mostly composed of linear saturates.

For tribo-performance evaluations, the steel balls were cleaned in hexane using ultrasonic vibrations. The balls were then arranged in a tetrahedral geometric pattern, where the lower three balls were placed in a ball pot filled with test lubricant sample. The upper fourth ball was attached to a rotating spindle and rotated over the bottom three stationary balls under a steady load for one hour. The standard test conditions of ASTM D-4172 B (Table 3.7) were maintained. The scars developed on the steel balls were studied using an optical microscope and their images captured.

3.3.1 Tribo-Performance of KE and MO

The friction behavior of KE and MO is shown in Figure 3.8. It is observed from the figure that the MO has higher friction values as compared to the KE over the entire test duration. The coefficient of kinetic friction recorded in KE and MO is 0.09 and 0.15 respectively. This suggests that the kinetic friction in KE is almost 35% less than the MO.

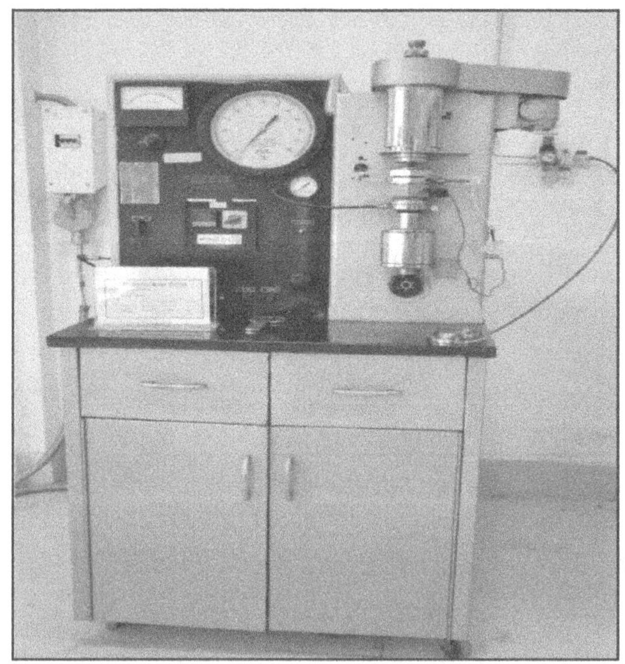

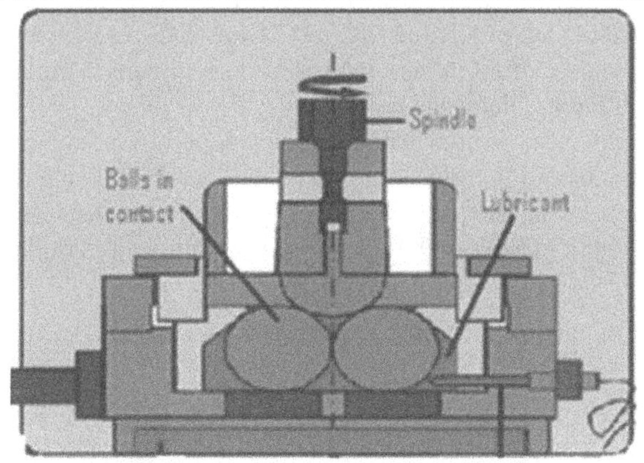

FIGURE 3.7 Four-ball tribo tester.

TABLE 3.7
Operating Parameters of Four-Ball Tester

S. No	Operating parameter	
1	Load (N)	392 N
2	Speed (RPM)	1200
3	Lubricant temperature (°C)	75°C
4	Test duration	3600 s

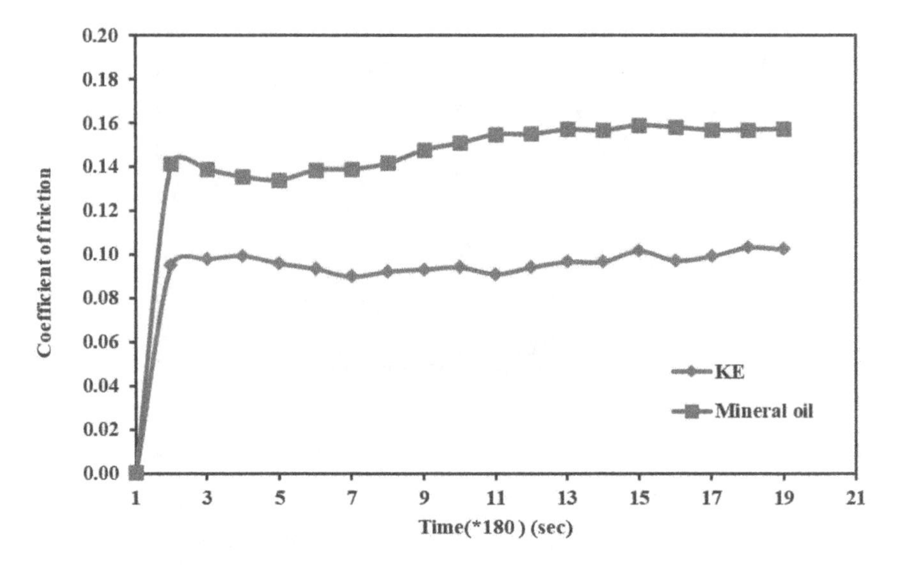

FIGURE 3.8 Friction behavior of KE and MO.

Similarly, lubricants anti-wear behavior in terms of WSD is shown in Figure 3.9. The WSD observed in case of KE and MO is 0.72 and 1.35, respectively. It is revealed that the WSD in the case of KE lubricated ball test specimens is almost 45% smaller than that of the MO lubricated specimens.

The friction and wear in lubricated contact depend on lubricant's structural and chemical aspects and the physical phenomenon on solid surfaces. Hence, the difference in the friction and wear behavior of the two oils can be attributed to the presence of polar groups in KE, leading to strong interactions between the

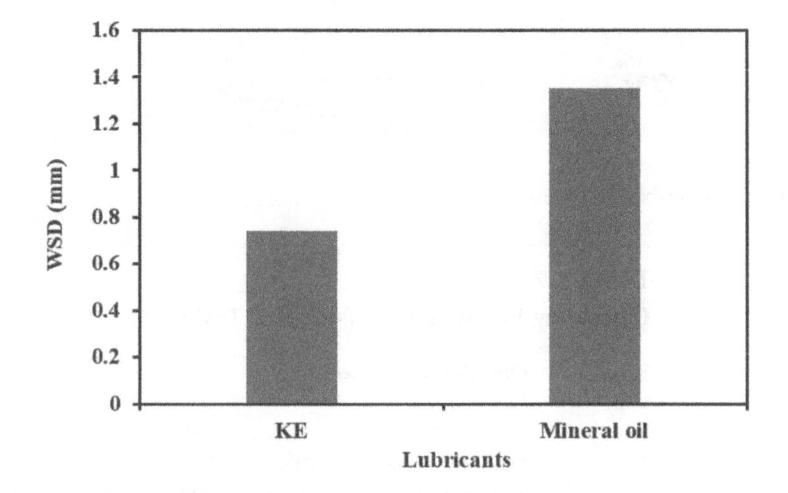

FIGURE 3.9 Wear behavior of KE and MO.

surface and twisted conformation, thereby, providing enhanced surface protection compared to MO.

3.3.2 Computational Studies of KE

The tribo-performance behavior of lubricants is greatly influenced by their ability to wet the surfaces and adsorb themselves, forming high strength films. Computational studies were undertaken to determine lubricants' adsorption and diffusion behavior on a steel substrate using Material Studio 6.0 software. Although oxides and hydroxides cover the metal surfaces, it is assumed that the Fe surface is clean to simplify the calculations. Characteristic properties such as interaction energy, diffusion coefficient, mean square displacement of atoms, and adsorption energy of the lubricants were determined using Forcite and Adsorption locator modules.

3.3.2.1 Geometry Optimization and Molecular Modeling of KE

The lubricant was modelled based on the constituent compounds, i.e. oleyl stearate, oleyl palmitate, oleyl linoleate, oleyl oleate oleyl linolineate of KE. The MO was modelled as n-hexadecane as a reasonable approximation discussed in the literature. The complete geometry optimization of oleyl esters of fatty acids was performed using the DMol3 method without any spin restrictions. Figure 3.10 shows the optimized structures for the lubricant's main constituents (KE and MO).

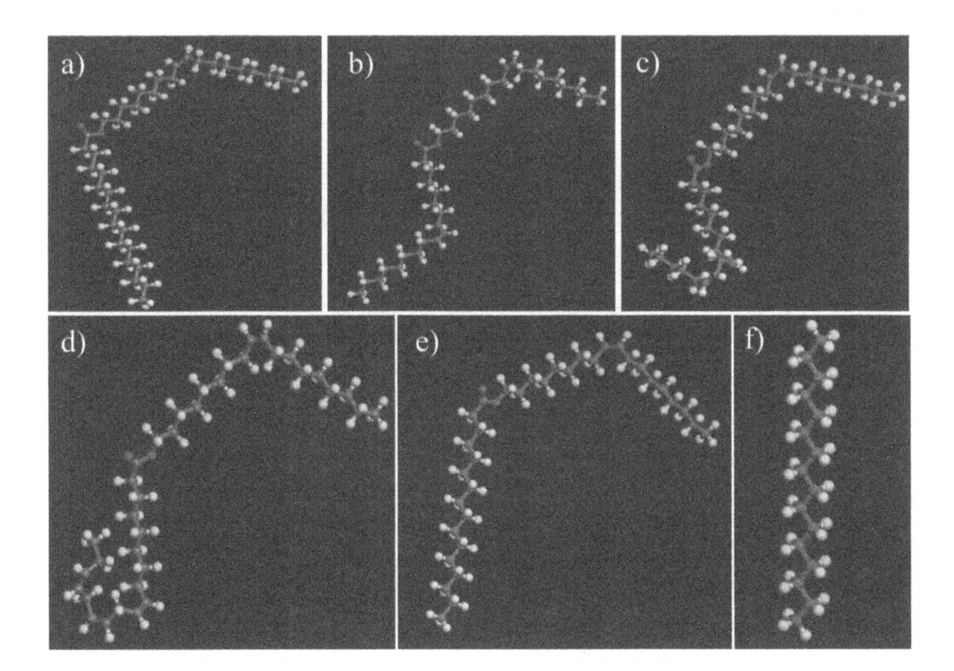

FIGURE 3.10 Optimized structures for the constituents of the lubricant (a) oleyl stearate, (b) oleyl oleate, (c) oleyl linoleate, (d) oleyl linolineate, (e) oleyl palmitate, and (f) n-hexadecane.

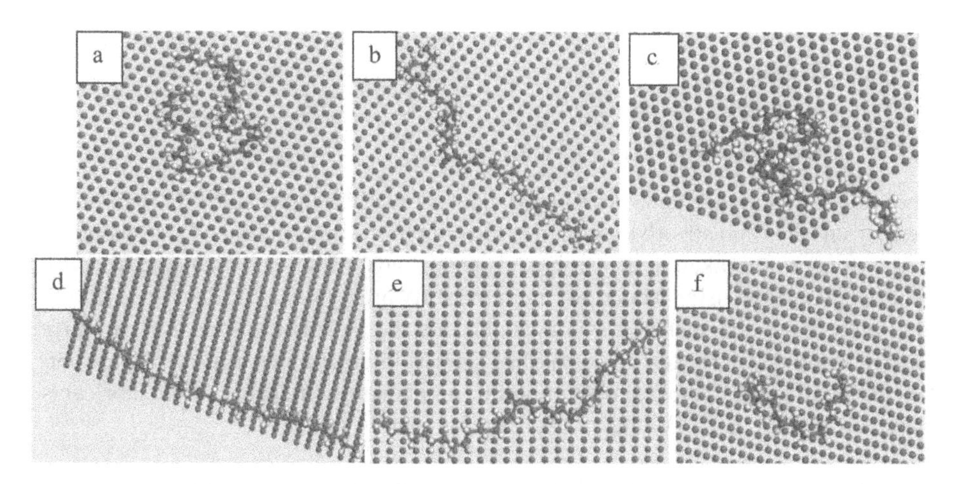

FIGURE 3.11 Mounting of the constituent compounds of the lubricant (a) oleyl linolineate (b) oleyl oleate, (c) oleyl linoleate, (d) oleyl stearate, (e) oleyl palmitate, and (f) n-hexadecane on the iron surface.

The oleyl ester molecules were mounted on the iron surface using the build layer section, as shown in Figure 3.11.

The VWN (Tvosko, Wilk, and Nusair) local density approximation functional was employed for the geometry optimization calculations of lubricant molecules. The optimization was carried out until the forces on all atoms were 0.1 eV/Å. The lubricant dynamics over the surface were carried out under N-V-T (canonical ensemble type of system) using a time step of 0.1 fs. The temperature was limited to 0.5ps and 298 K, respectively. The interaction energy over the metal surface was evaluated using,

$$\mathrm{E}_{interaction} = \mathrm{E}_{total} - \left(\mathrm{E}_{surface} + E_{oil\,component}\right) \tag{3.1}$$

where

E_{Total} = Total energy of the system comprising surface and component of oil
$E_{Surface}$ = Energy of the surface without any oil component over it
$E_{Component\,of\,oil}$ = Energy of the oil formulation (component)
$E_{interaction}$ = Affinity of an oil component over the surface in terms of the interaction energy

3.3.2.2 Quantum Chemical Calculations of KE

Quantum chemical calculations were performed to determine the adsorption energy of the lubricant. The lubricant component having the most substantial impact on the iron surface was selected, and its adsorption energy at different temperatures was determined using Adsorption Locator Module. Monte Carlo searches were performed to obtain the most stable adsorption configuration and the adsorption energy calculated. The adsorption studies were performed at constant volume and temperature with a time step of 0.1 fs and number of steps limited to 1,000,000.

The calculations to determine diffusion coefficients were performed considering SAM (Self Assembled Monolayer) of n-hexadecane to mimic MO and oleyl ester of Karanja oil, i.e. KE to quantify the spreading speed of the components. The simulations were run for 3 ns with 1 fs time step in N-V-T at 300 K. The Mean Square Displacement (MSD) is the sum of displacement of all the atoms in x, y, and z directions which provides information on the lubricant's spreading ability or its mobile nature over the surface.

3.3.2.3 Binding Energy

The MD simulations performed using one molecule of each component of chemically modified KE and n-hexadecane revealed individual component's interaction energy on the Fe (110) surface. Figure 3.12 shows a comparative assessment of their interaction energy at room temperature. The negative sign in interaction energy indicates lubricant component's affinity on the Fe surface [41].

It is revealed that oleyl oleate (OLOA) has the highest interaction energy of -118.462 kCal/mol, which suggests that OLOA has a strong affinity for the Fe(110) surface. OLOA constitutes 54% of the total composition of KE, which is the highest among all the constituents. Similarly, oleyl stearate (OLSA), which constitutes 2.92% of the total composition of KE, has the lowest interaction energy of -78.516 kCal/mol. On the contrary, n-hexadecane has the lowest interaction energy of the order of -18.987 kCal/mol on the Fe surface. Thus, due to the higher interaction energies, KE molecules show great attraction for Fe surface and form strong and stable lubricating films. As a result of this, KE exhibits superior tribological performance when compared with the MO.

The bond lengths of C=O and C-O of ester moieties determined before adsorption was 1.211 Å and 1.358 Å, respectively. It is observed that as the adsorption of KE proceeds, the bond lengths show an increase of 1.5% and 2.3% for C=O and C-O, respectively. This also reflects the type of bonding, i.e., monodentate or bidentate type between the surface and the ester. It is observed that as the adsorption

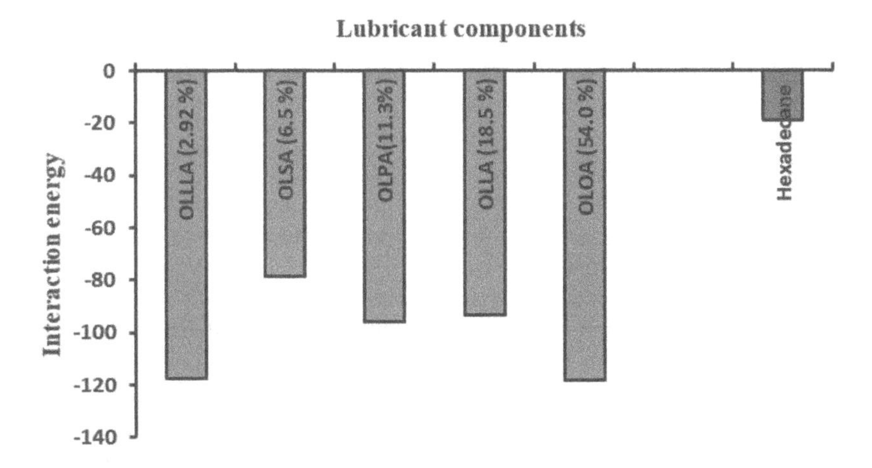

FIGURE 3.12 Interaction energies of the individual lubricant components.

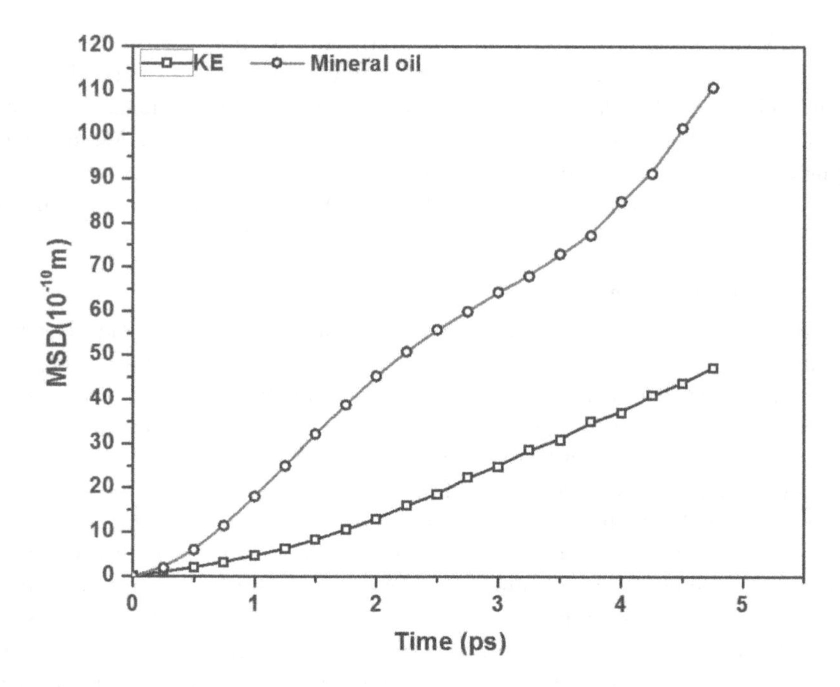

FIGURE 3.13 Variation in MSD with time

proceeds, synergistic bonding follows either monodentate (C-O-Fe) or bidentate type (C=O-Fe, C-O-Fe) with the substrate. In case of MO, there is no such considerable change in the hydrocarbon chain C-C or C-H bond lengths.

3.3.2.4 Diffusion Behavior

The results shown in Figure 3.13 suggest that the MSD values are relatively low for KE compared to MO. At the end of the simulation time, the MSD value for KE is 47.35 $Å^2$/ps compared to 110.76 $Å^2$/ps for MO. The lower values of MSD for KE can be attributed to the fact that i) It has strong interaction energy with the surface, which restricts its motion, and ii) In case of non-polar MO, the diffusion coefficient is high because of the weaker interaction energies and its almost twisted molecular conformation.

The value of the diffusion coefficient for MO is 3.39, which decreases to 2.01 in KE. Such behavior can be attributed to the fact that the polar interaction sites of polar esters demonstrate greater affinity to the metal surface and cause the molecules to adopt twisted conformations.

The diffusion is hampered by strong polar interactions in KE, while the trend gets reversed for MO as it has weak non-polar interactions. Furthermore, the atomic structure of model MO, n-hexadecane, contains all sp^3 hybridized carbon atoms that diffuse rapidly over the steel surface. TBOs, on the other hand, contain both saturated (sp^3 hybridized carbon atoms) and unsaturated (sp^2 hybridized carbon atoms), of which the latter bind through polar functional groups that lead to more stable lubricating films as compared to MO.

TABLE 3.8
Tribological Performance Results of PGPC

S. No	Sample	WSD (mm)	COF
1	PGPC	0.798	0.125
2	PGPC (Repeated)	0.792	0.095
3	PGPC (Average)	0.795	0.110

3.3.3 Tribo-Performance of PGPC

The average values of the COF and WSD of the synthesized ester observed were 0.110 and 0.795, reflecting a reduction in COF and WSD to about 20% and 42%, respectively to MO. As summarized in Table 3.8 and, Figures 3.14 and 3.15, the friction and wear results reveal that the synthesized ester yields improved tribological performance. The smoother surface of steel balls is observed in Figure 3.15. The PGPC adsorbs on the metal surfaces and forms boundary lubricating films. These adhered lubricant molecules protect the steel test ball's surfaces from further damage, resulting in smaller WSDs.

The presence of large molecular surface area in one plane enhances the anti-wear properties. In PGPC, the molecule has a branched and staggered structure with high molecular weight, resulting in improved lubricity. Therefore, the tribo-properties of PGPC are better than MO; however, they are slightly lower than those of TBO.

Meanwhile, in PGPC, the chemical modification retains their positive attributes such as exceptional lubricity, high viscosity, good corrosion behavior, and low

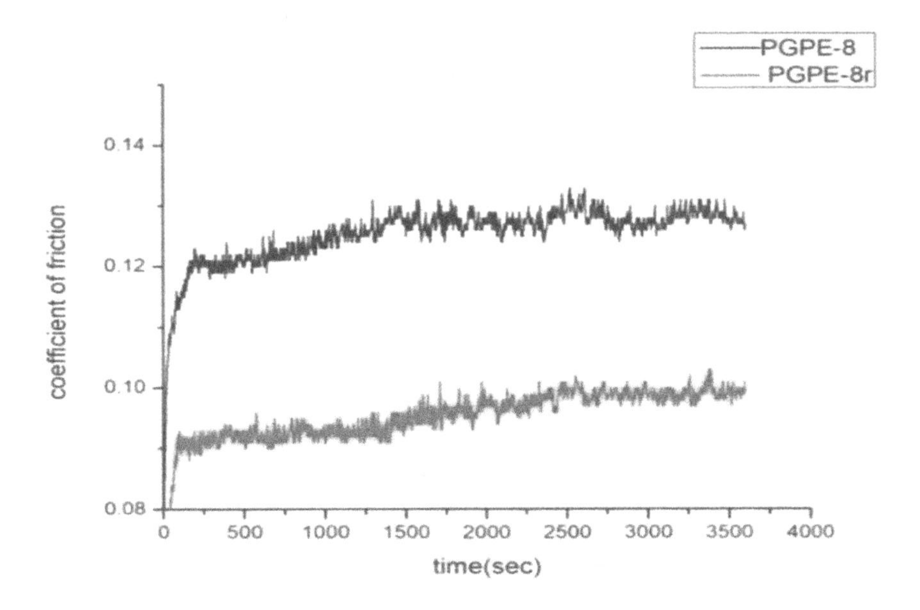

FIGURE 3.14 Friction behavior of PGPC ester.

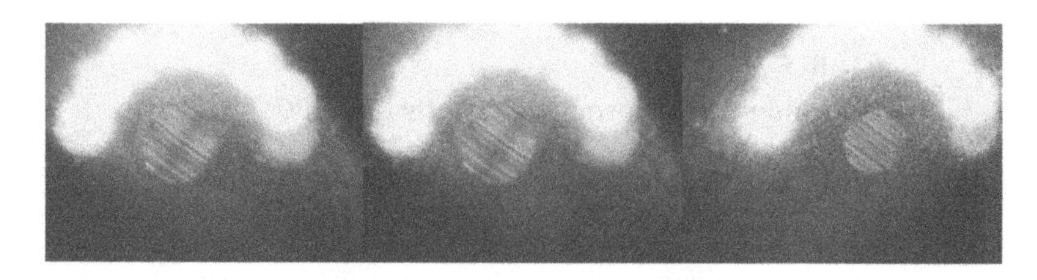

PGPE-8

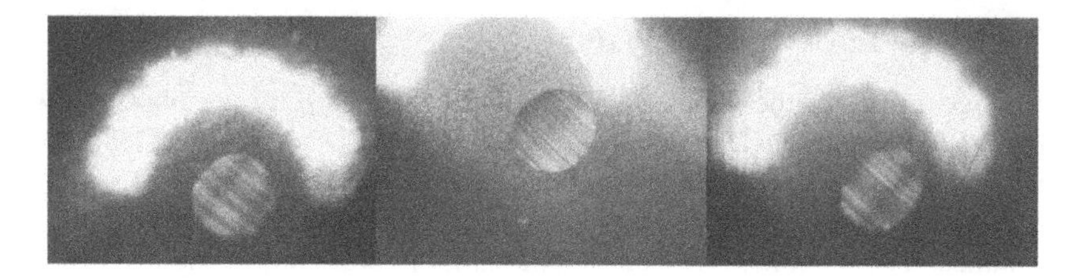

PGPE-8 repeated

FIGURE 3.15 Wear scar images for PGPC ester.

evaporation losses. The lower friction and wear of PGPC ester is due to stronger boundary film formation. Hence, the studied esters can be considered potential friction and wear reducers and can be used as industrial lubricants.

The post-experimental studies were performed using SEM to examine the surface morphology of wear scars on the used ball test specimens lubricated with PGPC. Figure 3.16 shows the SEM images of the used ball specimens lubricated with PGPC.

It can be observed from the figure that the wear scars have a smooth finish with minor abrasion marks along the sliding direction, further corroborating the data obtained for WSD and COF. The wear marks do not correspond to any severe wear damage in any test specimens. Rubbing wear with smoothening of surfaces is ascertained [42, 43].

However, due to software limitations, extensive simulations on the large and complex PGPC structures could not be carried out. It would be interesting to understand the atomic-scale behavior of these categories of novel cluster molecules and carry out such studies as future research.

3.4 BIODEGRADABILITY STUDIES OF LUBE BASE OILS

The demands placed on lubricants are changing rapidly from better performance throughout their lives. The lubricants are expected to be eco-friendly and eventually biodegradable. The use of biodegradable products in industrial and automobile sectors is increasing rapidly. A lubricant is considered to be biodegradable only when it has proven capability to decompose within a specific period through natural biological processes.

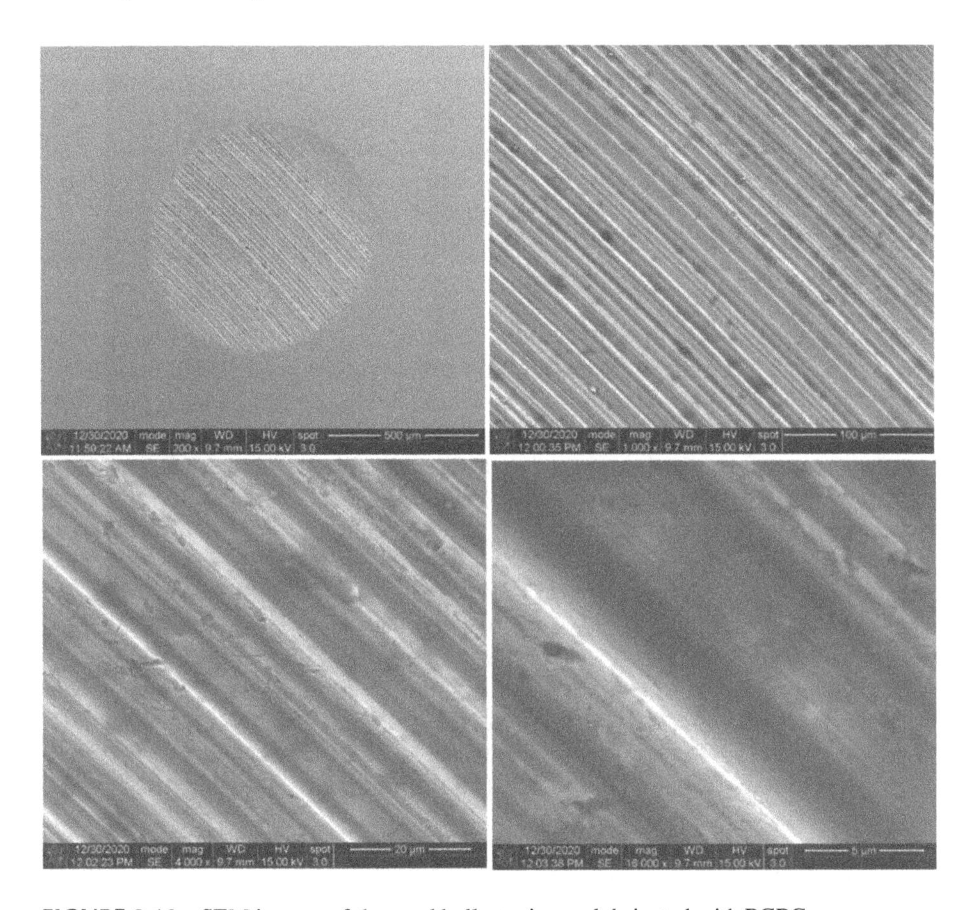

FIGURE 3.16 SEM images of the used ball specimens lubricated with PGPC.

The general biochemistry of microbial attack on esters involves ester hydrolysis, β- oxidation of lengthy hydrocarbon, and oxygen attack on aromatic nuclei [44–46]. The main factors which slow down the microbial breakdown are:

- Position and degree of branching;
- The degree to which ester hydrolysis is suppressed;
- Degree of saturation on the molecule; and
- Increase in molecular weight of esters.

The lube base stocks ultimate biodegradability characteristics were determined using a standard ASTM-D-5864-95 procedure in the present study. Biodegradability is assessed by considering the amount of CO_2 produced when the sample is under the attack of microorganisms in controlled aerobic aquatic conditions. This amount is then compared with the theoretical amount of CO_2 on the ester's complete oxidation [47].

The standard biodegradability test carried out shows that the vegetable oil and synthetic base stocks prepared in the laboratory range from fair to very good in biodegradability with 70–100% biodegradation in 28 days. However, the mineral-based lubricants show up to fair biodegradability, i.e., 15–35% biodegradation in 28 days [48].

Toxicity of the samples on natural sewage bacteria was determined by the modified method of Algal inhibition test No. L383A/179-185. The toxicity evaluation result indicates that the samples are non-toxic [49].

3.5 CONCLUSION

A comprehensive study to assess the lubrication behavior of chemically modified TBO (KE) and synthetic lubricant formulation developed by chemical transformation of glycerol to high-performance long service life green lubricants as potentially attractive alternatives to MO has been presented.

The characteristic anti-friction and anti-wear properties have been experimentally studied. The MD simulations and post-experimental studies have been presented to justify the tribo-performance behavior of the studied lubricants. We determine and deduce that:

- The oleyl ester of TBO Karanja oil (KE) has superior tribo-performance characteristics over MO of similar viscosity as constituents of KE are strongly adsorbed on the iron surface with interaction energies ranging between -78.516 kCal/mol to -118.462 kCal/mol forming strong chemical bonds as revealed from their interaction energies, whereas it is -18.987 kCal/mol in case of n-hexadecane (MO) which shows its weak affinity for the surface.
- The decrease in the spreading speed due to the strong affinity of KE justifies the adoption of flat conformation. It is almost twisted in n-hexadecane (MO), which shows weak interaction and high spreading speeds over the surface.
- The presence of polarity in esters causes them to get attached to the metal surface by forming chemical bonds with vacant d-orbitals of iron as indicated by increased bond lengths of carbonyl or ester moiety. Therefore, diffusion is hampered by strong polar interactions while the trend gets reversed for MO having weak non-polar interactions.
- A potential eco-friendly and biodegradable base oil developed using biocompatible PG and caprylic acid in the presence of an acidic catalyst results in product yields exceeding 90%. The synthesized PGPC is biodegradable and has excellent physico-chemical characteristics. The biodegradability and acceptance as food-grade lubricants in many countries make them particularly suitable for the applications in the food and agro-industry.
- Tribological evaluation of the synthesized PGPC showed good tribo-performance in terms of anti-wear and anti-friction properties.

Our initial studies indicate that PGPC possesses suitable properties as base stock for industrial applications. Further detailed studies of PGPC and correlations of experimental results with MD simulations are needed for the effective development of formulations for lubrication in industrial applications.

ACKNOWLEDGEMENTS

The authors are thankful to Director CSIR-IIP for permission to publish the research findings. We also thank CSIR, New Delhi, for providing the research fellowship. The Analytical Sciences and Tribology & Combustion Divisions of CSIR-IIP are acknowledged to support the present study.

REFERENCES

1. C. Lea, Energy savings through use of advanced biodegradable lubricants, *Industrial Lubrication and Tribology* 59 (2007) 132–136.
2. P. Dašić, F. Franek, E. Assenova, M. Radovanovic, International standardization and organizations in the field of tribology, *Industrial Lubrication and Tribology* 55(6) (2003) 287–291.
3. P. H. Jost, Tribology micro and macro economics: A road to economic savings. World Tribology Congress III, Washington, DC, *Tribology & Lubrication Technology* 61(10) (2005) 18–22.
4. G. Karmakar, P. Ghosh, B. Sharma, Chemically modifying vegetable oils to prepare green lubricants, *Lubricants* 5 (2017) 44.
5. R. Singh, S.K. Padhi, Characterization of jatropha oil for the preparation of biodiesel, *Natural Product Radiance* 8 (2009) 127–132.
6. A.B. Chhetri, M.S. Tango, S.M. Budge, K.C. Watts, M.R. Islam, Non-edible plant oils as new sources for biodiesel production, *International Journal of Molecular Sciences* 9 (2008) 169–180.
7. A. Murugesan, C. Umarani, T. Chinnusamy, M. Krishnan, R. Subramanian, N. Neduzchezhain, Production and analysis of biodiesel from non-edible oils—A review, *Renewable and Sustainable Energy Reviews* 13 (2009) 825–834.
8. F. Klocke, A. Schulz, K. Gerschwiler, Saubere Fertigungstechnologien–Ein Wettbewerbsvorteil von morgen, *Aachener Perspektiven*, VDI, Düsselorf (1996).
9. F. Ma, M.A. Hanna, Biodiesel production: A review, *Bioresource Technology* 70 (1999) 1–15.
10. L. Meher, D.V. Sagar, S. Naik, Technical aspects of biodiesel production by transesterification—A review, *Renewable and Sustainable Energy Reviews* 10 (2006) 248–268.
11. B. Fabiano, A. Reverberi, A. Del Borghi, V. Dovì, Biodiesel production via transesterification: Process safety insights from kinetic modelling, *Theoretical Foundations of Chemical Engineering* 46 (2012) 673–680.
12. S. Arumugam, G. Sriram, Effect of bio-lubricant, and biodiesel-contaminated lubricant on tribological behaviour of cylinder liner–piston ring combination, *Tribology Transactions* 55 (2012) 438–445.
13. W. Castro, D.E. Weller, K. Cheenkachorn, J.M. Perez, The effect of the chemical structure of base fluids on antiwear effectiveness of additives, *Tribology International* 38 (2005) 321–326.
14. M.T. Siniawski, N. Saniei, J. Pfaendtner, Tribological degradation of two vegetable-based lubricants at elevated temperatures, *Lubrication Science* 24 (2007) 167–179.
15. S. Boyde, Green lubricants. Environmental benefits and impacts of lubrication. *Green Chemistry* 4(4) (2002) 293–307.
16. T. Ushikusa, T. Maruyama, I. Nhya, M. Okada, Pyrolysis behaviours and thermostability of polyglycerols and polyglycerol fatty acid esters, *Journal of the Japan Oil Chemists' Society* 39 (1990) 314–320, 5.
17. D. Dermawan, Polyglycerol esters of oleic acid – estolide as renewable lubicating oil base stock, *Regional symposium of Chemical Engineering*, Singapore (2006).

18. B. De Meulenaer, B. Vanhoutte, A. Huyghebaert, Development of a chromatographic method to determine the degree of polymerization of polyglycerols and polyglycerol fatty acid esters, *Chromatographia* 51 (2000) 49–51.
19. J.B. Rodríguez, The food additive polyglycerol polyricinoleate (E-476): Structure, applications, and production methods, *ISRN Chemical Engineering* (2013) Article ID124767: 21.
20. K.S. Shikhaliev, N.V. Stolpovskaya, M.Y. Krysin, A.V. Zorina, V. Lyapun, F.I. Denis Zubkov, K.Y. Yankina, Production and emulsifying effect of polyglycerol and fatty acid esters with varying degrees of esterification, *Journal of the American Oil Chemists' Society* 93 (2016) 1429–1440.
21. V.K. Babayan, R.T. McIntyre. Preparation and properties of some polyglycerol esters of short and medium-chain length fatty acids, *Technical* 48 (1968) 307–309.
22. C.S. Miner, N.N. Dalton, Chemical Properties and Derivatives of Glycerine, Reinhold Publication Corp., New York, Printed in USA (1953) 4–21.
23. P. Nagendramma, Development of eco-friendly/biodegradable synthetic (polyol & complex) ester lube base stocks. Thesis. Srinagar, India: HNB Garhwal Central University (2004).
24. P. Nagendramma and S. Kaul, Development of eco-friendly/biodegradable lubricants: An overview, *Renewable and Sustainable Energy Reviews* 16 (2012) 764–774.
25. https://www.inovyn.com, products, organic chlorine derivatives. (2019).
26. P. Nagendramma, S. Goyal, A. Ray, J. Singh, N. Atrey, Studies on the sustainability of vegetable oil as a biocompatible multipurpose green lubes and additives, *Indian Journal of Chemical Technology* 26 (2019) 454–457.
27. E. Mirci, A. Patrut, Synthetic adipic complex tetra esters as eco-friendly lubricants, *Lubrication Science* 25 (2013) 339–350.
28. P. Nagendramma, Study of pentaerythritol tetraoleate ester as industrial gear oil, *Lubrication Science* 23 (2011) 355–362.
29. Daniel Wayne Lemke, Processes for preparing linear Polyglycerols and polyglycerol esters US6620904B2 United States (2003).
30. J. Nowicki, J. Mosio-Mosiewski, Esterification of fatty acids with C8–C9 alcohols over selected sulfonic heterogeneous catalysts. *Polish Journal of Chemical Technology Tech*, 15 (2013) 42–47.
31. J. McNutt, S. H. Quan, Development of biolubricants from vegetable oils via chemical modification, *Journal of Industrial and Engineering Chemistry* 36 (2016) 1–34.
32. F.S. Yan, W. He, Q. Jia, Q. Wang, S. Xia, P. Ma, Prediction of ionic liquids viscosity at variable temperatures and pressures, *Chemical Engineering Science* 184 (2018) 134–140.
33. T. Housel, Synthetic esters: Engineered to perform, *Machinery Lubrication Magazine*, (2014) 1–10.
34. A. N. Raof, R. Yunus, U. Rashid, N. Azis, Z. Yaakub, Effects of molecular structure on the physical, chemical, and electrical properties of ester-based transformer insulating liquids, *Journal of the American Oil Chemists' Society* 96 (2019) 607–616.
35. M. H. Ghatee, M. Zare, F. Moosavi, A R. Zolghadr, Temperature-dependent density and viscosity of ionic liquids 1-alkyl-3-methylimidazolium iodides: Experiment and molecular dynamics simulation, *Journal of Chemical & Engineering Data* 55(9) (2010) 3084–3088.
36. A. Wolak, Changes in lubricant properties of used synthetic oils based on the total acid number, *Measurement and Control* 51(3–4) (2018) 65–72.
37. P. Nagendramma, B.M Shukla, D.K. Adhikari, Synthesis, characterization and tribological evaluation of new generation materials for aluminum cold rolling oils, *Lubricants* 4 (2016) 23.
38. R. Nutiu, M. Maties, M. Nutiu, Correlation between structure and physical and rheological properties in the class of neopentanepolyol esters used as lubricating oils, *Journal of Synthetic Lubrication* 7(2) (1990) 145–154.

39. F. J. Flider, Polyglycerol esters as functional fluids and functional fluid modifiers, US5380469A United States (1993).
40. N. E. Rothfuss, M. D. Petters, Influence of functional groups on the viscosity of organic aerosol, *Environmental Science & Technology* 51(1) (2016) 271–279.
41. G. Bahlakeh, B. Ramezanzadeh, A detailed molecular dynamics simulation and experimental investigation on the interfacial bonding mechanism of an epoxy adhesive on carbon steel sheets decorated with a novel cerium-lanthanum nanofilm, *ACS Applied Materials & Interfaces* 9 (2017) 17536–17551.
42. M. Osama, A. Singh, R. Walvekar, M. Khalid, T. C. S. M. Gupta, W. W. Gupta, Recent developments and performance review of metalworking fluids, *Tribology International* 114 (2017) 389–401.
43. M. Anna, Gradkowski, Antiwear action of mineral lubricants modified by conventional and unconventional additives, *Tribology* 27 (2007) 177–180.
44. A. R. Macrae, R. C. Hammond, Present and future applications of lipases, *Biotechnology and Genetic Engineering Reviews* 3 (1982) 1093–217.
45. J.M Wyatt, The microbial degradation of two-stroke oils. Ph.D. Thesis. University of Kent at Canterbury, U.K. (1983).
46. C.E. Cerniglia, Microbial transformation of aromatic hydrocarbons. In: Atlas RM (Ed.) Petroleum Microbiology Macmillan Publishing Co., New York, (1984) 98–128.
47. ASTM-D-5864-95, Standard Test Method for Determining Aerobic Aquatic Biodegradation of Lubricants or Their Components, American National Standards Institute, New York, NY (1996).
48. A. Emmanuel, K. O. Obahiagbon, M. Ori-jesu, Biodegradation of vegetable oils: A review article, *Scientific Research and Essays* 4(6) (2009) 543–548.
49. K. Van Miert, Methods for the determination of ecotoxicity, C3—Algal inhibition test, *Official Journal of the European Communities* 383A (1993) 179–185.

4 Nano-Technology-Driven Interventions in Bio-Lubricant's Tribology for Sustainability

Rajeev Nayan Gupta
National Institute of Technology Silchar
Silchar, India

A.P. Harsha
Indian Institute of Technology (Banaras Hindu University)
Varanasi, India

Tej Pratap
Motilal Nehru National Institute of Technology Allahabad
Prayagraj, India

CONTENTS

DOI: 10.1201/9781003139386-4

4.1 INTRODUCTION

Lubricants are being used in approximately all industries to lubricate equipment and components. The aim is simply to keep the equipment in healthy condition to perform their intended function by improving anti-friction and anti-wear properties [1]. However, nowadays along with performance, the health of the environment is also a prime concern for the lubricant manufacturer. Around 85% of petroleum-based lubricants are being used worldwide but these not eco-friendly [2]. So, it is important to recognize the uncertainties, challenges and opportunities in relation to the demand and supply of lubricant. According to an OPEC (Organization of the Petroleum Exporting Countries) report, the world's oil requirement may increase from 95.4 to 111.1 mb/d (million barrels per day) up to 2040 [3]. Worldwide, one third of all lubricants are consumed in Europe, America and Asia alone [4]. The environment is negatively affected by this huge use of petroleum base stock that is linked with production, application, and disposal of lubricants, which result in contamination of soil, air, surface water and groundwater [5].

Worldwide, increasing industrialization and motorization have increased the demand for petroleum products. Petroleum base stock is limited and concentrated in certain regions of the world and is also non-biodegradable which affects our environment. Therefore, alternatives to this petroleum base stock need to be explored. On considering the environmental health aspect, the researcher focused on developing a universal biodegradable base stock that could replace mineral oil base stock [6, 7]. The environment is being negatively affected directly or indirectly by the enormous use of lubricants in industries and commercial applications. Lubricant loss contaminates the water and soil directly while volatile lubricant or lubricant miasma influence the air indirectly [8]. Therefore, all concerns regarding the best possible protection of our environment, nature in general and especially living beings must be considered during production, application, and disposal of lubricants. Vegetable oils are investigated as a potential source of lubricant in terms of biodegradability, renewability, non-toxicity, eco-friendliness and lubrication performance, except for the limitation of low oxidation and thermal stability and low temperature performance [9, 10]. A big question is, can vegetable oil-based bio-lubricants be a good base stock to replace petro-products? Many kinds of research are being explored to overcome its inferior thermo-oxidative properties, to fulfill all the requirements for ideal and economical lubricants [11]. The research includes chemical modification of the vegetable oils or the use of low/zero SAPS (sulfated ash, phosphorus, and sulfur) additives [12].

Vegetable oil-based bio-lubricants have a polar group along with a long hydrocarbon chain, which makes them amphiphilic surfactants. This helps in good adherence of the lubricant with the metal surface and formation of a monomolecular layer, which is thus good for both boundary and hydrodynamic lubrication situations [13, 14]. Various researchers reported the improved oxidation stability and low temperature performance of vegetable oils, i.e. rapeseed oil [15], soybean oil [16, 17], coconut oil [18], sunflower oil [19] and jojoba oil [20]. However, the tribological properties of the vegetable oil depend on the variation of the fatty acid compositions, which affect lubricating properties, the formation of protective film thickness and friction and wear [21]. It is also reported that vegetable oil alone is not capable to perform at par with mineral or synthetic oil without additives [22]. This is because additives may behave in different and/or complex ways between the mating surfaces, which results in protective film formation capability and thus affects friction and wear [21].

4.2 BASE OILS

4.2.1 Petroleum-Based Oils

Petroleum base stock is a natural substance produced by the decomposition of plants, animals and other living organisms inside the earth's crust with the passing of time. According to the crude oil source, oil composition varies distinctly, and oil can be categorized as paraffinic, aromatic and naphthenic compounds [23]. Crude petroleum base stock consists of a majority of covalently linked hydrocarbon molecules (e.g. paraffin, aromatic etc.) and a low amount of sulfur, nitrogen, metal trace etc. The carbon skeleton, i.e. the hydrocarbon chain, varies for different petroleum base stock according to its length, branching arrangement, as well as hydrogen saturation. A number of refining procedures are performed, aimed at removing the minor elements such as sulfur and nitrogen along with reducing average molecular weight (called cracking) of paraphenes to prevent wax deposition during lubrication. The most commonly used petro-products, e.g. mineral oils, are also produced by fractional distillation and refining. In addition, the behavior of the petro-product depends upon the cracking and hydrogenation. Mineral oil offers numerous positive properties, like availability, good oxidation stability and low cost, including some shortcomings like low temperature solidification and viscosity loss.

4.2.2 Synthetic Oils

Synthetic oils are artificially made, which leads to predicted lubricating properties with unique molecular structure. These oils consist of high molecular weight compounds formed by chemical modification of petroleum products rather than crude oil. Synthetic oil exhibits superior thermal and oxidation stability, viscosity index and biodegradability as compared to mineral oils. Therefore, the worldwide application of synthetic oil is increasing undeviatingly whereas mineral oil use is not being sustained. On the basis of chemical composition, synthetic oils are classified as synthetic hydrocarbons (polyalphaolefins, ester, polyalkylene glycols), silicon analogue hydrocarbons (silicones, silahydrocarbons), organohalogen, polyether oils etc. [23]. The properties of the synthetic oil can vary according to the application but production cost is very high.

4.2.3 Vegetable Oil-Based Bio-Lubricants

Environmental concern has focused on exploring oil that degrades after use with the passing of time. Bio-lubricants are actually lubricants based on natural resources like plants and animal fats. "Bio" means a material originating from a living organism, and "lubricant" means a substance (like oil) used to reduce the friction and wear. Biodegradability is defined as the susceptibility of a substance to undergo degradation under the influence of biological agents such as bacteria, fungi, yeast etc. Vegetable oils are considered in the category of bio-lubricants.

The fact that vegetable oils are renewable and non-toxic has concentrated attention on technologies that would enhance their usefulness as bio-fuels and industrial and commercial lubricants [24]. The biodegradability of different base oils is presented in Table 4.1, which shows that vegetable-based oils are more biodegradable than the other oils [6, 24].

Vegetable oil is produced from oil containing seeds, nuts or animal fats through different processing methods, solvent extraction or a combination of these. The most frequently used vegetable oils are soybean, olive, cottonseed, sunflower, castor, rapeseed, coconut, palm and canola; less commonly used vegetable oils are rice bran oil, tiger nut oil, niger seed, piririma oil, and much more [25]. Oils and fats consist mainly of a mixture of fatty acid esters of trihydroxy alcohol or glycerol [26]. In other words, the major content of vegetable oil is triglycerides, which are glycerol molecules with three long chain fatty acids attached at the hydroxyl group through ester linkages [27]. The minor contents are mono- and diacylglycerols, free fatty acid, phosphatides, sterols, fatty alcohol, fat soluble vitamins etc. Different fatty acids combined in the ester have different chemical structures causing variation in bonding forces, which are responsible for different melting points of fats. These differences are also responsible for different chain length, the presence or otherwise of unsaturation.

TABLE 4.1
Biodegradability of Different Base Oils [6, 9, 24]

Base Oil	Biodegradability (%)
Mineral oil	20–40
Trimelliates	0–70
Polyols	70–100
Poly alpha olefin	30–55
Synthetic esters	75–100
Vegetable oil	90–98

TABLE 4.2
Typical Composition of Common Vegetable Oil [7]

Base Oil	Saturates (%)	Mono-Unsaturates (%)	Poly-Unsaturates (%)
Canola	7.9	55.9	33.2
Corn	12.7	24.2	58.7
Cottonseed	25.8	17.8	52
Peanut	13.6	17.8	52
Olive	13.2	73.3	8.5
Safflower	8.5	12.1	74.5
Soybean oil	14.2	22.5	57.8
Sunflower	10.5	19.6	65.7

Fatty acids having a single carbon-to-carbon bond are termed as saturated, while those with one or more carbon-to-carbon double bond are termed as unsaturated. When fatty acid contains one double bond, it is called monounsaturated, while one with more than one double bond is called polyunsaturated. The compositions of most commonly used vegetable oils are presented in Table 4.2, which shows the percentage of mono-unsaturates, poly-unsaturates and saturates of base oil [7].

4.3 CHARACTERISTICS OF BIO-LUBRICANTS AND APPROACH TO IMPROVE THERMO-OXIDATIVE STABILITY

4.3.1 Effect of Fatty Acid Composition on Tribological Properties

Triglyceride structure plays a vital role in the value of vegetable oil as a lubricant. On one hand, a triglyceride has a long, polar fatty acid chain which provides high strength lubricant films that interact strongly with metallic surfaces to reduce friction and wear. The strong molecular interaction is also responsible for resiliency to temperature changes, providing a more stable viscosity or high viscosity coefficient. On the other hand, triglyceride structure limits the vegetable oil as a lubricant. In fatty acids, unsaturated double bonds act as active sites for many reactions, including oxidation which lowers the oxidation stability of vegetable oil and also causes hydrolysis. The strong intermolecular interactions also result in poor low temperature performance of vegetable oil [28], since fatty acids found in vegetable oil have different chain lengths and numbers of double bonds. So, the composition of fatty acids is evaluated by the ratio and position of the carbon-carbon double bond. One, two or three double bonds (oleic, linoleic and linolenic fatty acid components respectively) are found generally in a long carbon chain [29, 30]. Chemical compositional structure, physico-chemical properties and application of different vegetable-based oils are presented in Table 4.3.

Reeves et al. [31] studied the anti-wear and anti-friction behavior of different vegetable oils viz. safflower, corn, soybean, sesame, peanut, canola, olive and avocado oil. They found that higher oleic acid (C18:1)- containing vegetable oil has the lowest friction and wear. Also, it was speculated that dominant fatty acid C18:1 has a

TABLE 4.3

Structure of Fatty Acid, Physico-Chemical Properties and Application of Different Vegetable Oils [32–41]

Common Name	C:D	Base Oil							
		Sunflower	Castor	Rapeseed	Palm	Olive	Linseed	Soybean	Jatropha
Saturated									
Myristic acid	14:0	-	-	-	1.5	0.1	-	-	-
Palmitic acid	16:0	6.0	-	9.8	43.0	7.3	5.0	1.5	12–17
Stearic acid	18:0	3.0	2–3	1.6	5.0	2.7	3.0	4.3	6.7
Arachidic acid	20:0	Traces		9.2	0.5	-	-	-	-
Behenic acid	22:0	Traces	-	-	-	-	-	0.5	-
Unsaturated									
Palmitoleic	16:1	-	-	-	-	-	-	10.4	-
Oleic acid	18:1	17.0	3–5	18.4	40.0	60.7	22.0	24.4	37–63
Linoleic acid	18:2	74.0	3–5	16.8	10.0	4.4	17.0	51.6	19–41
α-Linolenic acid	18:3	6.0	80–90	6.5	-	0.5	52.0	7.7	-
Erucic acid	22:1	-	-	37.7	-	-	-	-	-
Physico-Chemical Properties									
Kinematic Viscosity (@40°C, mm^2/s)		4.45	15.25	4.45	5.72	4.52	3.74	4.05	4.82
Density (kg/m^3)		878	898	880	875	892	890	885	878
Oxidation stability 110°C (h)		0.9	1.2	7.5	4.0	3.4	0.2	4.05	2.3
Pour point (°C)		-12.0	-15.0	-12.0	9.0	-9.0	-30.0	-9.0	0.0
Flash point (°C)		252	286	240	260	207	222	240	240
Application									
		Grease, diesel-fuel substitute	Gear lubricant, greases	Chain saw bar lubricant, air com-pressor farm equipment	Greases, Rolling lubricants, Steel industries	Automotive lubricants	Paints, varnishes, coatings	Bio-diesel fuel, metal working, Lubricants, Hydraulic	Greases, lubricant application
Reference		[32]	[32–34]	[35]	[36, 37]	[38]	[27, 39]	[27, 33, 40]	[41]

tendency to form a denser mono-layer and is adsorbed on the mating surfaces to protect against wear.

4.3.2 Improvement in Bio-Lubricants' Stability

4.3.2.1 By Modification of Base Oil

4.3.2.1.1 Chemical Modification of Vegetable-Based Oil

- Modification of carboxyl group
  Esterification
  Transesterification
- Modification of fatty acid chain

Esterification is the reaction in which a carboxylic acid combines with an alcohol in the presence of an acid catalyst to form an ester. A small amount of concentrated sulfuric acid is usually selected as a catalyst. The esterification reaction is both slow and reversible, as is given in the equation:

$$RCO_2H + R'OH \leftrightarrow RCO_2R' + H_2O$$

Transesterification is a process in which there is an interexchange between organic group R″ of an ester and organic group R′ of an alcohol. In this organic reaction ester is transformed into another ester through an alcohol moiety interchange [42]. The general equation for the transesterification reaction is given by:

$$RCOOR' + R''OH \overset{catalyst}{\leftrightarrow} RCOOR'' + R'OH$$

Modification in the fatty acid chain improves the bio-lubricant properties. It leads to formation of C-C and C-O bonds to improve the oxidation property. Since a double bond is the most prone location in the fatty acid chain to act as the reactive site, therefore it can be functionalized by epoxidation. In this process a three-membered oxirane ring or epoxides are being formed by joining of an oxygen atom to an olefinically unsaturated molecule. By epoxidation oxygen stability of vegetable-based oil is improved [43].

4.3.2.1.2 Genetic Modification of Oilseed Crops

Genetic engineering helps to modify the oilseed crop by inserting additional DNA. Generally, one or two genes are inserted to modify the particular trait to improve their existing useful property. There are two types of genetic modification of a crop, first, to modify the input trait (such as herbicide tolerance and/or insect resistance), called a first generation genetically modified crop, and second, to modify the output traits (such as structure, fatty acid chain length etc.) resulting in different composition as compared to non-genetic modified crops. This is called a second generation genetically modified crop. In order to produce a wide range of modified oils for industrial application, new kinds of crop breeds are developed by various methods

[44, 45]. High oleic soybean oil and sunflower oil produced from genetically modified plants showed improved oxidation stability, higher load carrying capacity and good anti-wear properties [46].

4.3.2.2 By Using Chemical Reagent in Base Oil

Literature shows that the tribo-chemical properties of vegetable-based oil improved significantly with a small amount of chemical compound additive. In an internal combustion engine test rig, the blend of 2%w/v of zinc dialkyldithiophosphate (ZDDP) in base coconut oil improves the tribological property by reducing the friction and wear value [47]. Rapeseed oil with functional additives such as triethanolamine oleic acid and triethanolamineoleate was enhanced in both tribological and thermal stability [48]. Soybean oil in three modes, i.e. soybean 25% oleic, soybean 85% oleic and epoxidized soybean with organosulfur phosphorous, phosphorodithioate and amine phosphate, was used to study the effect on tribological properties. Organosulfur phosphorous and phosphorodithioate reveals good anti-wear behavior with soybean 25% and 85% oleic but is not very effective with epoxidized soybean oil, while amine phosphate shows improved properties in all conditions [49].

4.4 ADDITIVE EVOLUTION

4.4.1 Low SAPS Lubricant Additive

Modern engine lubricant limits the use of sulfur, phosphorus and sulfated ash either present as an active element or generated during the application. This is because sulfur and phosphorus oxides along with metallic ash produced from the additive are hazardous to the environment. Also, in the automobile application, these block the exhaust after a treatment filter is used to suppress the engine exhaust and reduce air pollution.

Zinc dialkyldithiophosphate (ZDDP) is one of the multi-functional low SAPS additives used in the lubricant mainly to improve the anti-wear property along with being an anti-oxidant and corrosion inhibitor. ZDDP contains four sulfur atoms, two phosphorus atoms and one zinc atom in each molecule. Zinc has a tendency to form ash. Although ZDDP has a huge negative impact on environmental health, in spite of this it is being used to date. This is because no other lubricant additive can perform in a multi-functional way with minimum addition in the lubricant. Researchers are exploring the alternative to ZDDP that can perform at least at par to ZDDP and still be environmentally friendly. Spikes [50] reported the alternatives to ZDDP with the presence and/or absence of sulfur, phosphorus and metal, and these are presented in Table 4.4.

In terms of the anti-wear property, only the environmentally hazardous metals lead and cadmium containing metal dialkyldithiophosphates (MDDP) are capable of performing like ZDDP [51–53]. Additives having a thiophosphate core, which excludes metal atoms by replacing them with an organic group (i.e. alkyl, amine, sulfides or more complex species like a hydroxyl group), were also studied [54, 55]. Sanin et al. [55] varied the organic group and number of the sulfur atom in thiophosphate and thiophosphites and reported decreased initial seizure load and lower

TABLE 4.4
Possible ZDDP Substitutes by Low- and Zero-SAPS Additives [50, 51]

Phosphorus	Sulfur	Metal	Lubricant Additive
Y	Y	Y	Metal dialkyldithiophosphate (here metal is not Zn)
Y	Y	N	Thiophosphate
Y	N	Y	Metal dialkylphosphate
Y	N	N	Phosphates, amine phosphates
N	N	Y	Organometallics (e.g. Ti, Sn compounds)
N	N	N	Organoboron compounds
N	N	N	Halide compounds
N	N	N	Zero SAPS
N	N	N	N and O Heterocyclics

Y: Present and N: Absent atoms

wear with an increase in the sulfur atom, along with higher initial seizure load with a shorter alkyl chain. Metal organophosphates containing phosphorus and metal (including Ca, Zn and Al) were also used as an anti-wear additive [56]. However, due to better solubility and tribo-performance of thiophosphates in lubricant, metal phosphates are rarely used as an anti-wear additive. Film-forming and load-carrying performance of metal- and sulfur-free and phosphorus-containing lauryl phosphates and phosphites additives were also studied [57]. It was reported that lauryl phosphates have better performance as compared to phosphite additives. Phosphorous- and sulfur-free organometallic additives also show good anti-wear performance. Although environmental concerns are associated with a few metal-based additives like lead and cadmium. Lead naphthenate and organocadmium additives have superior anti-wear and extreme pressure properties, but are not acceptable due to their hazardous exhaust impact [50, 58].

4.4.2 Zero-SAPS Lubricant Additive

Table 4.4 also represents SAPS-free lubricant additives. Various researchers have worked to investigate the zero-SAPS additive like boron-containing compounds in boundary lubrication [59]. The effect of the chemical structure and high temperature performance of these additives were also explored. Liu et al. [60] studied the tribological performance of a triborate ester series having $(R^1O)(R^2O)(R^3O)B$ structure with a different chain length. They found improved anti-wear and anti-friction properties with all the esters; however, superior performance was observed with the esters with longer chain length. Kapadia et al. [61] studied the film-forming capacity of ZDDP with a boron-containing compound at a higher temperature. They reported that the boron-containing compound has a tendency to form rapid and thicker film on the mating surfaces and some diffusion of boron into the surface also. Jaiswal et al. [62] and Rastogi et al. [63] investigated the tribo-performance of Schiff base compounds (i.e. 4-aminoantipyrine with benzaldehyde, salicylaldehyde, *p*-chlorobenzaldehyde and *p*-methoxybenzaldehyde) containing borate ester in paraffin oil. They reported that all the Schiff bases with borate ester show good tribological properties; however, *p*-methoxybenzaldehyde exhibits the best results. Also, the formation of the donor–acceptor complex between nitrogen–boron results in improved performance by forming a durable protective film. The interaction of novel SAPS-free organic anti-wear additive (OAW) was also compared with ZDDP and over-based calcium sulfonate in PAO6 oil [64]. The lowest friction coefficient and wear were reported with OAW tested on a pin-on-plate tester. This shows that SAPS-free lubricant additives also have potential to compete with the low SAPS additive.

4.4.3 Nano-Particles as Lubricant Additive

The nano-particle application in base oil improves the anti-wear property and reduces the friction as well. But the effectiveness of nano-particles in the base oil strongly depends upon parameters such as size, shape and concentration. The surface modification of nano-particles also provides good tribological properties. The nano-particles (e.g. ceramic, metallic or polymeric etc.) are blended in base oil by using various methods like mechanical or magnetic stirring or ultrasonication to formulate the test nano-lubricants.

4.4.3.1 Tribo-Performance of Nano-Lubricants

4.4.3.1.1 Metallic and Metal Oxide Nano-Particles

Numerous metallic [65–71] and metal oxide nano-particles [72–82] are being used, as an admixture, in different base lubricants to evaluate the tribo-performances. Nano-lubricant formation is processed with various variables like quantity and morphology of the particles. The tribo-performances of the nano-lubricants with some metals and oxides are summarized in Table 4.5. The major concern with the solid nano-particles is their suspension for a longer time in the base lubricant. This is because it directly reflects upon the nano-lubricant performance. For the prolonged nano-particles suspension, different surfactants or dispersants are used. In some cases, the nano-particles are capped first with the surfactant and then blended in the lubricant [68]. It is totally dependent on the nano-particles' intrinsic property as well as compatibility with the base oil. Various authors reported both improved and impaired properties for the nano-lubricants. Battez et al. [72] explored ZnO nano-particles up to 3 wt% in PAO6. They used two dispersing agents (OL100 and OL300; commercial name Octacare (R) DSP-OL100 and DSP-OL300) along with ZnO nano-particles and found that OL300 is the better dispersant. On the basis of experimental results, they argued that ZnO nano-particles do not function as an anti-wear additive; however, they are a good extreme pressure additive with PAO6 + 3% OL300 + 0.3% ZnO. Moreover, the rest of the other concentration of OL300 and ZnO impaired the tribo-properties. A comparative extreme pressure study with a four-ball tribotester was also performed by Battez et al. [73] for PAO6 containing CuO, ZnO and ZrO_2 nano-particles in 0.5, 1.0 and 2.0 wt% separately. CuO showed better tribological behavior while ZrO_2 showed worse. ZrO_2 nano-particles behave similarly regardless of the concentration. ZnO revealed an increasing trend of the load-wear index (LWI) with an increase in concentration,

TABLE 4.5
Summary of Nano-Lubricant Tribo-Performance with Various Metal and Metal Oxide Nano-Particles

Base Oil	Nano-Particle	Operating/Variable Parameter	Test Rig	Findings	Ref.
			Metallic Nano-Particles		
Coconut oil	Cu and Ag	Concentration: 0.1 to 1.0 wt%	Pin-on-Disk	AW and AF↑ at 0.25% concentration	Khan et al. [65]
Pongamia oil	Cu	Concentration: 0.025, 0.05, 0.075 & 0.1% Speed: 200–800 rpm	Pin-on-disk	• 0.075% was optimum • 1% impaired tribological property	Rajubhai et al. [66]
TBA	Pd	Load 1–20 N Rotative and reciprocative mode	Ball-on-disk	• 2% concentration gives good tribo-performance. • electrical resistance↓ • μ ↓	Abad et al. [67]
Glycerol	Al	0–1 wt% concentration,	Thrust collar	At 0.6667 wt% concentration, friction and wear minimum.	Le et al. [68]
Macs oil	Sn (30–60 nm) and Fe (20–70 nm)	Concentration: 0.1, 0.5, & 0.1%	Four-ball tribotester	μ ↓ effectively by Sn and AW ↓ with Fe 1% optimum.	Zhang et al. [69]
Ionic liquid-in-castor oil	In-situ synthesized Cu nano-particles	0.16 and 0.32 wt% concentration, Castor oil 25 mM Reaction temperature: 80°C	Four-ball tribotester	μ ↓ by 13.73%, wear↓ by about 40% 0.32 wt% concentration was optimum	Wang et al. [70]
SAE 10 and rapeseed	Fe	Size variation 50–340 nm 0.1 wt% concentration	Four-ball tribotester	At 0.1 wt% μ ↑ but AW↓; 50–140 nm was optimum.	Maliar et al. [71]
			Metal Oxide Nano-Particles		
Coconut oil, paraffin oil and engine oil	CeO_2 Zno	concentration: 0.46, 0.51 and 0.63 wt%	Pin-on-disk	• μ ↓ by 22% • Specific wear rate ↓ by 17%, for coconut oil at optimum concentration 0.51 wt%	Thottackkad et al. [74]

(Continued)

TABLE 4.5 (*Continued*)
Summary of Nano-Lubricant Tribo-Performance with Various Metal and Metal Oxide Nano-Particles

Base Oil	Nano-Particle	Operating/Variable Parameter	Test Rig	Findings	Ref.
Coconut oil	CeO_2/CuO hybrid nano-additives	0.1 to1.0 wt% concentration	Four-ball tribotester	0.25 wt% CeO_2/CuO: 50/50 optimum μ↓ by 15.7% Wear scar ↓ 23.4%	Sajeeb et al. [75]
Coconut oil	CuO	s: 1000 m speed: 1.4 to 5.6 m/s concentration 0.1, 0.2, 0.3, and 0.4%	Pin-on-disk	0.34% concentration and 3.7 m/s speed is optimum. AF↑	Thottackkad et al. [76]
Sunflower oil, Soybean oil, mineral oil and synthetic oil	CuO and ZnO	Concentration: 0.5 wt% Load: 10 N Time: 1 hr	High frequency reciprocating tribo-tester	CuO exhibits better tribo-properties with synthetic oil than others.	Alves et al. [77]
Palm oil	TiO_2	0.05,0.1 and 0.2 wt%	Four-ball tribotester	0.1 wt% optimum Higher concentration impairs the anti-wear property	Shaari et al. [78]
Modified palm oil	TiO_2	Time 10 min. RT L 40, 80, 120, 160 kg	Four-ball tribotester	At 160 kg AW↑ by 11% and μ↓ 15%	Zulkifli et al. [79]
Base oil	a. TiO_2 (P25) b. TiO_2 (anatase)	L 14.715 & 0.05 m/s (constant)	Pin-on-disc	Anatase phase TiO_2 is superior than commercial TiO_2 (b) AF↑	Ingole et al. [80]
Lubricating oil	Al_2O_3	0.05–1.0 wt% concentration	Four-ball tribotester& Thrust-ring	0.1 wt% optimum AF↑ 23.92 and AW↑ 41.75% (maximum)	Luo et al. [81]
Castor oil	Reduced graphene oxide (rGO)	10, 20, 30, 40, 50, 60 and 70 vol% blending of rGO in oil	Four-ball tribotester	40% blending improves the AW properties by 45.8%.	Bhaumik et al. [82]

↑: increase/improve, ↓: decrease/impair, μ: COF, AF: anti-friction, AW: anti-wear, L: load, RT: room temperature, T: temperature

while CuO improved LWI with 0.5 and 2.0 wt%. This means a minimum concentration (i.e. 0.5 wt%) of CuO showed the best LWI performance. A few researchers also reported synthesis of composite oxide nano-particles as an additive and improved tribo-performance [82].

4.4.3.1.2 Polymeric Nano-Particles

Some researchers used polymeric particles in base oils to evaluate the anti-wear and friction characteristics using various test rigs. For example, polytetrafluoroethylene (PTFE) is a well-known anti-friction material, especially in the case of coatings. However, its compatibility in lubricants and effect on tribo-performances have also been explored by various researchers. Gupta et al. [83] used PTFE (≈ 90.4 nm)), CuO (≈151.2 nm) and CeO_2 (≈ 80 nm) nano-particles in different concentrations (0.1, 0.25 and 0.5 w/v%) in raw rapeseed oil and epoxidized rapeseed oil having different viscosity. They examined wear scar, friction variation and weld point with different compositions of the oil using a four-ball tribotester. They reported that PTFE-based nano-lubricants have shown substantial improvement in anti-wear and anti-friction properties over the other CuO and CeO_2 nano-particles. The lowest concentration, i.e. 0.1%w/v, was considered as optimum, because of minimum wear scar size and friction coefficient. However, PTFE nano-particles have shown independent behavior from any concentration in base oils, which means the same weld point was obtained for both raw oil and PTFE-based nano-lubricant. Dubey et al. [84] studied PTFE-based nano-lubricants with a four-ball tribotester and Optimol-SRV III oscillating friction and wear tester. They used four different sizes, 50, 150, 400 nm and 12 μm, and three concentrations, 4, 8 and 12 wt% in 150 N API Group II base oil. They obtained reduction in wear and friction trend and reported that with lower size and increased PTFE concentration, tribo-performance improved. The maximum 40% reduction was observed with 12 wt% of PTFE while wear scar was reduced by 20%.

4.4.3.1.3 Carbon Allotropes

Carbon allotropes like diamond, carbon sphere, graphene, single and multiwall carbon nano-tubes etc. are also being used as a lubricant additive. For the last two decades, these allotropes in different forms and combinations have been used by various investigators who reported better tribological properties. Among of them, graphene is a nascent allotrope which has superior tribo-properties. The suspension properties of the carbon allotropes are different and also dependent on the morphology of the particles [85]. Song et al. [85] examined the suspension stability of different forms of carbon allotropes, i.e. spherical carbon sphere (CS) and flaky graphene oxide (GO) for both particles and found GO exhibits more stable and uniform suspension than CS. Gupta et al. [83] reported that the prolonged suspension potential of the different nano-particles varies with the physical properties of lubricant as well as the compatibility of the dispersant. They observed distinct suspension behavior for similar nano-particles in rapeseed oil and modified rapeseed oil, in the presence of sodium dodecyl sulfate dispersant (Figure 4.1). Therefore, it is important to optimize the appropriate dispersant in different base oils for proper suspension of the

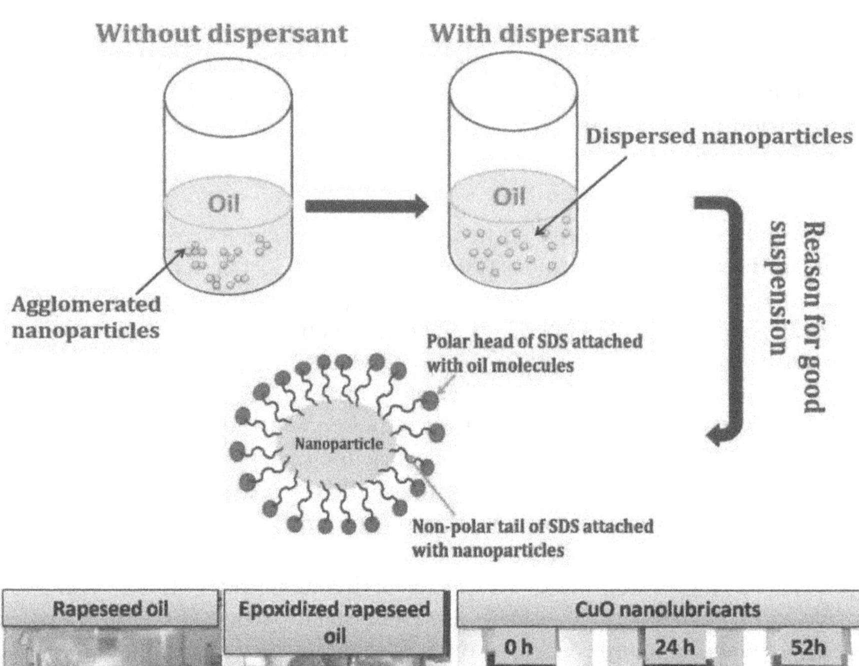

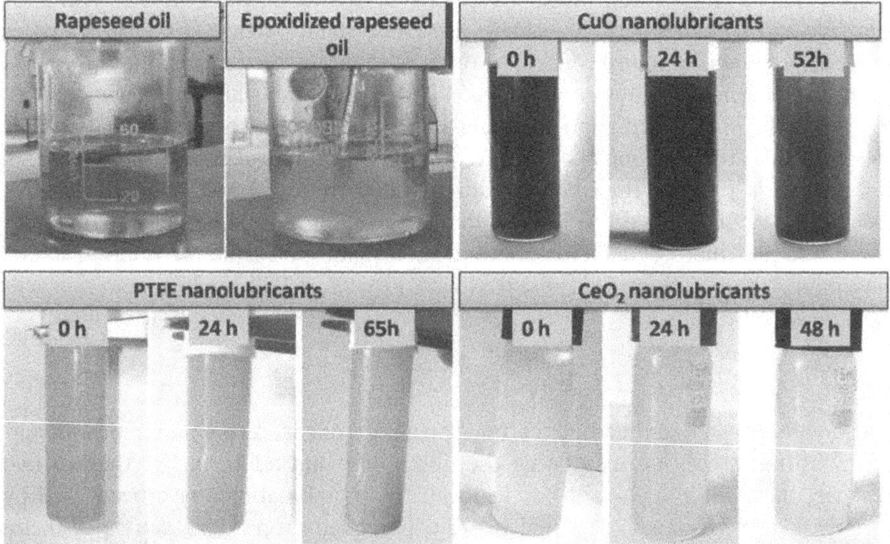

FIGURE 4.1 Suspension image of nano-particles in lubricating oil [83].

nano-particles. The friction and wear performance of some carbon allotropes are summarized in Table 4.6.

4.4.3.1.4 Other Nano-Particles in Various Lubricants

Excluding the metal, metal oxide, polymer and carbon allotropes, other nano-particles have also been explored by various researchers to evaluate the tribological performance of different base oils. In addition, a substantial improvement was reported in terms of wear resistance anti-friction properties of different lubricants as presented in Table 4.7.

TABLE 4.6
Summary of Tribo-Performance of Different Lubricants with Carbon Allotropes as an Additive

Base Oil	Nano-Particle	Operating/ Variable Parameter	Test Rig	Findings	Ref.
Sunflower oil	a. carbon sphere (1.8–2.2 μm) b. flaky graphene oxide	L 5 N and speed 300 rpm (constant) 0.5–3.0 wt% for a 0.1–1.0 wt% for b	Ball-on-disk	• 0.3 wt% optimum for graphene oxide. • AF↑ • Flaky structure better than sphere	Song et al. [85]
Palm oil-based lubricants	Graphene	L 392 N and speed 1200 rpm& 25, 50 and 100 ppm concentration	Four-ball tribotester	• 50 ppm optimum • AW and AF↓ at lower concentration • High concentration causes high wear	Kiong et al. [86]
Modified jojoba oil blended with SAE20W40	Graphene	Load 50 to 150 N Speed 1 to 5 m/s 0.05, 0.075, and 0.1 wt% concentration	Pin-on-disk	0.075 wt% shows minimum friction and superior anti-wear behavior for 20% jojoba oil blending	Kannan et al. [87]
LB2000 vegetable oil	Graphite; 35 and 80 nm	L 2, 10 N; Speed 100 rpm, 0.05 & 0.25 wt% concentration	Pin-on-disk	• 35 nm better than 80 nm • At both loads 0.25 wt% improved AW and AF↑	Su et al. [88]
Oil and water	single- and multi-walled CNT	p 0.8 and 1.1 GPa 0.01 and 0.05 wt% concentration	Twin-disk	0.01 wt% for multi-walled CNT in oil and 0.05 wt% for single-walled CNT optimum in water	Cornelio et al. [89]
Rapeseed oil and mineral oil	Eichhornia crassipes carbon nano-tubes (EC-CNTs)& ZDDP	Frequency 2 and 5 Hz Temp 75°C, 1.0 wt% concentration (const.) L 25 and 100 N	High frequency reciprocating test rig	At low frequency and load rapeseed oil shows better performance than mineral oil and at par to ZDDP	Opia et al. [90]

(Continued)

TABLE 4.6 (*Continued*)
Summary of Tribo-Performance of Different Lubricants with Carbon Allotropes as an Additive

Base Oil	Nano-Particle	Operating/ Variable Parameter	Test Rig	Findings	Ref.
Corn oil	Multi-wall CNTs	0.1, 0.2 and 0.3% concentration	4-S single cylinder IC engine	Thermal performance improved about 17%	Abdet al. [91]
Vegetable oil	Graphene nano-sheets (GN), Carbon nano-tubes (CNT), and Graphene oxide (GO)	Concentration 50 and 100 ppm Operating condition of 1200 rpm, 75°C, 392 N Load	Four-ball tribotester	50 ppm GN is optimum, AF↑ AW↑	Kiu et al. [92]
Groundnut oil	TiO_2/gC_3N_4 nano-composite	L 400 N; Speed 600 rpm, T: 29°C and 75°C Concentration: 0.00625, 0.0125, 0.0187, and 0.0250 mg/ml	Four-ball tribotester	μ↓ by 23.3% and AF↑ 17.1%	Ranjan et al. [93]

AW: anti-wear property, AF: anti-friction property, μ: COF, L: load, RT: room temperature, T: temperature, ↑: increase/improve, ↓: decrease

TABLE 4.7
Summary of Other Nano-Particles as a Lubricant Additive [94–103]

Base Oil	Nano-Additive	Concentration Size Shape (Optimum)	Test Rig	Operating Parameter	Tribological Results	[Ref]
500SN	zinc borate	0.5–2.5 wt% 20–50 nm Regular (0.92 wt%)	Four-ball tribotester Block-on-ring	L:294, t:30 L: 300, v: 3.85	Friction ↓ P_B↑	Dong and Hu [94]
500SN	CeF_3	25 nm Spherical and cylindrical	Four-ball tribotester	N: 1480	Friction ↓ P_B↑	Sunquing et al. [95]
Liquid paraffin	ZnS	0.05–0.4 wt% 4 nm (0.1 wt%)	Four-ball tribotester	L: 100,200,300,400 N:1450, t:30	Anti-wear↑ Load carrying capacity↑	Liu and Chen [96]
Paraffin oil	WS_2	100 nm	Pin-on-disk	v: 0.6 m/s L:100–500	Friction↓ Roughness↑	Rapoport et al. [97]
Poly-alpha-olefin	Mo–S–I nano-wires	0.5–2.0 wt% Wire (1.0 wt%)	Pin-on-flat	L: 1, 2, 5, 10 T: 25 v: 2.5	Friction↓	Joly-Pottuz et al. [98]
Diesel engine oil (50 CC)	Serpentine	0.5–2.0 wt% 1 μm (1.5 wt%)	MM-10W sliding friction tribotester	L: 100,200,300,400 v: 1.51 t: 120	Wear↓ Friction ↓	Yu et al. [99]
Base oil	Cerium borate	50 nm	Four ball tribotester	L: 390 N: 1450, t: 30	Friction ↓	Lingtong et al. [100]

(Continued)

TABLE 4.7 (*Continued*)
Summary of Other Nano-Particles as a Lubricant Additive [94–103]

Base Oil	Nano-Additive	Concentration Size Shape (Optimum)	Test Rig	Operating Parameter	Tribological Results	[Ref]
Engine oil (API SM grade 5W-30)	Talc	0.05–1.5 wt% 1 µm (100°C and 0.15 wt%)	Ball-on-plate	L: 5 v: 0.04 p: 0.75 s:2000 T: RT to 120	Friction and wear↓ at high temperature, not at RT.	Rudenko and Bandyopadhyay [101]
SAE 20W-40	MoS_2	0.25–1.0 wt% ≤ 150 nm Regular (0.25 to 0.5 wt%)	Four-ball tribotester	L: 392, N:1200 t: 60, T: 75	Load-wear index↑ Weld load↑ Friction and wear↓	Srinivas et al. [102]
SAE 20W50	h-BN	1.0–3.0 wt% 50 nm Lamellar (100 N, 3.0 wt%)	Four-ball tribotester Pin-on-block	L: 392, N:1200 t: 60, T: 75 L:50–150;v:0.03; t:10	Wear loss↓ by 30–70% Friction and wear is function of nano-additive concentration	Charoo and Wani [103]

L: load (in N), N: speed (rpm), p: contact pressure (GPa), P_B: maximum non-seizure load (N), RT: room temperature (°C), s: sliding distance (m), t: time (min), T: temperature (°C), v: velocity (m/s)

4.5 EFFECT OF NANO-PARTICLE PARAMETERS ON TRIBOLOGICAL BEHAVIOR

4.5.1 Effect of Nano-Particles' Size in Contact Zone

Two main characteristics are associated with the size of the nano-particle-based lubricants. First, the size of the nano-particles affects their physico-chemical and mechanical potential that directly influences the tribological properties. The decrease in particle size increases hardness due to increase in dislocation pileup for a crystal generally in the size range of 100 nm or higher [104, 105]. The harder nano-particle, when it comes in contact with comparatively softer mating surfaces, creates indents and scratches. Thus, nano-particle size-induced variations in performance (based on hardness) must be kept in mind during nano-lubricant formulation.

Second, lubricants containing nano-particles should stay at the contact interface during loading and shearing to protect against wear [106]. For example, if the roughness of the friction surface is less than the nano-particle size (Figure 4.2), the nano-particles tend to escape from the contact zone, which leads to poor lubrication. On the contrary, if the roughness of the friction surface is higher than the nano-particle size, then these nano-particles fill the surface dimples and valleys to smooth them, which improves the tribo-performance of nano-lubricants [107].

4.5.2 Effect of Nano-Particles' Shape in Contact Zone

Another parameter of consideration for nano-lubricants is their shape. The spherical shape of the nano-particles is more pronounced as compared to sheets, tubes and irregular shapes. This is because of the rolling action of the spherical particles. In other words, a nano-sphere acts as a nano-bearing by carrying fraction of actual applied load in the contact zone and keeping the mating surfaces away, thus improving tribological properties. However, due to the point contact geometry of nano-spheres, they experienced more contact pressure as compared to nano-sheets and irregular shapes [107].

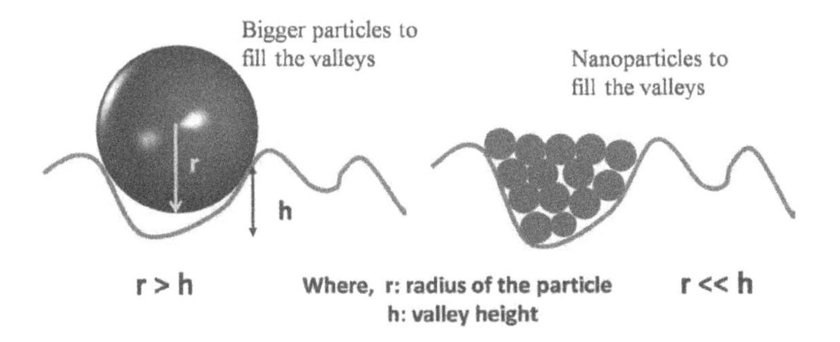

FIGURE 4.2 Effect of nano-particle size and surface roughness [107].

4.5.3 Effect of Nano-Structure in Contact Zone

The nano-structure of the nano-particle affects both mechanical and tribological properties. Vacancies within nano-sized particles are the pinpoint. Limited vacancies in nano-particles improve their tribo-mechanical properties. However, excessive vacancies inside the nano-material lower the mechanical and tribological properties and vice versa [107].

4.5.4 Effect of Nano-Particles' Concentration in Contact Zone

The concentration of nano-particles plays a vital role to influence the tribological properties of the nano-lubricant [107]. For a particular range of concentration, there may be an optimum concentration at which the minimum coefficient of friction is obtained. With low concentration there may be asperity-asperity collision under higher contact stress as well as an insufficient number of particles at the interface; on the contrary, at high concentration particle agglomeration and abrasion take place [75, 107, 108]. Past studies exhibit that the optimum concentration of the nano-particles in the lubricating oil is strongly system-dependent, and it needs to be investigated for each different operation condition.

4.6 EFFECT OF NANO-LUBRICANT ON THERMAL CONDUCTIVITY AND TRIBOLOGICAL BEHAVIOR

Thermal properties of the base oil improve with addition of the nano-particles. Wang et al. [109] used Al_2O_3 and CuO nano-particles in water, vacuum pump fluid, engine oil and ethylene glycol and found that thermal conductivity of the nano-lubricant was better than base fluids. They also observed that the trend of the increase in thermal conductivity was linear with increase in concentration. Taha-Tijerina et al. [110] correlated experimentally the relation among thermal conductivity, nano-particle concentration and tribological behavior. They used h-BN and graphene as nano-additives in mineral oil with varying concentrations as 0.01, 0.05 and 0.1 wt%. As the concentration of the nano-particles increases the thermal conductivity of the nano-lubricant increases; on the contrary, wear scar diameter and friction coefficient decrease.

4.7 HYPOTHETICAL MECHANISMS OF NANO-LUBRICANTS TO IMPROVE TRIBO-PERFORMANCE

The nano-particles' behavior in lubricating oil varies in different ways under different contact situations. Especially in boundary lubrication, when mating surfaces are too close, the role of these ultrafine particles is crucial in oil. Some of the hypothetical mechanisms are as follow and are also depicted in Figure 4.3:

- *Reduction in real area of contact*
  At the atomic level, it is obvious that no surface is perfectly smooth. It contains numerous asperities and valleys. For close contact and dynamic situations, the chance of asperity-asperity locking, asperities collision and material loss is more prone. In this situation the nano-particles inside the oil come into

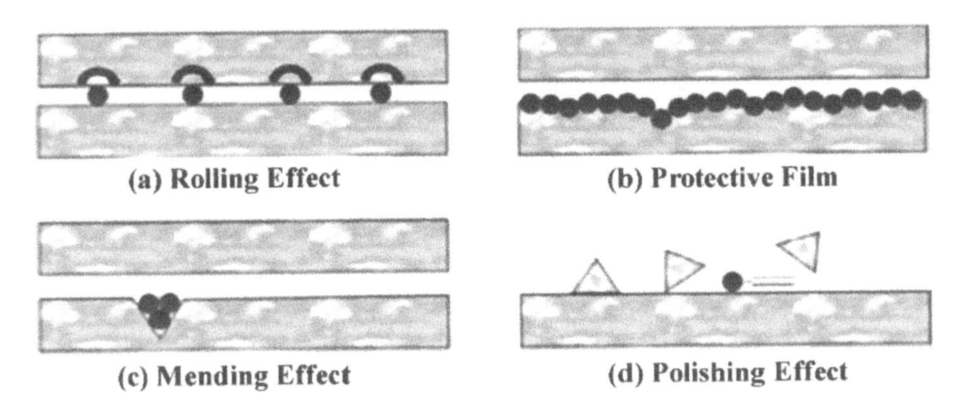

FIGURE 4.3 Hypothetical mechanism of nano-lubricant action under tribological contact showing: (a) reduction in contact area with spherical particles under rolling-sliding; (b) formation of nano-particle film as protective layer; (c) surface mending and (d) surface polishing with irregular shaped particles [108].

the contact zone and reduce the real area of contact (Figure 4.3(a)) [108, 111, 112]. This reduces the wear loss.

- *Nano-bearing*
  If the nano-particles have spherical, almost spherical, tube or capsule-like morphology, during surface sliding they roll over the friction surfaces keeping them separate. They act as nano-bearings and carry a fraction of the applied load as in Figure 4.3(a) [108, 111].
- *Formation of protective film and surface mending*
  The nano-particles have the ability to form a thin film that reduces the interfacial shear stress between tribo-pairs and prevents micro-damage and scratches on the surfaces. In addition, the nano-particles fill the surface valleys and dimples to make them smooth as in Figure 4.3(b). These protective layers, formed by the nano-particles, are also called deposition film [113]. Nano-particles may also repair the rough surfaces by filling the micro-valleys and pits at the interface, as presented in Figure 4.3(c), for enhanced tribo-performance.
- *Surface polishing*
  If the nano-particles have irregular morphology as presented in Figure 4.3(d), it breaks the micro- asperities and material is transferred away from the contact zone by the lubricants. This phenomenon helps smooth the friction surfaces[114].
- *Tribo-sintering of the nano-particles*
  Under severe operating conditions, i.e. higher loading and localized temperature, the nano-particles may melt or attain a semi-molten state that causes tribo-sintering. In other words, in tribo-sinterization nano-particles accumulate in the valleys and/or dimples which are surrounded by the asperities. Moreover, under high temperature and pressure, these nano-particles form compact tribo-film and reduce direct metal-to-metal contact [115].

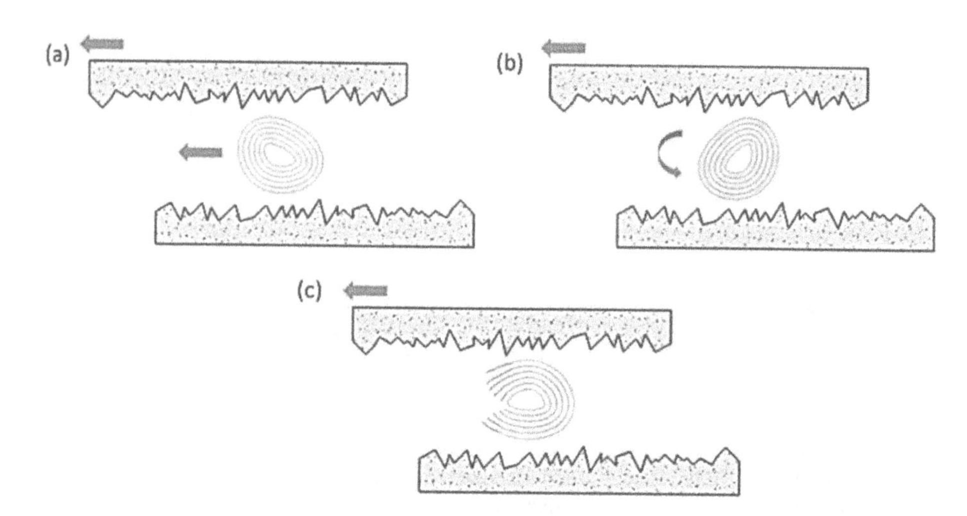

FIGURE 4.4 Exfoliation mechanism of multi-layer particles [117].

- *Shearing of trapped nano-particles at the interface*
  Wäsche et al. [116] reported that under continuous high loading the nano-particles experience something like fatigue which induces micro-shearing of the top surface of the nano-particles, which generates secondary particles. These secondary particles help the nano-additive in surface mending, thus improving tribological properties.
- Exfoliation
  This mechanism is associated with layered particles when used in the base oil, as presented in Figure 4.4. When the tribo-pairs are in dynamic condition, the layered particles may : (i) slide or roll at the interface (Figure 4.4(a) and (b)) and (ii) start to peel out and propagate from the top layer to the core under high contact stress (Figure 4.4(c)) [117]. These peeled out parts come in contact with friction surfaces and form a protective layer to reduce friction and wear.
- *Hardening the contacting surfaces*
  Continuous rolling and sliding of the nano-particle in the contact zone make the friction surface work harden up to a certain extent. This also may responsible for the improvement in the tribological properties for the nano-lubricants [118].
- Synergistic lubrication
  Synergistic in-situ lubrication mechanism of surface interaction with nano-lubricants is presented in Figure 4.5. It is speculated that more than one phenomenon may occur simultaneously under sliding conditions. Also, Figure 4.5 shows that apart from the surface separation, nanoparticles forms protective layer and nano-bearing action may proceed simultaneously. With nanolubrication and heterogeneous nanoparticle deposition at interface causes alteration of lubrication mechanism. Somewhere it may be traditional lubrication or interface separation due to nano-particles or synergistic effect of nano-particles with lubricant's tribo-film.

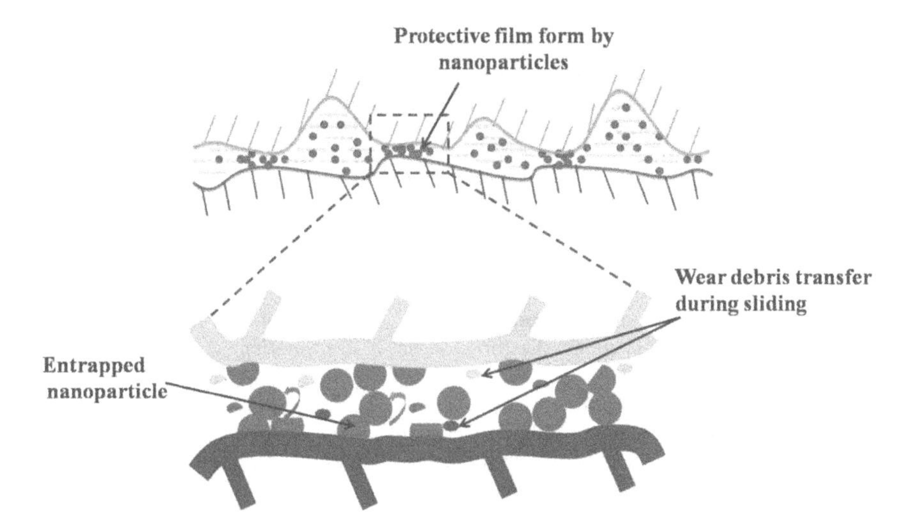

FIGURE 4.5 Typical model for synergistic interaction of mating surface under nano-lubrication.

4.8 NANO-LUBRICANTS' LIMITATION

In spite of superior positive characteristic (like high chemical reactivity, high absorption, wide diffusivity), nano-lubricants have some limitations including suspension stability, the exact role of the micro-rolling mechanism and correlation of the nano-particle's physical properties with tribo-performance. Nano-particles have high specific area, which is suitable for protective film formation but inferior dispersion in oil. Nano-particles in the oil tend to aggregate and settle down at the interface, which may cause higher abrasion and thus higher material loss. So it became mandatory for the nano-lubricants to keep the ultrafine particles uniformly suspended for a prolonged time. The use of dispersant or surface modification of the nano-particles are two popular methods. However, to optimize the appropriate concentration of the nano-particles as well as compatible dispersant is a big challenge so far.

4.9 CONCLUSION

Lubrication is a primary tool to minimize the material and energy loss of any mechanical system. All categories of the base lubricant, i.e. petro-product, synthetic and bio-lubricant, have some shortcomings either in terms of cost, physical properties, environmental concerns or performance. In tribological contact situations these oils in raw condition, under higher contact and speed, fail to perform their intended function. Additives in small amounts show a significant change in tribo-performance of the base oils. However, additive-based lubricants depend on the compatibility of the additives with the base oils. Researchers are exploring SAPS-free additives considering environmental issues. In this sense, nano-particles are getting much attention as a lubricant additive. In spite of the numerous positive characteristics of the nano-particle, it also presents the biggest challenge to formulate a homogeneous mixture and maintain it for a prolonged period during operation. Therefore, modification of the nano-particles' surface is an important and challenging task. The mechanisms of the nano-particles in lubricating oil vary distinctly according to their size, shape, morphology and physical nature.

REFERENCES

1. Hamrock BJ, Schmid SR, Jacobson BO. Fundamentals of fluid film lubrication, 2nd edition, 2004, Marcel Dekker, New York.
2. Pop L, Puşcaş C, Bandur G, Vlase G, Nuţiu R. Basestock oils for lubricants from mixtures of corn oil and synthetic diesters. Journal of the American Oil Chemists' Society, 2008, 85(1), 71–76.
3. Griffin J, Fantini A-M. Organization of the Petroleum Exporting Countries. OPEC World Oil Outlook, October 2017. Helferstorferstrasse 17 A-1010 Vienna, Austria.
4. Bartz WJ. Lubricants and the environment, Tribology International, 1998, 31(1–3), 35–47.
5. Abdalla HS, Patel S. The performance and oxidation stability of sustainable metalworking fluid derived from vegetable extracts. Proceedings of the Institution of Mechanical Engineers, Part B: Journal of Engineering Manufacture, 2006, 220(12), 2027–2040.
6. Gnanasekaran D, Chavidi VP. Chapter 2: Biodegradable, renewable, and eco-friendly vegetable oil: Lubricants, Vegetable oil based bio-lubricants and transformer fluids, 2018, 29–47, Springer, Singapore. https://doi.org/10.1007/978-981-10-4870-8_2.
7. Rafiq M, Lv YZ, Zhou Y, Ma KB, Wang W, Li CR, Wang Q. Use of vegetable oils as transformer oils–a review. Renewable and Sustainable Energy Reviews, 2015, 52, 308–324.
8. Rizvi SQA. Chapter 13: Lubricants and the environment, A comprehensive review of lubricant chemistry, technology, selection, and design, 2009, 579–600. MNL11473M, ASTM International, West Conshohocken, PA.
9. Anand ON, Chhibber VK. Vegetable oil derivatives: environment-friendly lubricants and fuels. Journal of Synthetic Lubrication, 2006, 23, 91–107.
10. Kumar BS, Padmanabhan G, Krishna PV. Performance assessment of vegetable oil based cutting fluids with extreme pressure additive in machining. Journal of Advanced Research in Materials Science, 2016, 19(1), 1–13.
11. Kashyap A, Harsha AP. Tribological studies on chemically modified rapeseed oil with CuO and CeO_2 nanoparticles. Proceedings of the Institution of Mechanical Engineers, Part J: Journal of Engineering Tribology, 2016, 230(12), 1562–1571.
12. Rastogi RB, Maurya JL, Jaiswal V. Low sulfur, phosphorus and metal free antiwear additives: Synergistic action of salicylaldehyde *N*(4)-phenylthiosemicarbazones and its different derivatives with Vanlube 289 additive, Wear, 2013, 297 (1–2), 849–859.
13. Jahanmir S, Beltzer M. Effect of additive molecular structure on friction coefficient and adsorption. Transactions of the ASME, Journal of Tribology, 1986, 108, 109–116.
14. Jain AK, Suhane A. Research approach & prospects of non edible vegetable oil as a potential resource for biolubricant – a review. Advanced Engineering and Applied Sciences: An International Journal, 2012, 1(1), 23–32.
15. Wu X, Zhang X, Yang S, Chen H, Wang D. The study of epoxidized rapeseed oil used as a potential biodegradable lubricant. Journal of the American Oil Chemists' Society, 2000, 77(5), 561–563.
16. Doll KM, Sharma BK. Physical properties study on partially bio-based lubricant blends: Thermally modified soybean oil with popular commercial esters. International Journal of Sustainable Engineering, 2012, 5(1), 33–37.

17. Castro W, Perez JM, Erhan SZ, Caputo F. A study of the oxidation and wear properties of vegetable oils: Soybean oil without additives. Journal of the American Oil Chemists' Society, 2006, 83(1), 47–52.
18. Mannekote JK, Kailas SV. Experimental investigation of coconut and palm oils as lubricants in four-stroke engine. Tribology Online, 2011, 6, 76–82.
19. Campanella A, Rustoy E, Baldessari A, Baltanás MA. Lubricants from chemically modified vegetable oils. Bioresource Technology, 2010, 101(1), 245–254.
20. Bisht RPS, Sivasankaran GA, Bhatia VK. Additive properties of jojoba oil for lubricating oil formulations, Wear, 1993, 161(1), 193–197.
21. Biresaw G, Bantchev G. Effect of chemical structure on film-forming properties of seed oils. Journal of Synthetic Lubrication, 2008, 25(4), 159–183.
22. Hsien WLY. Towards green lubrication in machining – Utilization of vegetable oil as bio-lubricant and additive. Springer Briefs in Green Chemistry for Sustainability, 2014, 7–17. DOI 10.1007/978-981-287-266-1_2.
23. Shahnazar S, Bagheri S, Hamid SBA. Enhancing lubricant properties by nanoparticle additives. International Journal of Hydrogen Energy, 2016, 41(4), 3153–3170.
24. Aleksandar RAC, Aleksandar VENCL. Ecological and technical aspects of the waste oils influence on environment. The Annals of University "Dunărea De Jos" of Galaţi fascicle VIII, Tribology, 2012, 18(1), 5–11.
25. Aluyor EO, Obahiagbon KO, Ori-Jesu M. Biodegradation of vegetable oils: A review. Scientific Research and Essays, 2009, 4(6), 543–548.
26. Nwobi BE, Ofoegbu OO, Adesina OB. Extraction of qualitative assessment of African sweet orange seed oil. African Journal of Food Agriculture Nutrition and Development, 2006, 6(2), 1–11.
27. Shashidhara YM, Jayaram SR. Vegetable oil as a potential cutting fluid--An evolution. Tribology International, 2010, 43(5–6), 1073–1081.
28. Fox NJ, Stachowiak GW. Vegetable oil-based lubricants – A review of oxidation, Tribology International, 2007, 40(7), 1035–1046.
29. Mongkolwongrojn M, Arunmetta P. Theoretical characteristics of hydrodynamic journal bearings lubricated with soybean-based oil. Journal of Synthetic Lubrication, 2002, 19(3), 213–228.
30. Orsavova J, Misurcova L, Ambrozova JV, Vicha R, Mlcek J. Fatty acids composition of vegetable oils and its contribution to dietary energy intake and dependence of cardiovascular mortality on dietary intake of fatty acids. International Journal of Molecular Science, 2015, 16(6), 12871–12890.
31. Reeves CJ, Menezes PL. Evaluation of boron nitride particles on the tribological performance of avocado and canola oil for energy conservation and sustainability. International Journal of Advanced Manufacturing Technology, 2017, 89(9), 3475–3486.
32. Asadauskas S, Erhan SZ. Thin film test to investigate liquid oxypolymerization of nonvolatile analyses: Assessment of vegetable oils and biodegradable lubricants. Journal of American Oil Chemists' Society, 2001, 78(10), 1029–1035.
33. Quinchia LA, Delgado MA, Reddyhoff T, Gallegos C, Spikes HA. Tribological studies of potential vegetable oil based lubricants containing environmentally friendly viscosity modifiers. Tribology International, 2014, 69, 110–117.
34. Luna FMT, Rocha BS, Rola Jr. EM, Albuquerque MCG, Azevedo DCS, Cavalcante Jr CL. Assessment of biodegradability and oxidation stability of mineral, vegetable and synthetic oil samples. Industrial Crops and Products, 2011, 33(3), 579–583.
35. Joseph PV, Saxena D, Sharma DK. Study of some non-edible vegetable oils of Indian origin for lubricant application. Journal of Synthetic Lubrication, 2007, 24, 181–197.
36. Krishna Reddy KSV, Kabra N, Kunchum U, Vijayakumar T. Experimental investigation on usage of palm oil as a lubricant to substitute mineral oil in CI engines. Chinese Journal of Engineering, 2014, Article ID 643521, 2014, 1–5. DOI:10.1155/2014/643521.

37. Norhaizan ME, Hosseini S, Gangadaran S, Lee ST, Kapourchali FR and Moghadasian MH. Palm oil: Features and applications. Lipid Technology, 2013, 25(2), 39–42.
38. Choi US, Ahn BG, Kwon OK, Chun YJ. Tribological behavior of some antiwear additives in vegetable oils. Tribology International 1997, 30(9), 677–683.
39. Popa V-M, Gruia A, Raba D-N, Dumbrava D, Moldovan C, Bordean D, Mateescu C. Fatty acids composition and oil characteristics of linseed (LinumUsitatissimum L.) from Romania, Journal of Agroalimentary Processes and Technologies, 2012, 18 (2), 136–140.
40. Siniawski MT, Saniei N, Adhikari B, Doezema LA. Influence of fatty acid composition on the tribological performance of two vegetable-based lubricants. Journal of Synthetic Lubrication, 2007, 24, 101–110.
41. Pramanik K. Properties and use of jatrophacurcas oil and diesel fuel blends in compression ignition engine. Renewable Energy, 2003, 28(2), 239–248.
42. Schuchardt U, Sercheli R, Vargas R.M, Transesterification of vegetable oils: A review, Journal of the Brazilian Chemical Society, 1998, 9, 199–210.
43. Saurabh T, Patnaik M, Bhagt SL, Renge VC. Epoxidation of vegetable oils: A review. International Journal of Advanced Engineering Technology, 2011, 2(4), 491–501.
44. Cahoon EB, Kinney AJ. Production of vegetable oils with novel properties: Using genomic tools to probe and manipulate fatty acid metabolism. European Journal of Lipid Science and Technology, 2005, 107(4), 239–243.
45. McKeon TA. Chapter 1: Genetic modification of seed oils for industrial applications, Industrial uses of vegetable oils, 2005, USDA, ARS, WRRC, Albany, CA. https://doi.org/10.1201/9781439822388.ch1.
46. Adhvaryu A, Erhan SZ. Epoxidized soyabean oil as a potential source of high temperature lubricants. Industrial Crops and Products, 2002, 15(3), 247–254.
47. Jayadas NH, Nair KP, Ajithkumar G. Tribological evaluation of coconut oil as an environment-friendly lubricant. Tribology International, 2007, 40(2), 350–354.
48. Dmitrieva TV, Sirovatka LA, Bortnitskii VI. Composites based on rapeseed oil and functional additives. TrenieIznos, 2001, 22(6), 693–698.
49. Castro W, Weller DE, Cheenkachorn K, Perez JM. The effect of chemical structure of basefluids on antiwear effectiveness of additives. Tribology International, 2005, 38(3), 321–326.
50. Spikes H. Low- and zero-sulphated ash, phosphorus and sulphur anti-wear additives for engine oils. Lubrication Science, 2008, 20(2), 103–136.
51. Gallopoulos NE. Thermal decomposition of metal dialkyldithiophosphate oil blends. ASLE Transactions, 1964, 7, 55–63.
52. Born M, Hipeaux JC, Marchand P, Parc G. Relationship between chemical structure and effectiveness of some metallic dialkyl and diaryldithiophosphates in different lubricated mechanisms. Lubrication Science, 1992, 4, 93–116.
53. Allum KG, Forbes ES. The load-carrying properties of metal dialkyldithiophosphates: The effect of chemical structure. Proceedings of the Institution of Mechanical Engineering: Conference Proceedings, 1968, 183, 7–14.
54. Rowe CN, Dickert Jr. JJ. The relation of antiwear function to thermal stability and structure for metal O,O-dialkylphosphorodithioates. ASLE Transactions, 1967, 10(1), 85–90.
55. Sanin PI, Shepeleva ES, Ulyanova AV, Kleimenov BV. The effect of synthetic additives in lubricating oil on wear under friction. Wear, 1960, 3(3), 200–218.
56. Farrington BB, Clayton JO. Compounded mineral oil. US Patent 2228658, 1941.
57. Sakurai T, Sato K. Chemical reactivity and load carrying capacity of lubricating oils containing organic phosphorus compounds. ASLE Transactions, 1970, 13(4), 252–261.
58. Didziulis SV, Fleischauer PD. Chemistry of the extreme-pressure lubricant additive lead naphthenate on steel surfaces. Langmuir, 1991, 7(12), 2981–2990.

59. Choudhary RB, Pande PP. Lubrication potential of boron compounds: an overview. Lubrication Science, 2002, 14(2), 211–222.
60. Liu W, Xue Q, Zhang X, Wang H. Effect of molecular structure of organic borates on their friction and wear properties. Lubrication Science, 1993, 6(1), 41–49.
61. Kapadia R, Glyde R, Wu Y. In situ observation of phosphorous and non-phosphorous antiwear films using a mini traction machine with spacer layer image mapping. Tribology International, 2007, 40(10–12), 1667–1679.
62. Jaiswal V, Kalyani, Rastogi RB, Kumar R. Tribological studies of some SAPS-free Schiff bases derived from 4-aminoantipyrine and aromatic aldehydes and their synergistic interaction with borate ester. Journal of Materials Chemistry A, 2014, 2(27), 10424–10434.
63. Rastogi RB, Maurya JL, Jaiswal V. Zero SAPs and ash free antiwear additives: Schiff bases of salicylaldehyde with 1, 2-phenylenediamine, 1, 4-phenylenediamine, and 4, 4′-diaminodiphenylenemethane and their synergistic interactions with borate ester. Tribology Transactions, 2013, 56(4), 592–606.
64. Greenall A, Neville A, Morina A, Sutton M. Investigation of the interactions between a novel, organic anti-wear additive, ZDDP and overbased calcium sulphonate. Tribology International, 2012, 46(1), 52–61.
65. Khan MS, Sisodia MS, Gupta S, Feroskhan M, Kannan S, Krishnasamy K. Measurement of tribological properties of Cu and Ag blended coconut oil nanofluids for metal cutting. Engineering Science and Technology, an International Journal, 2019, 22(6), 1187–1192.
66. Rajubhai VH, Singh Y, Suthar K., Surana AR. Friction and wear behavior of Al-7% Si alloy pin under pongamia oil with copper nanoparticles as additives. Materials Today Proceeding, 2019, 25, 695–698.
67. Abad MD, Sánchez-López JC. Tribological properties of surface-modified Pd nanoparticles for electrical contacts. Wear, 2013, 297, 943–951.
68. Le VN, Lin J-W. Tribological properties of aluminum nanoparticles as additives in an aqueous glycerol solution. Applied Science, 2017, 7(1), 80, 1–15.
69. Zhang S, Hu L, Feng D, Wang H. Anti-wear and friction-reduction mechanism of Sn and Fe nanoparticles as additives of multialkylatedcyclopentanes under vacuum condition. Vacuum, 2013, 87, 75–80.
70. Wang A, Chen L, Xu F, Yan Z. In situ synthesis of copper nanoparticles within ionic liquid-in-vegetable oil microemulsions and their direct use as high efficient nanolubricants. RSC Advances, 2014, 4, 45251–45257.
71. Maliar T, Achanta S, Cesiulis H, Drees D. Tribological behaviour of mineral and rapeseed oils containing iron particles. Industrial Lubrication and Tribology, 2015, 67(4), 308–314.
72. Battez AH, Rico JEF, Arias AN, Rodriguez JLV, Rodriguez RC, Fernandez JMD. The tribological behaviour of ZnO nanoparticles as an additive to PAO6. Wear, 2006, 261(3–4), 256–263.
73. Battez AH, González R, Felgueroso D, Fernández JE, Fernández MR, García MA, Peñuelas I. Wear prevention behaviour of nanoparticle suspension under extreme pressure conditions. Wear, 2007, 263(7–12), 1568–1574.
74. Thottackkad MV, Rajendrakumar PK, Prabhakaran NK, Tribological analysis of surfactant modified nanolubricants containing CeO_2 nanoparticles. Tribology-Materials, Surfaces & Interfaces, 2014, 8(3), 125–130.
75. Sajeeb A, Rajendrakumar PK. Tribological assessment of vegetable oil based CeO_2/CuO hybrid nano-lubricant. Proceedings of the Institution of Mechanical Engineers, Part J: Journal of Engineering Tribology, 2020, 234(12), 1940–1956.
76. Thottackkad MV, Perikinalil RK, Kumarapillai PN. Experimental evaluation on the tribological properties of coconut oil by the addition of CuO nanoparticles. International Journal of Precision Engineering and Manufacturing, 2012, 13(1), 111–116.

77. Alves SM, Barros BS, Trajano MF, Ribeiro KSB, Moura E. Tribological behavior of vegetable oil-based lubricants with nanoparticles of oxides in boundary lubrication conditions. Tribology International, 2013, 65, 28–36.
78. Shaari MZ, Roselina NRN, Kasolang S, Hyie KM, Murad MC, Abu Bakar MA. Investigation of tribological properties of palm oil biolubricant modified nanoparticles. Jurnal Teknolgi, 2015, 76, 69–73.
79. Zulkifli NWM, Kalam MA, Masjuki HH, Yunus R. Experimental analysis of tribological properties of biolubricant with nanoparticle additive. Procedia Engineering, 2013, 68, 152–157.
80. Ingole S, Charanpahari A, Kakade A, Umare SS, Bhatt DV, Menghani J. Tribological behavior of nano TiO_2 as an additive in base oil. Wear, 2013, 301(1–2), 776–785.
81. Luo T, Wei X, Huang X, Huang L, Yang F. Tribological properties of Al_2O_3 nanoparticles as lubricating oil additives. Ceramics International, 2014, 40(5), 7143–7149.
82. Bhaumik S, Kamaraj M, Paleu V. Tribological analyses of a new optimized gearbox biodegradable lubricant blended with reduced graphene oxide nanoparticles. Proceedings of the Institution of Mechanical Engineers, Part J: Journal of Engineering Tribology, 2020, 1–15. DOI: 10.1177/1350650120925590.
83. Gupta RN, Harsha AP, Singh S. Tribological study on rapeseed oil with nano-additives in close contact sliding situation. Applied Nanoscience, 2018, 8(4), 567–580.
84. Dubey MK, Bijwe J, Ramakumar SSV. PTFE based nano-lubricants. Wear, 2013, 306(1–2), 80–88.
85. Song H, Wang Z, Yang J. Tribological properties of graphene oxide and carbon spheres as lubrication additives. Applied Physics A, 2016, 122, 933/1–9.
86. Kiong KSS, Yusup S, Soon CV, Arpin T, Samion S, Kamil RNM. Tribological investigation of graphene as lubricant additive in vegetable oil. Journal of Physical Science, 2017, 28, 257–267.
87. Kannan KT, Rameshbabu S. Tribological behavior of modified jojoba oil with graphene nanoparticle as additive in SAE20W40 oil using pin on disc tribometer. Energy Sources, Part A: Recovery, Utilization and Environmental Effects, 2017, 39, 1842–1848.
88. Su Y, Gong L, Chen D. An investigation on tribological properties and lubrication mechanism of graphite nanoparticles as vegetable based oil additive. Journal of Nanomaterials, 2015, 2015, 1–7.
89. Cornelio JAC, Cuervo PA, Hoyos-Palacio LM, Lara-Romero J, Toro A. Tribological properties of carbon nanotubes as lubricant additive in oil and water for a wheel–rail system. Journal of Materials Research and Technology, 2016, 5(1), 68–76.
90. Opia AC, Kameil AHM, Daud ZHC, Mamah SC, Izmi MI, Rahim AB. Tribological properties enhancement through organic carbon nanotubes as nanoparticle additives in boundary lubrication conditions. JurnalTribologi, 2020, 27, 116–131.
91. Abd HS, Abdulmunem AR, Jabal MH, Samin PM, Rahman HA. Working features evaluation of the diesel engine lubricated with blends of renewable corn oil and carbon nanotubes. Journal of Mechanical Engineering Research and Developments, 2020, 43(2), 384–395.
92. Kiu SSK, Yusup S, Chok VS, Taufiq A, Kamil RNM, Syahrullail S, Chin BLF. Comparison on tribological properties of vegetable oil upon addition of carbon based nanoparticles. IOP Conf. Series: Materials Science and Engineering, 2017, 206, 012043, doi:10.1088/1757-899X/206/1/012043.
93. Ranjan N, Shende RC, Kamaraj M, Ramaprabhu S. Utilization of TiO_2/gC_3N_4 nanoadditive to boost oxidative properties of vegetable oil for tribological application. Friction, 2021, 9, 273–287.
94. Dong JX, Hu ZS. A study of the anti-wear and friction-reducing properties of the lubricant additive, nanometer zinc borate. Tribology International, 1998, 31(5), 219–223.

95. Sunqing Q, Junxiu D, Guoxu C. Tribological properties of CeF_3 nanoparticles as additives in lubricating oils. Wear, 1999, 230, 35–38.
96. Liu W, Chen S. An investigation of the tribological behaviour of surface-modified ZnS nanoparticles in liquid paraffin. Wear, 2000, 238, 120–124.
97. Rapoport L, Leshchinsky V, Lapsker I, Volovik Y, Nepomnyashchy O, Lvovsky M, Popovitz-Biro R, Feldman Y, Tenne R. Tribological properties of WS_2 nanoparticles under mixed lubrication. Wear, 2003, 255, 785–793.
98. Joly-Pottuz L, Dassenoy F, Martin JM, Vrbanic D, Mrzel A, Mihailovic D, Vogel W, Montagnac G. Tribological properties of Mo–S–I nanowires as additive in oil. Tribology Letters, 2005, 18(3), 385–393.
99. Yu HL, Xu Y, Shi PJ, Wang HM, Zhao Y, Xu BS, Bai ZM. Tribological behaviors of surface-coated serpentine ultrafine powders as lubricant additive. Tribology International, 2010, 43, 667–675.
100. Lingtong K, Hua H, Tianyou W, Dinghai H, Jianjian F. Synthesis and surface modification of the nanoscale cerium borate as lubricant additive. Journal of Rare Earths, 2011, 29(11), 1095–1099.
101. Rudenko P, Bandyopadhyay A. Talc as friction reducing additive to lubricating oil. Applied Surface Science, 2013, 276, 383–389.
102. Srinivas V, Thakur RN, Jain AK. Antiwear, antifriction and extreme pressure properties of motor bike engine oil dispersed with molybdenum disulfide nanoparticles. Tribology Transactions, 2017, 60(1), 12–19.
103. Charoo MS, Wani MF. Tribological properties of h-BN nanoparticles as lubricant additive on cylinder liner and piston ring. Lubrication Science, 2017, 29, 241–254.
104. Weertman JR. Hall-Petch strengthening in nanocrystalline metals. Material Science and Engineering A, 1993, 166(1–2), 161–167.
105. Yamakov V, Wolf D, Phillpot SR, Mukherjee AK, Gleiter H. Deformation mechanism map for nanocrystalline metals by molecular-dynamics simulation. Nature Materials, 2004, 3, 43–47.
106. Narayanunni V, Kheireddin BA, Akbulut M. Influence of surface topography on frictional properties of Cu surfaces under different lubrication conditions: Comparison of dry, base oil, and ZnS nanowire-based lubrication system. Tribology International, 2011, 44, 1720–1725.
107. Akbulut M. Nanoparticle-based lubrication systems. Journal of Powder Metallurgy and Mining, 2012, 1, 1–3.
108. Kotia A, Chowdary K, Srivastava I, Ghosh SK, Ali MKA. Carbon nanomaterials as friction modifiers in automotive engines: Recent progress and perspectives. Journal of Molecular Liquids, 2020, 310, 113200.
109. Wang X, Xu X, Choi SUS. Thermal conductivity of nanoparticle-fluid mixture. Journal of Thermophysics and Heat Transfer, 1999, 13(4), 474–480.
110. Taha-Tijerina J, Peña-Paras L, Narayanan TN, Garza L, Lapray C, Gonzalez J, Palacios E, Molina D, García A, Maldonado D, Ajayan PM. Multifunctional nanofluids with 2D nanosheets for thermal and tribological management. Wear, 2013, 302, 1241–1248.
111. Lee K, Hwang Y, Cheong S, Choi Y, Kwon L, Lee J, Kim SH. Understanding the role of nanoparticles in nano-oil lubrication. Tribology Letters, 2009, 35, 127–131.
112. Ghaednia H, Jackson RL. The effect of nanoparticles on the real area of contact, friction, and wear. Transactions of ASME, Journal of Tribology, 2013, 135, 041603.
113. Liu G, Li X, Qin B, Xing D, Guo Y, Fan R. Investigation of the mending effect and mechanism of copper nano-particles on a tribologically stressed surface. Tribology Letters, 2004, 17, 961–966.
114. Tao X, Jiazheng Z, Kang X. The ball-bearing effect of diamond nanoparticles as an oil additive. Journal of Physics D: Applied Physics, 1996, 29, 2932–2937.

115. Jaiswal V, Rastogi RB, Kumar R, Singh L, Mandal KD. Tribological studies of stearic acid-modified $CaCu_{2.9}Zn_{0.1}Ti_4O_{12}$ nanoparticles as effective zero SAPS antiwear lubricant additives in paraffin oil. Journal of Materials Chemistry A, 2014, 2, 375–386.
116. Wäsche R, Hartelt M, Hodoroaba V-D. Analysis of nanoscale wear particles from lubricated steel-steel contacts. Tribology Letters 2015, 58(3), Article number 49, 1–10.
117. Dai W, Kheireddin B, Gao H, Liang H. Roles of nanoparticles in oil lubrication. Tribology International, 2016, 102, 88–98.
118. Chou C-C, Lee S-H. Tribological behavior of nanodiamond-dispersed lubricants on carbon steels and aluminum alloy. Wear, 2010, 269, 757–762.

5 Tribology of Polymer Composites with Green Nano-Materials

Guoxin Xie, X.H. Sun, H.J. Gong, Y.L. Ren, H. Chen, M.Y. Li, Y.B. Li, and L. Zhang
Tsinghua University
Beijing, China

Z.J. Ji
Qinghai University
Xining, China

L.N. Si
North China University of Technology
Beijing, China

CONTENTS

5.1 INTRODUCTION

Friction consumes more than 30% of the primary energy all over the world [1], and the wear between moving parts is one of the main failure reasons for materials and mechanical equipment [2]. The economic loss caused by friction and wear is up to hundreds of billions of USD every year, and lubrication is the most effective way to diminish friction and wear [3].

Polymer composites have good self-lubricity and wear resistance, and can provide extremely low coefficients of friction and wear rates under specific or highly controlled test conditions. Polymer composites with green nano-materials have a combination of properties that are not found in other materials [4–8]. They have

DOI: 10.1201/9781003139386-5

been extensively employed in various applications as tribological components such as bearings, gears, and seals due to their light weight, self-lubrication, chemical stability, and bio-compatibility [9–12].

To obtain a deep understanding on the tribological behavior of the polymer composites, this chapter systematically reviews the tribological properties of polymer composites with green nano-materials and the underlying tribological mechanism. This chapter is divided into Section 1 (Introduction), Section 2 (Polymer with micro-/nano-particles), Section 3 (Polymer with micro-/nano-fibers), Section 4 (Polymer with 2D materials), and Section 5 (Conclusion).

5.2 POLYMER WITH MICRO-/NANO-PARTICLES (0D)

The addition of nano-/micro-particles into solid materials can improve the tribological properties and can obviously influence the structure and dispersion of nano-/micro-particles. The traditional mechanical mixing of fillers and the matrix can lead to phase separation or uneven dispersion of the particles resulting in a decrease in the lubricating properties of the composites. In contrast to ordinary particles, core-shell nano-/micro-particles have shell materials that can interlock the matrix and particles, as well as improve the dispersion of particles in the matrix. Meanwhile, the core material can play a role in wear resistance, friction reduction, and fortification.

Core-shell micro-/nano-particles with characteristic structural advantages are of great significance to improve the tribological properties of lubricating materials. In general, there are two ways for core-shell particles to improve the tribological properties of the matrix: in the form of micro-/nano-containers with liquid lubricants and core-shell particles with solid cores.

5.2.1 Polymer Composites with Liquid Core-shell Particles

Micro-/nano-containers with stimuli-responsive properties have been used to fabricate lubricating composites such as porous materials and capsules in Figure 5.1 The micro-/nano-containers store and hold liquid lubricants when not in use. When needed, they continuously and steadily release lubricants to the friction surface. Micro-/nano-containers optimize the existing state of liquid lubricants in composites and make it possible to achieve the effect of fluid lubrication under the condition of dry friction without leakage problems.

The porous PEEK composites are a typical matrix material of porous lubricating composites. They can be fabricated by mold-leaching. Liquid lubricants are then incorporated under a vacuum condition [13, 14]. An ultralow coefficient of friction (COF) of 0.0197 was obtained when porous sweating PEEK composites containing ionic liquids slid against the steel counterpart with an average roughness of 0.15–0.3 μm [15]. Wang et al. prepared oil-containing porous polyimide (PI) as a retainer of thrust ball bearings by cold pressing, sintering, and vacuum immersion processes [16]. The results showed that the porous PI with higher porosity possessed higher oil supply content leading to more effective anti-friction effect [16]. The bearings with oil-containing PI retainers operated smoothly for 12 h at 800 rpm and 5 h at 1200 rpm with an axial load of 500 N [16]. Only some slight wear was found on the raceway

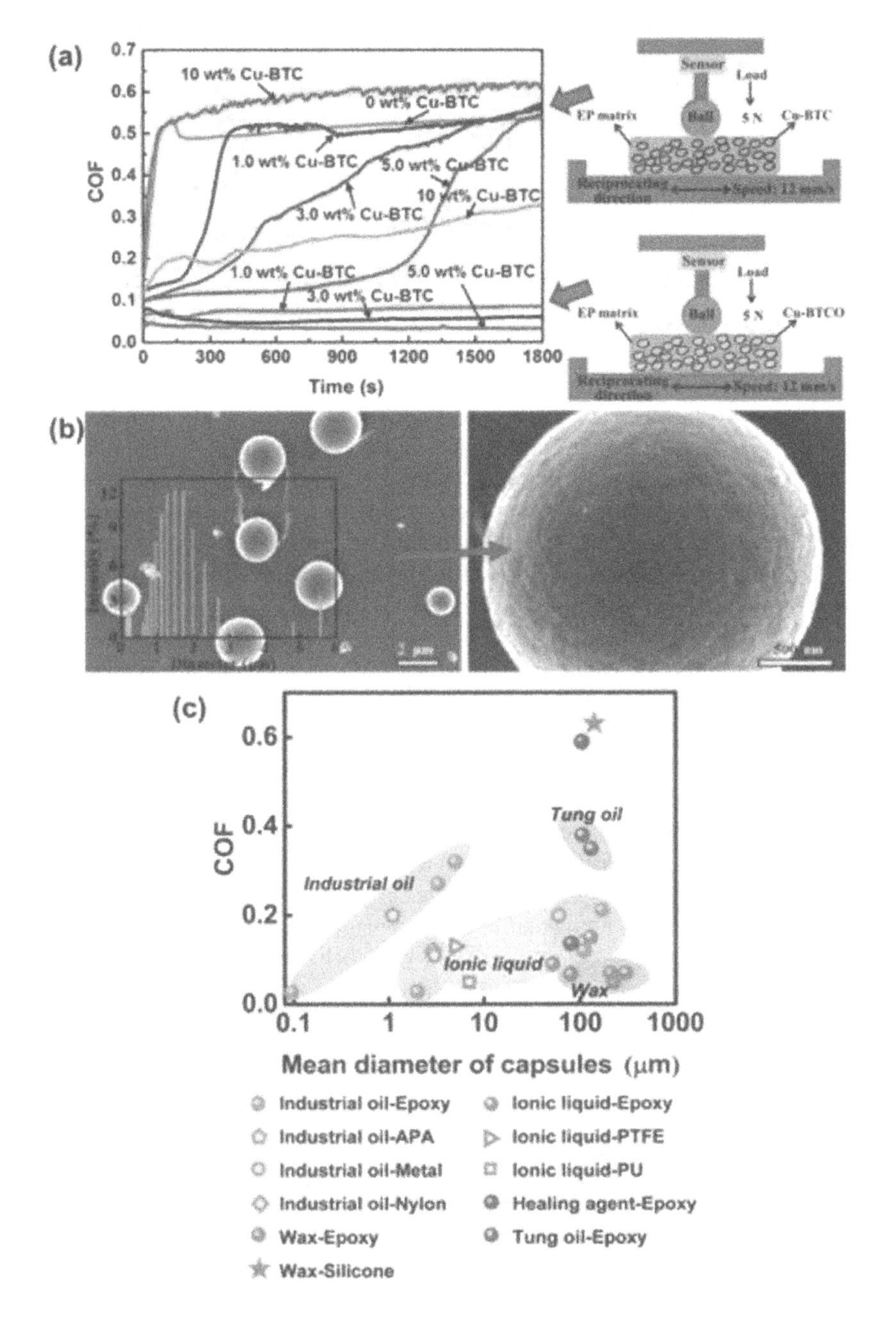

FIGURE 5.1 Tribological properties of lubricating materials containing liquid lubricants [2]. (a) Variation curves for COF as a function of the test time for the pure epoxy resin, epoxy composites with different contents of Cu-BTC, and epoxy composites with different contents of Cu-BTC storing oleylamine (Cu-BTCO) [17]; (b) SEM images of mono-dispersed polystyrene/poly alpha olefin micro-capsules [20]; (c) An overview of COFs and average capsule diameters of lubricating materials with different cores of capsules and matrices. APA, PTFE, and PU are aromatic polyamide, polyetheretherketone, and polyurethane, respectively [2].

surface while the bearings without oil-loaded retainers failed quickly due to the severe adhesion wear [16].

In addition, Zhang et al. [17] reported self-lubricating nano-composites with mesoporous Cu-BTC metal-organic frameworks (MOFs) storing oleylamine as smart nano-containers. The nano-composites with 5 wt% Cu-BTC storing oleylamine showed an ultralow COF of 0.03 (Figure 5.1(a)) at a speed of 12 mm/s and an applied load of 5 N [17]. The low friction resulted from the lubricating film that was produced in situ by the polymerization reaction between the oleylamine and the epoxy matrix [17]. The worn surface of Cu-BTC nano-composite storing oleylamine was smooth without cracks while the worn surface of the neat matrix and its composites only filled with Cu-BTC exhibited layer detachment and abrasive wear [17].

Capsules containing liquid lubricants have also been used to make lubricating composites. The capsule shell is damaged during the friction process, and the lubricant is released to the friction interface for lubrication. At the same time, the cavity formed after the capsule ruptures collected abrasive debris, weakening the abrasive effect and reducing friction. Guo et al. [18] first proposed capsule-based self-lubricating composites in 2009. Oil-containing poly (melamine formaldehyde) micro-capsules with a particle size of 110 μm were synthesized via in-situ polymerization and embedded into the epoxy matrix [18]. The COF and the wear rate of composites with 10 wt% micro-capsules were reduced by 75% and 98.3%, respectively [18]. Poly (urea-formaldehyde) micro-capsules containing a mixture of wax lubricants and multi-wall carbon nano-tubes with a diameter of 150–300 μm were synthesized by in-situ polymerization in an oil-in-water emulsion [19]. The synergistic effect of wax lubricants and multi-wall carbon nano-tubes further improved the lubricating performance of epoxy composites; the optimal COF was 0.049 with only 5 wt% micro-capsules when the sample slid against the steel counterpart [19].

The lower friction of capsule-based polymer lubricating materials (Figure 5.1(b) and (c)) was reported by Zhang et al. [20]. The epoxy composites with 15 wt% mono-dispersed polystyrene/poly alpha olefin micro-capsules exhibited extremely low friction and wear (COF of 0.028 and wear rate of 6.7×10^{-7} mm^3/Nm) under 6 N and 32 mm/s when sliding against a GCr45 bearing steel ball with a surface roughness of 5 nm [20]. The superior lubricating performance was attributed to the uniform polymer transfer films and dense lubricant films on the contact surface [20].

The lubricating composites with inorganic shell capsules exhibit superior chemical stability, thermal stability, flame retardancy, and mechanical strength. Yang et al. [21] synthesized silica (SiO_2) shell micro-capsules containing ionic liquids and the tribological properties of polyurethane composites with SiO_2 capsules were optimized. The COF of the composite with 20 wt% SiO_2/ionic liquid micro-capsules decreased to about 0.05 due to the formation of an orderly adsorption film on the friction surface [21]. This was related to the molecular polarity of the ionic liquid.

To expand the application temperature range of micro-/nano-capsules, ionic liquid-loaded micro-capsules were synthesized by Bandeira et al. [22] using the simple solvent evaporation method and incorporated into a PTFE top-coat by spraying followed by a coating cure with the base-coat at 380°C for 30 min. Polysulfone has excellent thermal stability and was selected as the shell material. Thermogravimetric analysis showed that the initial thermal decomposition temperature of the micro-capsules was 420°C, which could resist high temperature curing and spraying during the preparation of the PTFE coating [22]. This maintained capsule integrity. In addition, the lubricating material exhibited excellent anti-friction effect under high load conditions due to the outstanding mechanical properties of polysulfone [22]. Another type of thermally stable polysulfone micro-capsules with an average diameter and shell thickness of about 128 μm and 10 μm, respectively, were obtained similarly [23]. The results indicated that micro-capsules began to break down at 440°C, which greatly expanded the application field of micro-capsules as lubricating additives [23].

However, the tribological properties of composites seriously declined with the increasing capsule size or content as a result of the propagation of cracks. Moreover, the large size of micro-capsules limits their applications in the coating field. The performance loss caused by capsules can be recovered via the introduction of reinforcing fillers, the reduction in the capsule size, the improvement in the capsule strength, and the enhancement in the interaction with the matrix (Figure 5.2) [2].

SiO_2 nano-particles demonstrated an effective improvement in the mechanical properties of capsule-based lubricating materials [24]. The hardness and flexural strength of composites with 23 wt% SiO_2 nano-particles and 10 wt% micro-capsules increased by 72% and 100% as compared with those with only 10 wt% micro-capsules [24]. Moreover, ternary composites containing grafting SiO_2 nano-particles, carbon fibers, and oil-containing micro-capsules were fabricated by Guo et al. [25]. The results showed that SiO_2 nano-particles and carbon fibers both played a dual role as solid lubricants and reinforced fillers, inhibiting the molecular structure degradation of the friction surface and greatly improving the tribological properties of the materials [25].

In most capsule-based lubricating composites, the diameter of the capsules is hundreds of micro-meters. A systematic study evaluated the influence of the micro-capsules with diameters of 38–63 μm and 63–90 μm on the tribological properties of composites; the hardness and Young's modulus of composites gradually declined while the lubricating properties improved with increasing capsules' size and content [26]. Furthermore, a tough and ultra-low friction resin nano-composite was reported recently that incorporated shell-designed nano-capsules filled with liquid lubricants [27]. The tensile strength and toughness of the nano-composites with a capsule diameter of 72 nm were all 1.4-fold that of micro-capsule-based composites. This proved that reducing the capsule size into the nano-scale is a promising solution to improving the mechanical performance without introducing any fillers [27].

In most cases, brittle polymers are used as shell materials for lubricating capsules. A novel Ni shell capsule was prepared by Yang et al. through the electroless plating [28]. The strength of the metal capsules increased by about two orders of magnitude in contrast to conventional micro-capsules [28]. The introduction of conventional micro-capsules significantly diminished the compressive strength of the resin matrix while the Ni shell micro-capsule exhibited a comparable compressive strength to the pure epoxy resin at different strain rates (Figure 5.2(f)) [28].

Surface functionalization can also provide enhanced or additional properties in composites and realize synergistic effects of multiple functionalities [29]. Surface functionalization of capsules plays a key role in minimizing the aggregation of capsules and improving the mechanical properties of composites by reducing the

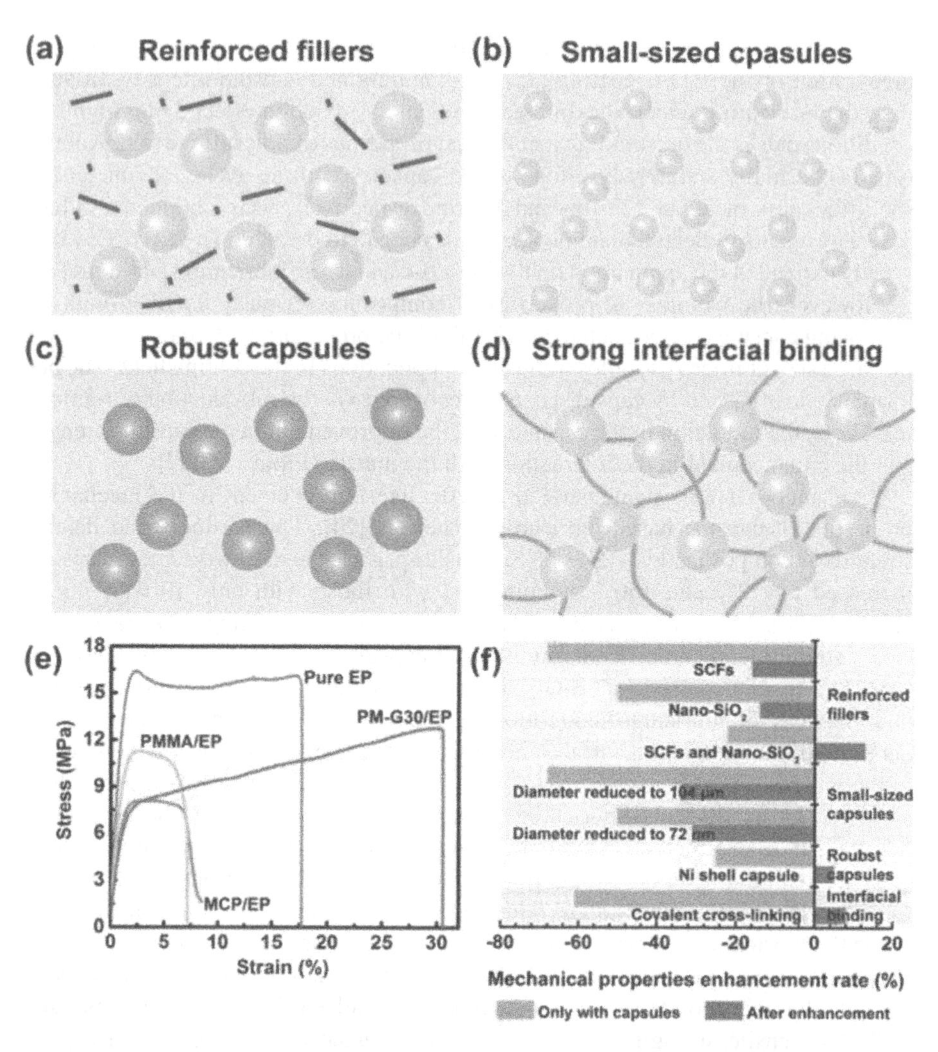

FIGURE 5.2 Four strategies for compensating mechanical property loss caused by capsules in lubricating materials [2]. (a) The introduction of reinforcing fillers; (b) The reduction of the capsule size; (c) The improvement of the capsule strength; (d) The enhancement of the interaction between capsules and the matrix; (e) The stress-strain curves of pure epoxy (pure EP), the epoxy composite with 10 wt% polymethyl methacrylate (PMMA) micro-particles (MPC/EP), the epoxy composite with 10 wt% PMMA nano-capsules (PMMA/EP), and PMMA nano-capsules copolymerized with 30% glycidyl methacrylate (PM-G30/EP) [27]; and (f) The mechanical property enhancement rates of different reinforcement strategies.

chemical dissimilarity between the matrix and capsules. Gong et al. [27] reported the highly tough resin composite with cross-linkable nano-capsules. The co-monomer of glycidyl methacrylate with epoxy groups in the capsule shell reacted with the curing agent in the matrix and increased the crosslinking point between nano-capsules and the matrix [27]. In contrast to the non-crosslinked samples, the crosslinked nano-capsules gave the resin nano-composite remarkable toughness that was 2.7-fold larger due to the contribution of strong interfacial bonding (Figure 5.2(e)) [27].

5.2.2 Polymer Composites with Solid Core-Shell Particles

The parameters of some core-shell particles that enhance the lubrication properties of the matrix are summarized in Table 5.1 [12]. In contrast to nano-/micro-capsules, solid core-shell particles can reduce the COF of the matrix while maintaining or improving the mechanical properties. Conventionally, MoS_2 is an ideal solid lubrication additive in the polymer matrix owing to its outstanding self-lubricating ability [30]. However, the lubrication properties of MoS_2 would deteriorate after being exposure to humid atmosphere, and therefore the encapsulation of MoS_2 into urea formaldehyde (UF) resin is used to improve the chemical stability of MoS_2 in a humid environment. The friction properties of composites were studied when the MoS_2@UF resin core-shell particles were used as fillers [31]. The results showed that the composites incorporated with the MoS_2@UF resin exhibited lower sliding friction and smoother surface morphology because of the protection provided by shells, which delayed the oxidation of the MoS_2 crystals [31]. MoS_2 can also be used as a shell material to improve the interfacial adhesion between the core-shell particles and the matrix. CNF@MoS_2 core-shell particles were prepared via a one-step hydrothermal method to enhance the anti-friction and anti-wear properties of epoxy resin [32].

TABLE 5.1
Parameters of Enhanced Tribological Properties [12]

Composites		Friction Condition		Friction Properties (Compared with Pure Matrix)		Ref.
Filler (Content)	Matrix	Load	Sliding Speed	COF	Wear Rate mm³/(N·m)	
MoS_2@UF resin (10 wt%)	HDPE	1.05 MPa	150 rpm	0.01 (89.5% lower)		[31]
CNF@MoS_2	EP	4 N	200 rpm	0.075 (82.1% lower)	8.6×10^{-5} (87.5% lower)	[32]
h-BN@Ni (5 vol%)	Al_2O_3/C	20 N	200 rpm	0.45 (6.25% lower)	10^{-6} (26.2% lower)	[33]
Ni@NiO (5 wt%)	EP	1.0 MPa	1.0 m/s	1.05 (36.4% lower)	10^{-6}	[34]
m-Si_3N_4@PANI (2.0 wt%)	Phenolic resin			0.1681 (48.48% lower)	1.13×10^{-8} (87.5% lower)	[35]
SiC@GNSs (5 vol%)	Al_2O_3	90 N	0.1 m/s	0.45 (29.4% lower)	2.6×10^{-5} (90.1% lower)	[36]
SiC@CNT (5 vol%)	Al6061 alloy	98.1 N	200 rpm	0.45 (31% lower)	3.25×10^{-8} (45% lower)	[37]
Cu@GO (0.8 wt%)	Cu/Ti_3AlC_2	4 N	200 rpm	0.2 (about 50% lower)	2.0×10^{-8}	[38]

Furthermore, the MoS_2 debris generated during the sliding process and could form a dense and uniform lubricating film to improve the lubrication performance [32]. In addition to MoS_2, h-BN@Ni particles were synthesized by an electroless plating technique as fillers in the ceramic [33]. By adding h-BN@Ni particles, the micro-structures of the ceramic composites were more homogeneous than those of the composite containing uncoated h-BN particles [33].

Epoxy resin incorporated with Ni@NiO magnetic nano-particles were prepared by Wang et al [34]. The mechanical and tribological properties of the composites were both improved greatly (the COF was reduced by 36.4%, the wear resistance increased by 22.2-fold, the hardness increased by 37.8%, and the elastic modulus increased by 16.3%) [34]. The reason for the reinforcement of the Ni@NiO particles is that they filled up the cracks and defects produced during the friction process and formed a smooth transfer film, which in turn improved the tribological properties of the composites.

m-Si_3N_4@PANI core-shell particles were synthesized to improve the poor thermal conductivity of the phenolic resin/carbon fiber (PF/CF) composites [35]. The results indicated that the addition of m-Si_3N_4@PANI improved the thermal conductivity, electrical conductivity, COF, and the wear rate of PF/CF composites effectively [35]. Zhang et al. [36] fabricated SiC@GNSs by a wet ball milling process as the fillers in alumina matrix composites, and the composites showed outstanding tribological properties owing to the graphene nano-sheets (GNSs)-rich tribofilms. In a similar way, SiC was coated on the surface of carbon nano-tubes (CNT) to obtain SiC@CNT nano-particles using the chemical vapor deposition (CVD) method, and the particles were mixed with Al6061 alloy to fabricate SiC@CNT/Al6061 composites via spark plasma sintering [37]. The COF and wear rate of the composites decreased owing to the transition of the wear mode from adhesive wear to abrasive wear because of the uniform dispersion of SiC@CNT debris at the friction interface [37]. A novel Cu matrix composite with a low COF (0.2) was prepared by hot-press sintering the core-shell Cu@GO particles and Cu-decorated Ti_3AlC_2 powders [38]. The GO and Ti_3AlC_2 could form continuous and compact transfer films on the friction surface, and thus reducing the COF and wear rate [38].

In summary, the friction mechanism of core-shell nano-/micro-particles in a solid matrix is the formation of transfer films or lubricating layers at the friction interface to separate the samples and the counterparts to avoid direct contact [12], as shown in Figure 5.3. However, for nano-/micro-capsules, the transfer films or lubricating

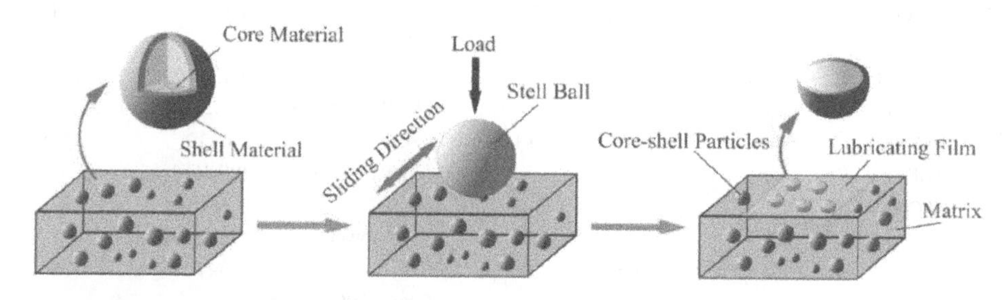

FIGURE 5.3 Formation of smooth agent film during sliding wear [12].

layers are mainly generated by organic cores (lubricant oil) and shells. Neither matrix (most of them are EP and another polymer) nor capsules can withstand high temperatures and high loads [12]. These factors will limit the applications of nano-/micro-capsules as lubricating materials in mechanical devices that need to operate under high temperatures or for a long time. Most of the core-shell solid particles are inorganic and can maintain the desired mechanical and tribological properties under high temperatures; however, the tribological properties of the composites still need to be enhanced prior to practical applications. Consequently, it is necessary to further develop core-shell self-lubricating materials which can withstand high temperatures or high loads to meet the requirements of the mechanical components in aerospace, transportation, etc.

5.3 POLYMER WITH MICRO-/NANO-FIBER MATERIALS (1D)

One-dimensional polymer nano-materials are mainly divided into carbon nano-tubes, carbon fibers (CF), glass fibers, and other materials.

CF is a special fiber made of carbon. It offers high temperature resistance, friction resistance, electrical conductivity, thermal conductivity, and corrosion resistance [39]. Due to the preferred orientation of graphite micro-crystalline structure along the fiber axis, it has high strength and modulus along the fiber axis. The main purpose of CF is to be a composite with resin, metals, ceramics, or carbon as reinforcing materials to make advanced composite materials [39]. Future work will improve the process, expand production, reduce costs, and develop new applications. Some special carbon fibers include oxidation-resistant carbon fibers (to improve the service temperature of composite materials), low fineness carbon fibers (for 0.035 mm ultra-thin prepreg), high thermal conductivity and low resistance carbon fibers (for shielding electromagnetic and radio frequency interference and emitting excess heat), low thermal expansion coefficient carbon fibers (for products such as satellite antenna systems and reflectors), and hollow carbon fiber (for high temperature and low resistance carbon fiber) [40].

Carbon nano-tubes (CNTs) are one-dimensional quantum materials with special structures (radial size is nano-meter, axial size is micron, and both ends of the tube are basically sealed). Carbon nano-tubes are mainly composed of hexagonal carbon atoms arranged as coaxial tubes with several to dozens of layers; The distance between layers is about 0.34 nm, and the diameter is generally 2–20 nm [41]. According to the different orientations of the carbon hexagon along the axis, it can be divided into three types: serrated, armchair, and spiral [41].

Carbon nano-tubes are graphene sheet layers that have been curled. They are classified according to the number of graphene sheet layers [41, 42]: single-walled carbon nano-tubes (SWCNTs) and multi-walled carbon nano-tubes (SWCNTs). When MWCNTs are first formed, they can easily become trap centers between layers to capture various defects, and thus, the wall of MWCNTs is usually full of small hole-like defects [42]. In contrast to multi-wall tubes, single wall tubes have a smaller diameter distribution, fewer defects, and higher uniformity [42].

Common preparation methods of carbon nano-tubes include arc discharge, laser ablation, chemical vapor deposition (hydrocarbon gas pyrolysis), solid phase pyrolysis, glow discharge, gas combustion, and polymerization synthesis [42]. Glass fibers are an inorganic non-metal material with excellent performance. It has many advantages such as good insulation, strong heat resistance, good corrosion resistance, and high mechanical strength [43–45]. It is made of pyrophyllite, quartz sand, limestone, dolomite, bornite, and brucite by high temperature melting, wire drawing, winding and weaving [43]. The diameter of the mono-filament ranges from several microns to more than 20 microns, which is equivalent to 1/20–1/5 of a hair, and each bundle of fiber precursor consists of hundreds or even thousands of mono-filaments [43].

When glass fibers are used as the reinforcing material of reinforced plastics, the biggest characteristic is the high tensile strength [46–48]. The tensile strength is 6.3–6.9 g/D in standard state and 5.4–5.8 g/D when wet [47]. The glass fibers have good heat resistance, and there is no effect on the strength when the temperature reaches 300°C [47]. The mechanical properties of the polymers are characterized by three basic indexes: stiffness (elastic modulus), strength (tensile strength), and toughness (elongation at break). Although some polymers have outstanding mechanical properties, some key properties such as impact strength or toughness still need to improve. The mechanical properties of polymers are usually improved by adding fillers with excellent mechanical properties. According to the morphology, the reinforced fillers are divided into two-dimensional (2D) fillers, one-dimensional (1D) fillers, and zero-dimensional (0D) fillers (Figure 5.4) [10]. The mechanical properties of the materials can be optimized by fully understanding the mechanism of reinforcement and toughening.

The enhancements of the stiffness and strength of polymer composites can be described by many theories, such as strain field distortion [49], dynamic changes [50], polymer bridges [51, 52] and stress transfer [53]. Stress transfer between the matrix and filler and the load bearing of fillers are the main mechanisms. When the composite is subjected to external stress, the matrix stress is transferred to the fillers through the matrix-filler interface, and the fillers become the main load-bearing phase. The elastic modulus and tensile strength of the filler are higher than those of the matrix so the polymer material is reinforced [53, 54]. 0.2 wt% carbon nano-tubes as 1D fillers were added into a UHMWPE coating, and the hardness of the coating was increased by 66% and the elastic modulus by 58% [55].

The stress transfer mechanism is applicable to all three types of reinforcing fillers, and the size, content, and dispersion state of the fillers also have a significant impact on the reinforcing effect [56, 57].

A large number of experimental studies on fiber-reinforced polymer composites show that short fibers are not as effective as long fibers or continuous fibers, and the phenomenon is explained in detail by shear lag theory [57]. Excellent mechanical properties can be obtained only when the fiber length exceeds the critical length. The critical length is mainly related to the interface interaction between the filler and the matrix. In addition, the spatial orientation of fillers will also significantly affect the reinforcement effect [52, 58]. Mortazavian and Fatemi [59] confirmed that

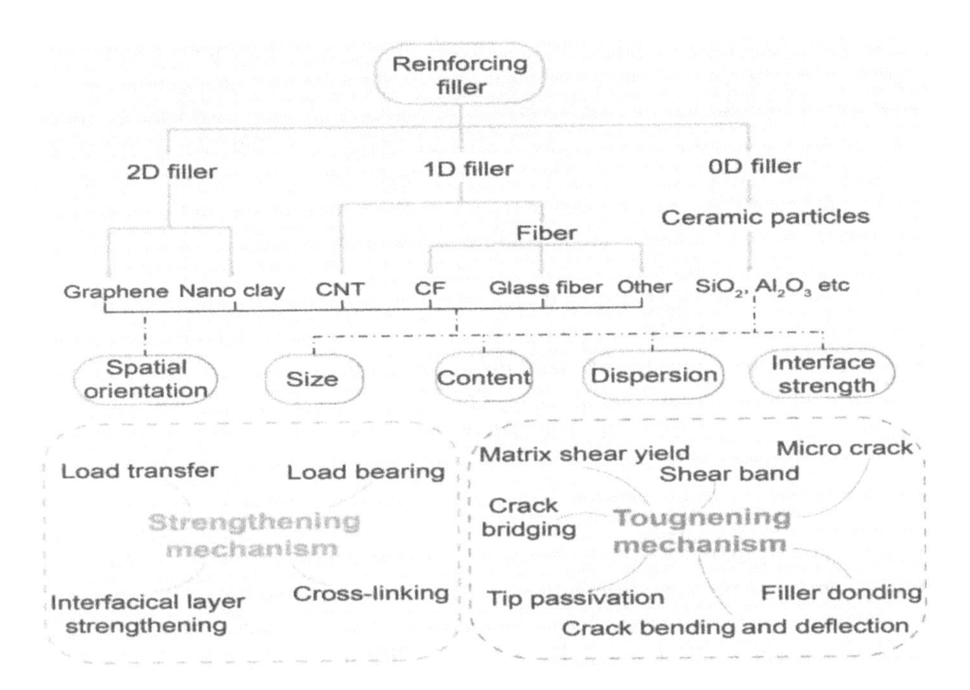

FIGURE 5.4 Classification of reinforcing fillers and influencing factors in the reinforcement effect [10].

the tensile strength of the composite changes nonlinearly with the angle of orientation in a given plane of the glass fiber in the sample.

Based on the analyses above, the reinforcement performance of polymer composites is directly related to the interfacial bonding strength between polymer matrix and filler [60, 61]. The good interface bonding property is helpful to stress transfer and restrain the generation of crack, and therefore the mechanical strength can be improved [60]. Surface modification such as physical adsorption or chemical grafting could improve the combination of filler and polymer matrix [62]. In addition, good dispersion is also an effective way to improve the mechanical properties of polymer [63, 64].

Because of its unique spatial morphology, one-dimensional materials have different enhancement effects from zero-dimensional materials and two-dimensional materials [65–69]: Due to the high aspect ratio (L/d), 1D fillers can enhance the mechanical properties more effectively—especially the tensile strength; Many researches suggested that the reinforcement effect of a 1D filler was better than that of a 2D material [??]. The reinforcing effects of 1D and 2D materials are greater than those of the 3D filler [67–69]. Nadiv et al. [69] introduced a robustness factor to measure the filler concentration range necessary for achieving a significant reinforcing effect, and the robustness factor increased with filler dimensionality. Here, 1D and 2D materials show better reinforcing effects as fillers, which may be related to the larger filler/polymer interfacial area caused by the high aspect ratio [53, 65–68].

Toughness is another important property of polymer materials. Toughness is closely related to crack growth. Increasing crack propagation path, decreasing crack propagation speed, and increasing energy dissipation are effective ways to toughening composites [70, 71]. The toughening mechanisms of polymer materials are reviewed, including crack bridging and filler extraction, shear yield of diffusion matrix, formation of shear band, micro-crack, crack pinning, crack tip passivation, crack deflection, and interfacial debonding of filler matrix [56].

The main toughening mechanism of carbon nano-tubes, or carbon fibers is crack bridging and pulling out [72, 73]: When cracks propagate, certain fibers hinder their expansion; With increasing applied energy, cracks grow around the fiber, i.e., crack bridging; This mechanism works until the matrix around the fiber breaks completely, and the fiber loses its reinforcing effect; The fiber is then pulled out from the matrix by the continued applied force. The size of the fibers is also a critical factor in the toughening process. When the size is smaller than the critical size, the matrix material will fail around the filler, and the filler is pulled out [73]. However, the stress will be completely transferred to the filler when the size is greater than the critical size, and the filler is more likely to be broken, eliminating the toughening effect [59, 72–74]. Zero-dimensional materials are different from one-dimensional materials. The toughening mechanism of zero-dimensional materials is mainly the formation of shear bands [75], crack bending and deflection [76–78], and particle debonding [79–81].

The improvement of the mechanical properties of polymer composites can increase the load-bearing capacity of composites [82]. The higher tensile strength and Young's modulus result in a higher load-bearing capacity of the composite. The enhancement in the bearing capacity of the composites means that plastic deformation is not easy to occur in the material during the friction process. Meanwhile the structural integrity of the composite material under a high load is maintained, and therefore the wear resistance is enhanced [83].

In addition, outstanding mechanical properties can dramatically inhibit the generation and propagation of cracks on the worn surface, and thereby improving the wear resistance [84]. Surface hardness is another critical factor that determines the wear resistance of a material. The harder surface has a higher wear resistance.

Self-lubricating composites based on polyphenylene sulfide/polytetrafluoroethylene (PPS/PTFE) reinforced with short CFs were prepared by melt-blending [85]. The results indicated that the incorporation of CFs improved the tensile strength, flexural modulus, and hardness of PPS/PTFE blends, as shown in Figure 5.5 [85]. Meanwhile, the specific wear rate and average friction coefficient of PPS/PTFE reinforced by 15 vol% CFs reached 5.2×10^{-6} mm^3/Nm and 0.085, which are 88% and 47% lower than that of a mono-PPS/PTFE blend under the same sliding condition [85].

The polyimide (PI)/carboxyl-functionalized multi-walled carbon nano-tube (MWCNTs-COOH) nano-composite films were synthesized by in-situ polymerization [86]. The results showed that the incorporation of MWCNTs-COOH greatly enhanced the thermal stability and mechanical property of PI [86]. Besides, the wear resistance of PI under seawater lubrication had been greatly improved by filling MWCNTs-COOH because the strong interfacial adhesion between PI matrix and MWCNTs-COOH nano-fillers could effectively transfer the load between contact surfaces, as shown in Figure 5.6 [86].

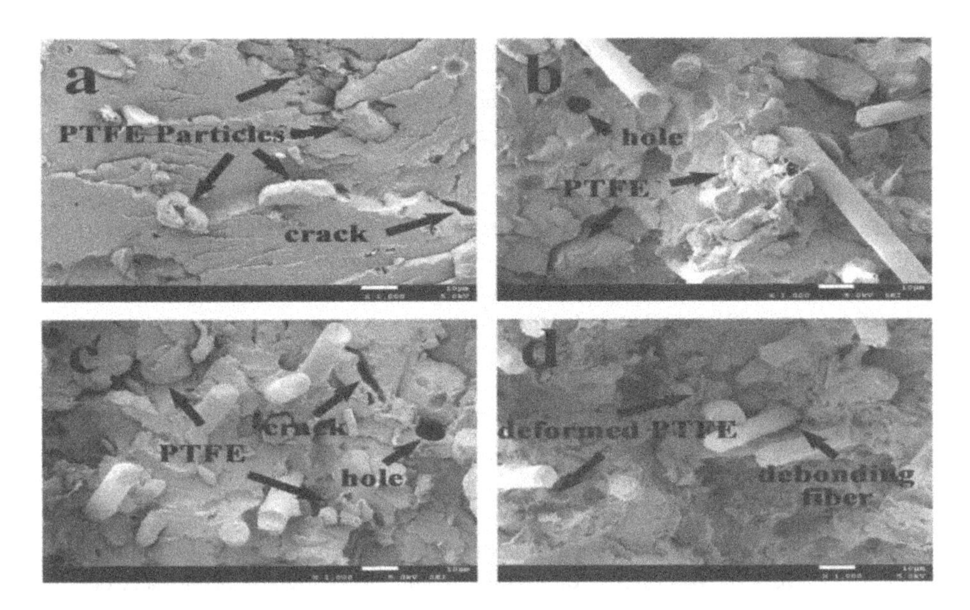

FIGURE 5.5 Tensile fractured surface of PPS/PTFE/CF composites [85]: (a) PPS/PTFE, (b) PPS/PTFE/5%CF, (c) PPS/PTFE10%CF, and (d) PPS/PTFE/15%CF.

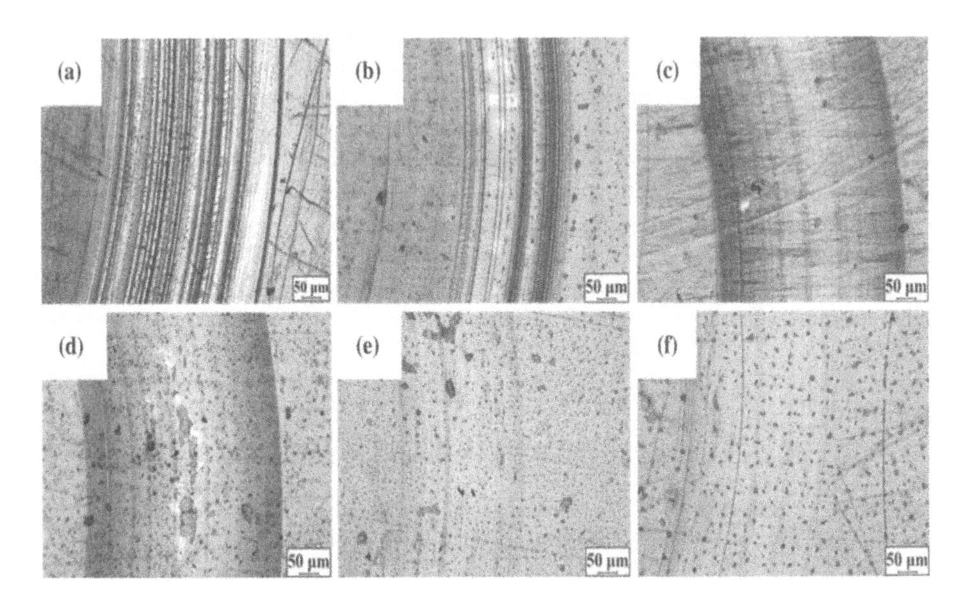

FIGURE 5.6 Optical photographs of worn surfaces of PI/MWCNTs COOH nano-composites under seawater lubrication [86]: (a) PI, (b) 0.1 wt% PI/MWCNTs-COOH, (c) 0.3 wt%PI/MWCNTs-COOH, (d) 0.5 wt% PI/MWCNTs-COOH, (e) 0.7 wt% PI/MWCNTs-COOH, and (f) 1 wt% PI/MWCNTs-COOH (load: 3 N, sliding speed: 0.1569 m/s, duration: 30 min).

5.4 POLYMER WITH TWO-DIMENSIONAL (2D) MATERIALS

In recent years, two-dimensional (2D) nano-materials having a film thickness of only a single or few atoms have attracted great attention [87–92]. Many studies have shown that 2D nano-materials have excellent properties and can be incorporated in composites to adapt to different application requirements. Many researchers are attempting to understand the reinforced composites with new 2D materials to improve their tribological properties, and achieve longer durability in various engineering applications.

In terms of their chemical elements and atomic arrangements, all 2D nano-materials can be classified into five categories [11, 93–98]: (1) Xenes, (2) transition metal carbides and nitrides (MXenes), (3) transition metal dichalcogenides (TMDs), (4) nitrides, and (5) organic frameworks. All 2D nano-materials have high in-plane strength due to the in-plane covalent bonding [98].

The hottest focus in the 2D nanomaterial-based composites (NBCs) field is graphene-based nano-composites [87, 99–102], but reviews regarding their tribological properties are rare. This section intends to discuss the tribological properties of composites reinforced with 2D nano-materials at different scales. More importantly, these 2D NBCs are promising for various applications such as aerospace, energy, and bio-medicine, and they could be used as lubricating materials due to their excellent tribological properties.

To achieve outstanding lubrication properties, many methods have been developed for the preparation of 2D NBCs such as cold or hot sintering, and sintering. For polymer composites, most typical fabrication methods are in-situ polymerization, solution mixing, and melt blending [103, 104]. In the case of in-situ polymerization, 2D nano-materials and the monomers or pre-polymers are swollen within the solvent, and polymerization is then initiated by adjusting the conditions [103]. In the case of solution mixing, the 2D nano-material and polymer can be easily dispersed in a solvent by mechanical mixing or ultrasonication, and the solvent is then removed to obtain the composite [104]. Melt blending is performed by high temperature melting without a solvent, and this method is generally used to fabricate thermoplastic composites [11]. In recent years, the resin transfer molding method has been presented to improve the dispersion of graphene in composites [105].

Based on the summary above (Table 5.2) [11], the dispersion of 2D nano-materials in a matrix is a crucial step in the preparation process of composites. Recently,

TABLE 5.2
Summary of Preparation Methods for Polymer-2D NBCs [11, 110–114]

Materials	Method	Advantages	Disadvantages	Ref.
Polymer matrix	In-situ polymerization	Excellent adhesion and dispersion	Complicated to operate, high cost, and only used in lab	[110]
	Solution mixing	Flexibility in various materials	Hard to remove solution	[111–113]
	Melt blending	Simple process, Eco-friendly	Unsuitable for thermosetting resin	[114]

various processing routes for dispersing 2D nano-material fillers into matrices such as liquid phase blending, melt mixing, and the freeze-dried masterbatch strategy have been reported [106–109].

The tribological properties of composites have drawn much attention and need to be considered [115, 116].

Since 2D NBCs consist of various matrices and 2D nano-materials, their tribological properties can be significantly modified by adjusting the chemical composition, micro-structure, and content of 2D nano-materials. The addition of very low amounts of 2D nano-materials significantly improves the tribological properties of a matrix.

To date, polymer composites incorporated with 2D nano-materials such as graphene, MoS_2, BN, MXene, and MOFs [117] have been widely researched. This section describes the recent developments on both the mechanical and tribological properties of 2D NBCs. The 2D nano-materials can improve the tribological performance of polymers because of their in-plane mechanical isotropy and a weak interlayer interaction [118]. The good tribological performance of the polymer composites achieved by incorporating 2D nano-materials is manifested in the reductions of the COF and the wear rate. For comparison, various 2D nano-materials added in the polymer matrix are presented [17, 119–121]. All of these results mentioned for the COF and wear rates are summarized in Table 5.3 [11].

The COF of PTFE is significantly reduced when the added graphene in the matrix is 4.0 wt%, and however, a higher content of graphene may compromise the lubrication property [132]. The addition of graphene to a PTFE matrix decreases the wear rate by three or four orders of magnitude [131]. The good self-lubrication and dispersion of 2D nano-fillers reduces friction and improves wear resistance of the composite [118]. Ultra-high molecular weight polyethylene (UHMWPE) had been reinforced by Ti_3C_2 [130] or graphene [132], and the results showed that the 2D nano-material increased the wear resistance and reduced the effective lateral force. Furthermore, 2D nano-materials can significantly contribute to excellent lubricity, tribological properties, and thermal conductivity [125].

2D nano-materials can also potentially solve the frictional heat problems of polymer composites. Qiu et al. [128] employed an infrared thermal imager to investigate the melting behavior of composites, and the results showed that the addition of MoS_2 reduced the frictional heat and retarded the melting wear. A 2D nano-material was also added in a base oil due to its good self-lubricating property [132]. Thus, 2D nano-materials are very attractive fillers and can significantly enhance the lubrication performance of polymer composites owing to their outstanding lubricating properties under harsh conditions [133, 134].

To understand the underlying lubrication mechanism of polymer composites incorporating 2D nano-materials, the analysis of the worn surface of composites is needed (Figure 5.7) [129]. The mechanism reported includes two points: (i) melting wear [128] and (ii) slightly abrasive wear [129]. A variety of studies have demonstrated that the formation of a load-carrying transfer layer resulted in the improvement of the tribological properties of polymer composites (Figure 5.8) [123, 124].

The functionalization of 2D nano-materials is an important approach to strengthening the interfacial bonding, and several problems relevant to the interfacial bonding must be solved to further enhance the lubrication performance [112].

TABLE 5.3
Tribological Properties of Polymer Composites Reinforced with 2D Materials [11, 102, 122–131]

Matrix	2D Material	Method	Operating Condition	COF	Wear Rate (mm³/(N·m))	Ref.
PTFE	4.0 wt% graphene	Cold compression and sintering	20 N, 0.1 m/s	0.18	7.5×10^{-6}	[122]
PTFE	10 wt% MoS_2	Cold press and sintering	5 N, 0.12 m/s	0.15	2.5×10^{-5}	[123]
PTFE	10 wt% g-C_3N_4	Cold press and sintering	5 N, 0.12 m/s	0.18	1.0×10^{-5}	[124]
PTFE	5 wt% phosphorene	Ball milling and SPS	3 N, 20 mm/s	0.041	6.9×10^{-6}	[125]
EP	0.5 wt% BN	Mechanical mixing	5 N, 20 mm/s	0.5	16×10^{-5}	[126]
PI	2 wt% BN	Mechanical mixing and spin-coated	3 N, 20 mm/s	0.10	2.79×10^{-6}	[127]
PEEK	10 vol% g-C_3N_4	Mechanical mixing and hot pressing	50 N, 1 m/s	0.6	4.0×10^{-7}	[102]
Phenol formaldehyde (PF)	0.3 wt% graphene oxide	Solution mixing	320 N, 2.24 m/s	0.12	-	[128]
Polyurethane	3 wt% MoS_2	Solution mixing	3 N, 60 mm/s	0.10	9×10^{-5}	[129]
EP	1.0 wt% MoS_2 @PPN	Mechanical mixing	80 N, 0.05 m/s	0.58	22.3×10^{-5}	[128]
Bismaleimide (BMI)	0.6 wt% PHbP @rGO/WS_2	Solution mixing	196 N, 0.42 m/s	0.13	1.22×10^{-6}	[129]
UHMWPE	2.0 wt% Ti_3C_2	Hot compression mold	200 N, 0.4 m/s	0.128	-	[130]
EP	2.0 wt% Ti_2CT_x	Mechanical mixing	98 N, 0.3 m/s	0.228	-	[131]

To further improve the performance of the composite, there have been many researches that reinforces them with 2D NBCs in recent years. Among them, graphene-based nano-composites are one of the most promising materials owing to the good wear resistance and outstanding specific strength. In addition, the composites incorporating MoS_2 and BN also possess excellent lubrication properties. The different types of applications in aviation and aerospace can be categorized into structural components, including fuselages, wings, and moving components such as bearings, gears, and hatch seals [134].

Due to the weak van der Waals interlayer interaction of 2D nano-materials, the composites possessing outstanding lubrication performance such as ultralow friction, ultralow wear rate, and self-lubricating properties can be realized [11]. 2D nano-materials can effectively improve the properties of composites on account of their distinctive characteristics that can give rise to optimized orientation, uniform dispersion, and strong interface bonding [11]. As a result, 2D nano-materials are

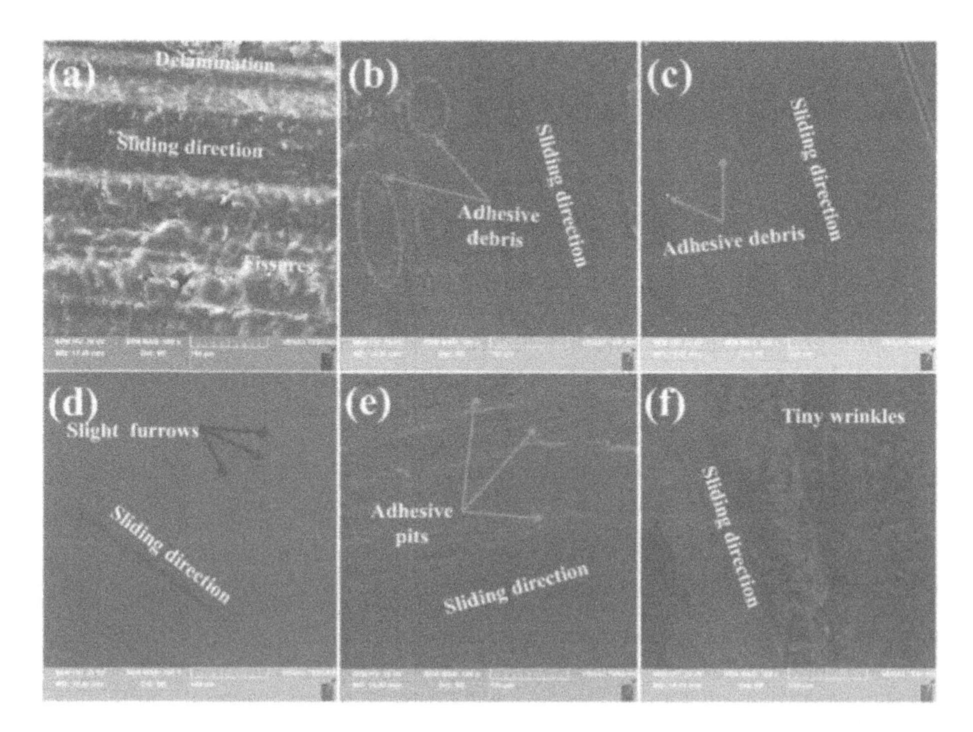

FIGURE 5.7 Scanning electron microscopy images [129] of worn surfaces of (a) bismaleimide resin and its composites incorporating (b) 0.2 wt%, (c) 0.4 wt%, (d) 0.6 wt%, (e) 0.8 wt%, and (f) 1.0 wt% PHbP@rGO/WS_2.

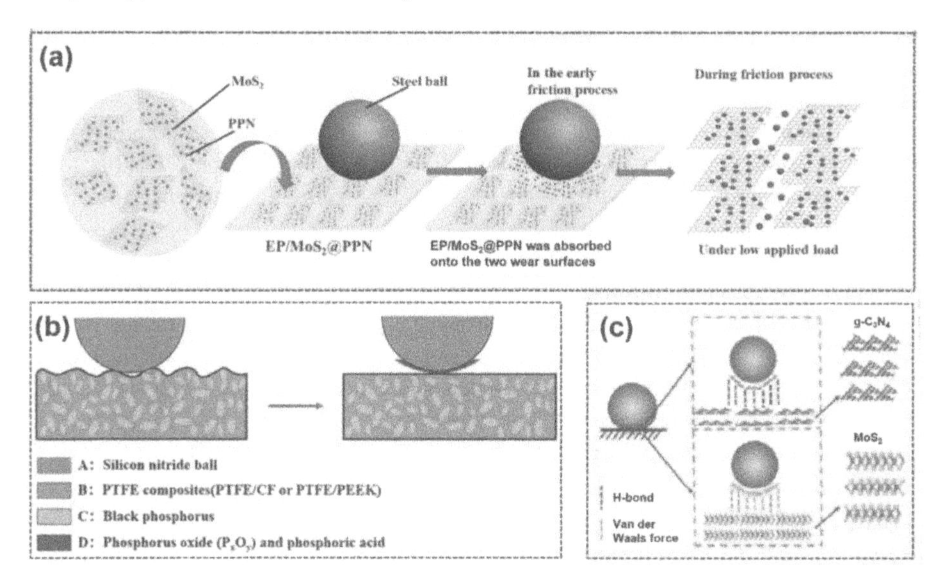

FIGURE 5.8 Summary of friction mechanism of the load-carrying transport layer formation [11]. Schematic of friction mechanism for (a) EP/MoS_2@PPN composites [128]. (b) PTFE/phosphorene [124], and (c) PTFE/MoS_2 and PTFE/g-C_3N_4 [123].

considered the best reinforcement for polymers. Despite the enormous advances, several critical points of 2D NBCs have not yet been fully resolved yet, and the challenges include the following [11]:

1. Preparation of tailored 2D nano-materials: High-quality and single-layer 2D nano-materials except for graphene are still lacking.
2. Uniform distribution of 2D nano-materials in matrices: Some easy and effective approaches are needed. Moreover, optimal orientation of the 2D nano-material needs to be kept as much as possible by an appropriate technique such as external electric fields or magnetic fields.
3. Fundamental mechanism of strengthening and lubrication for various composite systems: Some new systems such as phosphorene/polymer and MXene/polymer have not yet been investigated.
4. Low-cost applications of 2DNBCs. The cost and scalability should also be considered for real applications.

5.5 CONCLUSION

In recent years, there has been rapid development in the field of polymer tribology especially in the preparation methods and the tribological tests of polymer composites with green nano-materials. This review reports the recent progresses pertaining to the lubrication properties of composites reinforced with green nano-materials.

Core-shell micro-/nano-particles with characteristic structural advantages are of great significance to improve the tribological properties of lubricating materials. In contrast to ordinary particles, core-shell nano-/micro-particles have a shell material that can interlock the matrix and particles while also improving the dispersion of particles in the matrix. Meanwhile, the core materials can offer wear resistance, friction reduction, and fortification [2].

One-dimensional nano-materials such as carbon fibers, glass fibers, and carbon nano-tubes are very important for reinforcing polymer composite materials because of their one-dimensional structures. They have a low friction coefficient and excellent anti-friction properties [10].

This chapter also discusses the effect of incorporating 2D nano-materials in a polymer matrix. The 2D nano-materials can improve the lubrication performance of polymers because they exhibit in-plane mechanical isotropy and a weak interlayer interaction [135].

In consideration of the significance of polymer composites in various areas, effectively designing and fabricating polymer composites with green nano-materials needs to be addressed in the future. This review offers a new trail toward the design of advanced polymer composites with potential applications in a variety of fields.

REFERENCES

1. Holmberg K, Andersson P, Erdemir A. Global energy consumption due to friction in passenger cars. Tribol Int 47: 221–34 (2012).
2. Gong H J, Yu C C, Zhang L, Xie G X, Guo D, Luo J B. Intelligent lubricating materials: A review. Compos. B Eng 202: 108450 (2020).

3. Perry SS, Tysoe WT. Frontiers of fundamental tribological research. Tribol Lett 19(3): 151–61 (2005).
4. Wu S, Li J, Zhang G, Yao Y, Li G, Sun R, Wong C. Ultrafast self-healing nanocomposites via infrared laser and their application in flexible electronics. ACS Appl Mater Interfaces 9(3):3040–9 (2017).
5. Petrova P N, Fedorov A L. Polytetrafluoroethylene-based polymer composites with increased wear resistance in dry friction. Russ Eng Res 30: 895–899 (2010).
6. Podsiadlo P, Kaushik A K, Arruda E M, Waas A M, Shim B S, Xu J D, Nandivada H, Pumplin B G, Lahann J, Ramamoorthy A, Kotov N. Ultrastrong and stiff layered polymer nanocomposites. Science 318(5847): 80–83 (2008).
7. Panin S V, Duc Anh N, Kornienko L A, Ivanova L R, Ovechkin B B. Comparison on efficiency of solid-lubricant fillers for polyetheretherketone-based composites. AIP Conf Proc 2051: 020232 (2018).
8. Peng S G, Zhang L, Xie G X, Guo Y, Si L N, Luo J B. Friction and wear behavior of PTFE coatings modified with poly (methyl methacrylate). Compos. B Eng 172: 316–322 (2019).
9. Donnet C, Erdemir A. Solid lubricant coatings: Recent developments and future trends. Tribol Lett 17(3): 389–397 (2004).
10. Ren Y L, Zhang L, Xie G X, Li Z B, Chen H, Gong H J, Xu W H, Guo D, Luo J B. A review on tribology of polymer composite coatings. Friction 9: 429–470 (2021).
11. Ji Z J, Zhang L, Xie G X, Xu W H, Guo D, Luo J B, Prakash B. Mechanical and tribological properties of nanocomposites incorporated with two-dimensional materials. Friction 8(5): 813–846 (2020).
12. Chen H, Zhang L, Li M Y, Xie G X. Synthesis of core–shell micro/nanoparticles and their tribological application: a review. Materials 13(20): 4590 (2020).
13. Wang H, Zhang S, Wang G, Yang S, Zhu Y. Tribological behaviors of hierarchical porous PEEK composites with mesoporous titanium oxide whisker. Wear 297(1): 736–41 (2013).
14. Huang X, Wu J, Lu X, Feng X, Shi Y. Tribological properties of porous PEEK composites containing ionic liquid under dry friction condition. Lubricants 5(2): 19 (2017).
15. Zhu Y, Lin S, Wang H, Liu D. Study on the tribological properties of porous sweating PEEK composites under ionic liquid lubricated condition. J Appl Polym Sci 131(21): 1284–87 (2014).
16. Wang J, Zhao H, Huang W, Wang X. Investigation of porous polyimide lubricant retainers to improve the performance of rolling bearings under conditions of starved lubrication. Wear 380–381: 52–8 (2017).
17. Zhang G L, Xie G X, Si L N, Wen S Z, Guo D. Ultralow friction self-lubricating nanocomposites with mesoporous metal-organic frameworks as smart nanocontainers for lubricants. ACS Appl Mater Inter 9(43): 38146–52 (2017).
18. Guo Q B, Lau K T, Zheng B F, Rong M Z, Zhang M Q. Imparting ultra-low friction and wear rate to epoxy by the incorporation of microencapsulated lubricant? Macromol Mater Eng 294(1): 20–4 (2009).
19. Khun N W, Zhang H, Yang J, Liu E. Mechanical and tribological properties of epoxy matrix composites modified with microencapsulated mixture of wax lubricant and multi-walled carbon nanotubes. Friction 1(4):341–9 (2013).
20. Zhang L, Xie G X, Wu S, Peng S G, Zhang X Q, Guo D, Wen S Z, Luo J B. Ultralow friction polymer composites incorporated with monodispersed oil microcapsules. Friction 9(1): 29–40 (2019).
21. Yang M, Zhu X, Ren G, Men X, Guo F, Li P, Zhang Z. Tribological behaviors of polyurethane composite coatings filled with ionic liquid core/silica gel shell microcapsules. Tribol Lett 58(1): 9 (2015).

22. Bandeira P, Monteiro J, Baptista AM, Magalhães FD. Tribological performance of PTFE-based coating modified with microencapsulated [HMIM][NTf2] ionic liquid. Tribol Lett 59(1): 13 (2015).
23. Li H, Wang Q, Wang H, Cui Y, Zhu Y, Wang B. Fabrication of thermally stable polysulfone microcapsules containing [EMIm][NTf_2] ionic liquid for enhancement of in situ self-lubrication effect of epoxy. Macromol Mater Eng 301(12): 1473–81 (2016).
24. Imani A, Zhang H, Owais M, Zhao J, Chu P, Yang J, Zhang Z. Wear and friction of epoxy based nanocomposites with silica nanoparticles and wax-containing microcapsules. Compos Part A Appl Sci Manuf 107: 607–15 (2018).
25. Guo Q B, Lau K T, Rong M Z, Zhang M Q. Optimization of tribological and mechanical properties of epoxy through hybrid filling. Wear 269(1–2): 13–20 (2010).
26. Khun N W, Zhang H, Yue C Y, Yang J L. Self-lubricating and wear resistant epoxy composites incorporated with microencapsulated wax. J Appl Mech 81(7): (2014).
27. Gong H J, Song Y, Li G L, Xie G, Luo J B. A highly tough and ultralow friction resin nanocomposite with crosslinkable polymer-encapsulated nanoparticles. Compos B Eng 197:108157 (2020).
28. Zhang X, Wang P, Sun D, Li X, An J, Yu T, Yang E-H, Yang J. Dynamic plastic deformation and failure mechanisms of individual microcapsule and its polymeric composites. J Mech Phys Solids 139: 103933 (2020).
29. Shchukin D, Mohwald H. A coat of many functions. Science 341(6153): 1458–9 (2013).
30. Tang G G, Zhang J, Liu C C, Zhang D, Wang Y Q, Tang H, Li C. Synthesis and tribological properties of flower-like MoS_2 microspheres. Ceram Int 40: 11575–11580 (2014).
31. Yang Z R, Guo Z W, Yuan C Q. Effects of MoS_2 microcapsulation on the tribological properties of a composite material in a water-lubricant condition. Wear 432: 102919 (2019).
32. Chen B B, Jia Y H, Zhang M J, Liang H Y, Li X, Yang J, Yan F, Li C. Tribological properties of epoxy lubricating composite coatings reinforced with core-shell structure of CNF/MoS_2 hybrid. Compos Part A Appl Sci Manuf 122: 85–95 (2019).
33. Wu G Y, Xu C H, Xiao G C, Yi M D, Chen Z Q, Chen H. An advanced self-lubricating ceramic composite with the addition of core-shell structured h-BN@Ni powders. Int J Refract Met Hard Mater 72: 276–285 (2018).
34. Wang H Y, Lei Y, Liu D J, Wang C, Zhu Y J, Zhu J H. Investigation of the tribological properties: Core-shell structured magnetic Ni@NiO nanoparticles reinforced epoxy nanocomposites. Tribol Int 83: 139–145 (2015).
35. Fan C L, Li H L, Jin L, Zhang M J, Xiao L H, Li M, Ao Y. Improving tribological properties of phenolic resin/carbon fiber composites using m-Si3N4@PANI core-shell particles. J Appl Polym Sci 136: 47785 (2019).
36. Zhang J S, Yang S F, Chen Z X, Wyszomirska M, Zhao J W, Jiang Z Y. Microstructure and tribological behavior of alumina composites reinforced with SiC-graphene core-shell nanoparticles. Tribol Int 131: 94–101 (2019).
37. Yoo, S C, Kang B, Van Trinh P, Doan D P, Soon H H. Enhanced mechanical and wear properties of Al6061 alloy nanocomposite reinforced by CNT-template-grown core-shell CNT/SiC nanotubes. Sci Rep 10: 12896 (2020).
38. Lian W Q, Mai Y J, Wang J, Zhang L Y, Liu C S, Jie X H. Fabrication of graphene oxide-Ti_3AlC_2 synergistically reinforced copper matrix composites with enhanced tribological performance. Ceram Int 45: 18592–18598 (2019).
39. Xiong K Y, Wang Y, Tang B, Yi M, Wu Z Z, Wu X F. Review of activated carbon fiber. Synth. Fiber 49(10): 15–19 (2020).
40. Yuan N, Wang Y Z, Duan J K, Wang Y, Shi D D, Wu X F. Review of high performance fiber reinforced concrete composites. Synth Fiber 49(11): 52–56 (2020).

41. Liu A Y, Liu J C, Xu J L, Zhang J Q, Xing W F, Zhang H Y, Liu F B, Wang D Y, Li S H, Chen Y W. Research progress of carbon nanotube/carbon fiber composites. Eng Plast Appl 48(8): 158–162 (2020).
42. Li X M, Xu X C, Yu W, Gu Y J, Shi Y, Cheng R M, Chen Y W. Conductivity of carbon nanotubes/polyacetylene composites. Acta Comp Sin 23 (5): 70–74 (2006).
43. Sathishkumar T P, Satheeshkumar S, Naveen J. Glass fiber-reinforced polymer composites–a review. J Reinf Plast Compos 33(13): 1258–1275 (2014).
44. Adekomaya O, Adama K, Naveen J. Glass-fiber reinforced composites: The effect of fiber loading and orientation on tensile and impact strength. Nigerian J Technol 36(3): 782–787 (2017).
45. Harizi W, Chaki S, Bourse M, Ourak. Mechanical damage assessment of glass fiber-reinforced polymer composites using passive infrared thermography. Compos B Eng 59: 74–79 (2014).
46. Vyas S, Goli E, Zhang X, Geubelle P H. Manufacturing of unidirectional glass-fiber-reinforced composites via frontal polymerization: A numerical study. Compos Sci Technol 184: 107832 (2019).
47. Balachandar M, Ramnath Vijaya B, Jagadeeshwar P, Yokesh R. Mechanical behaviour of natural and glass fiber reinforced with polymer matrix composite. Mater Today Proc 16(2): 1297–1303 (2019).
48. Smaranika N, Nayak R K, Panigrahi I, Sahoo A K. Tribo-mechanical responses of glass fiber reinforced polymer hybrid nanocomposites. Mater Today Proc 18(7): 4042–4047 (2019).
49. Beckford S, Mathurin L, Chen J Y, Fleming R A, Zou M. The effects of polydopamine coated Cu nanoparticles on the tribological properties of polydopamine/PTFE coatings. Tribol Int 103: 87–94 (2016).
50. Pryamitsyn V, Ganesan V. Origins of linear viscoelastic behavior of polymer-nanoparticle composites. Macromolecules 39(2): 844–856 (2006).
51. Mujtaba A, Keller M, Ilisch S, Radusch H J, Beiner M, Thurn-Albrecht T, Saalwächter K. Detection of surface-immobilized components and their role in viscoelastic reinforcement of rubber-silica nanocomposites. ACS Macro Lett 3(5): 481–485 (2014).
52. Chen Q, Gong S S, Moll J, Zhao D, Kumar S K, Colby R H. Mechanical reinforcement of polymer nanocomposites from percolation of a nanoparticle network. ACS Macro Lett 4(4): 398–402 (2015).
53. Papon A, Montes H, Lequeux F, Oberdisse J, Saalwächter K, Guy L. Solid particles in an elastomer matrix: Impact of colloid dispersion and polymer mobility modification on the mechanical properties. Soft Matter 8(15): 4090–4096 (2012).
54. Papageorgiou D G, Li Z L, Liu M F, Kinloch I A, Young R J. Mechanisms of mechanical reinforcement by graphene and carbon nanotubes in polymer nanocomposites. Nanoscale 12(4): 2228–2267 (2020).
55. Samad M A, Sinha S K. Nanocomposite UHMWPE-CNT polymer coatings for boundary lubrication on aluminium substrates. Tribol Lett 38(3): 301–311 (2010).
56. Chih A, Ansón-Casaos A, Puértolas J A. Frictional and mechanical behaviour of graphene/UHMWPE composite coatings. Tribol Int 116: 295–302 (2017).
57. Ning H B, Lu N, Hassen A A, Chawla K, Selim M, Pillay S. A review of long fibre thermoplastic (LFT) composites. Int Mater Rev 65(3): 164–188 (2020).
58. Maillard D, Kumar S K, Fragneaud B, Kysar J W, Rungta A, Benicewicz B C, Deng H, Brinson L C, Douglas J F. Mechanical properties of thin glassy polymer films filled with spherical polymer-grafted nanoparticles. Nano Lett 12(8): 3909–3914 (2012).
59. Mortazavian S, Fatemi A. Effects of fiber orientation and anisotropy on tensile strength and elastic modulus of short fiber reinforced polymer composites. Compos B Eng 72: 116–129 (2015).

60. Zhang B, Jia L H, Tian M, Ning N Y, Zhang L Q, Wang W C. Surface and interface modification of aramid fiber and its reinforcement for polymer composites: A review. Eur Polym J 14: 110352 (2021).
61. Seretis G V, Manolakos D E, Provatidis C G. On the stainless steel flakes reinforcement of polymer matrix particulate composites. Compos B Eng 162: 80–88 (2019).
62. Shubham S K, Purohit R, Yadav P S, Rana R S. Study of nano-fillers embedded in polymer matrix composites to enhance its properties – A review. Mater Today Proc 26 (2): 3024–3029 (2020).
63. Li X, Kang H L, Shen J X. Effects of graft locations on dispersion behavior of polymer-grafted nanorods: A molecular dynamics simulation study. Polymer 211: 123077 (2020).
64. Broughton W R, Koukoulas T, Woolliams P. Assessment of nanoparticle loading and dispersion in polymeric materials using oscillatory photon correlation spectroscopy. Polym Test 49: 107–114 (2016).
65. Cheng S W, Bocharova V, Belianinov A, Xiong S M, Kisliuk A, Somnath S, Holt A P, Ovchinnikova O S, Jesse S, Martin H, Etampawala T, Dadmun M, Sokolov A P. Unraveling the mechanism of nanoscale mechanical reinforcement in glassy polymer nanocomposites. Nano Lett 16(6): 3630–3637 (2016).
66. Okumura T, Sonobe K, Ohashi A, Watanabe H, Watanabe K, Oyamada H, Aramaki M, Ougizawa T. Synthesis of polyamide-hydroxyapatite nanocomposites. Polym Eng Sci 60(7): 1699–1711 (2020).
67. Scotti R, Conzatti L, D'Arienzo M, Di Credico B, Giannini L, Hanel T, Stagnaro P, Susanna A, Tadiello L, Morazzoni F. Shape controlled spherical (0D) and rod-like (1D) silica nanoparticles in silica/styrene butadiene rubber nanocomposites: Role of the particle morphology on the filler reinforcing effect. Polymer 55(6): 1497–1506 (2014).
68. Pradhan S, Lach R, Le H H, Grellmann W, Radusch H J, Adhikari R. Effect of filler dimensionality on mechanical properties of nanofiller reinforced polyolefin elastomers. Int Sch Res Not 2013: 284504 (2013).
69. Nadiv R, Shachar G, Peretz-Damari S, Varenik M, Levy I, Buzaglo M, Ruse E, Regev O. Performance of nano-carbon loaded polymer composites: Dimensionality matters. Carbon 126: 410–418 (2018).
70. Thirumurugan R, Gnanasekar N. Influence of finite element model, load-sharing and load distribution on crack propagation path in spur gear drive. Eng Fail Anal 110: 104383 (2020).
71. Emdadi A, Zaeem M A. Phase-field modeling of crack propagation in polycrystalline materials. Comput Mater Sci 186: 110057 (2021).
72. Bhandari N L, Lach R, Grellmann W, Adhikari R. Depth-dependent indentation micro-hardness studies of different polymer nanocomposites. Macromol Symp 315(1): 44–51 (2012).
73. Opelt C V, Becker D, Lepienski C M, Coelho L A F. Reinforcement and toughening mechanisms in polymer nanocomposites--Carbon nanotubes and aluminum oxide. Compos B Eng 75: 119–126 (2015).
74. Boåsen M, Dahlberg C F O, Efsing P, Faleskog J. A weakest link model for multiple mechanism brittle fracture—Model development and application. J Mech Phys Solids 147: 104224 (2021).
75. Fu S Y, Sun Z, Pei H, Li Y Q, Hu N. Some basic aspects of polymer nanocomposites: A critical review. Nano Mater Sci 1 (1): 2–13 (2019).
76. Vasoya M, Unni A B, Leblond J B, Lazarus V, Ponson L. Finite size and geometrical non-linear effects during crack pinning by heterogeneities: An analytical and experimental study. J Mech Phys Solids 89: 211–230 (2016).

77. Patinet S, Alzate L, Barthel E, Dalmas D, Vandembroucq D, Lazarus V. Finite size and geometrical non-linear effects during crack pinning by heterogeneities: An analytical and experimental study. J Mech Phys Solids 61(2): 311–324 (2013).
78. Narducci F, Pinho S T. Exploiting nacre-inspired crack deflection mechanisms in CFRP via micro-structural design. Compos Sci Technol 153: 178–189 (2017).
79. Bonfoh N, Lipinski P. Ductile damage micromodeling by particles' debonding in metal matrix composites. Int J Mech Sci 49(2): 151–160 (2007).
80. Veluri B, Jensen H M. Steady-state propagation of interface corner crack. Int J Solids Struct 50(10): 1613–1620 (2013).
81. Steffensen S, Kibsgaard R L, Jensen H M. Debonding of particles in thin films. Int J Solids Struct 51 (15-16): 2850–2856 (2014).
82. Wei Y Z, Wang G S, Wu Y, Yue Y H, Wu J T, Lu C, Guo L. Bioinspired design and assembly of platelet reinforced polymer films with enhanced absorption properties. J Mater Chem A 2(15): 5516–5524 (2014).
83. He Y, Farokhzadeh K, Edrisy A. Characterization of thermal, mechanical and tribological properties of fluoropolymer composite coatings. J Mater Eng Perform 26(6): 2520–2534 (2017).
84. Su C, Xue F, Li T S, Xin Y S, Wang M M. Study on the tribological properties of carbon fabric/polyimide composites filled with SiC nanoparticles. J Macromol Sci Part B 55(6): 627–641 (2016).
85. Luo W, Qi Liu, Li Y, Zhou S T, Zou H W, Liang M. Enhanced mechanical and tribological properties in polyphenylene sulfide/polytetrafluoroethylene composites reinforced by short carbon fiber. Compos B Eng 91: 579–588 (2016).
86. Nie P, Min C Y, Song H J, Chen X H, Zhang Z Z, Zhao K L. Preparation and tribological properties of polyimide/carboxyl functionalized multi-walled carbon nanotube nanocomposite films under seawater lubrication. Tribol Lett 58: 7(2015).
87. Stankovich S, Dikin D A, Dommett G H B, Kohlhaas K M, Zimney E J, Stach E A, Piner R D, Nguyen S B T, Ruoff R S. Graphene-based composite materials. Nature 442(7100): 282–286 (2006).
88. Ramanathan T, Abdala A A, Stankovich S, Dikin D A, Herrera-Alonso M, Piner R D, Adamson D H, Schniepp H C, Chen X, Ruoff R S, Nguyen S T, Aksay I A, Prud'Homme R K, Brinson L C. Functionalized graphene sheets for polymer nanocomposites. Nat Nanotechnol 3: 327–331 (2008).
89. Huang X, Qi X Y, Boey F, Zhang H. Graphene-based composites. Chem Soc Rev 41(2): 666–686 (2012).
90. Kim S, Wang H T, Lee Y M. 2D nanosheets and their composite membranes for water, gas, and ion separation. Angew Chem Int Ed Engl 58(49): 17512–17527 (2019).
91. Li X, Sun M, Shan C X, Chen Q, Wei X L. Mechanical properties of 2D materials studied by in situ microscopy techniques. Adv Mater Interfaces 5(5): 1701246 (2018).
92. Baig Z, Mamat O, Mustapha M. Recent progress on the dispersion and the strengthening effect of carbon nanotubes and graphene-reinforced metal nanocomposites: A review. Crit Rev Solid State Mater Sci 43(1): 1–46 (2018).
93. Wang W, Xie G X, Luo J B. Black phosphorus as a new lubricant. Friction 6(1): 116–142 (2018).
94. Naguib M, Kurtoglu M, Presser V, Lu J, Niu J J, Heon M, Hultman L, Gogotsi Y, Barsoum M W. Two-dimensional nanocrystals produced by exfoliation of Ti_3AlC_2. Adv Mater 23(37): 4248–4253 (2011).
95. Halim J, Kota S, Lukatskaya M R, Naguib M, Zhao M Q, Moon E J, Pitock J, Nanda J, May S J, Gogotsi Y, Barsoum M. Synthesis and characterization of 2D molybdenum carbide (MXene). Adv Funct Mater 26(18): 3118–3127 (2016).

96. Hong Ng V M, Huang H, Zhou K, Lee P S, Que W X, Xu J Z, Kong L B. Recent progress in layered transition metal carbides and/or nitrides (MXenes) and their composites: Synthesis and applications. J Mater Chem A 5(7): 3039–3068 (2017).
97. Dai W Y, Shao F, Szczerbiński J, McCaffrey R, Zenobi R, Jin Y H, Schlüter A D, Zhang W. Synthesis of a two-dimensional covalent organic monolayer through dynamic imine chemistry at the air/water interface. Angew Chem Int Ed Engl 55(1): 213–217 (2016).
98. Liu L C, Zhou M, Jin L, Li L C, Mo Y T, Su G S, Li X, Zhu H W, Tian Y. Recent advances in friction and lubrication of graphene and other 2D materials: Mechanisms and applications. Friction 7(3): 199–216 (2019).
99. Li D, Kaner R B. Graphene-based materials. Science 320(5880): 1170–1171 (2008).
100. Kim H, Abdala A A, Macosko C W. Graphene/polymer nanocomposites. Macromolecules 43(16): 6515–6530 (2010).
101. Bartolucci S F, Paras J, Rafiee M A, Rafiee J, Lee S, Kapoor D, Koratkar N. Graphene–aluminum nanocomposites. Mater Sci Eng A 528(27): 7933–7937 (2011).
102. Papageorgiou D G, Kinloch I A, Young R J. Mechanical properties of graphene and graphene-based nanocomposites. Prog Mater Sci 90: 75–127 (2017).
103. Chieng B W, Ibrahim N A, Yunus W M Z W, Hussein M Z, Giita Silverajah V S. Graphene nanoplatelets as novel reinforcement filler in poly(lactic acid)/epoxidized palm oil green nanocomposites: Mechanical properties. Int J Mol Sci 13(9): 10920–10934 (2012).
104. El Achaby M, Arrakhiz F E, Vaudreuil S, El Kacem Qaiss A, Bousmina M, Fassi-Fehri O. Mechanical, thermal, and rheological properties of graphene-based polypropylene nanocomposites prepared by melt mixing. Polym Compos 33(5): 733–744 (2012).
105. Ni Y, Chen L, Teng K, Shi J, Qian X M, Xu Z W, Tian X, Hu C S, Ma M J. Superior mechanical properties of epoxy composites reinforced by 3D interconnected graphene skeleton. ACS Appl Mater Interfaces 7(21): 11583–11591 (2015).
106. Lee B, Koo M Y, Jin S H, Kim K T, Hong S H. Simultaneous strengthening and toughening of reduced graphene oxide/alumina composites fabricated by molecular-level mixing process. Carbon 78: 212–219 (2014).
107. Ye H Z, Liu X Y, Hong H P. Fabrication of metal matrix composites by metal injection molding—A review. J Mater Process Technol 200(1–3): 12–24 (2008).
108. Song Y, He G Y, Wang Y G, Chen Y. Tribological behavior of boron nitride nanoplatelet reinforced Ni_3Al intermetallic matrix composite fabricated by selective laser melting. Mater Design 165: 107579 (2019).
109. Hwang J, Yoon T, Jin S H, Lee J, Kim T S, Hong S H, Jeon S. Enhanced mechanical properties of graphene/copper nanocomposites using a molecular-level mixing process. Adv Mater 25(46): 6724–6729 (2013).
110. Zhang H, Wang L B, Zhou A G, Shen C L, Dai Y H, Liu F F, Chen J F, Li P, Hu Q K. Effects of 2-D transition metal carbide Ti_2CT_x on properties of epoxy composites. RSC Adv 6(90): 87341–87352 (2016).
111. Sorrentino A, Altavilla C, Merola M, Senatore A, Ciambelli P, Iannace S. Nanosheets of MoS_2-oleylamine as hybrid filler for self-lubricating polymer composites: Thermal, tribological, and mechanical properties. Polym Compos 36(6): 1124–1134 (2015).
112. Xiao Q, Han W H, Yang R, You Y, Wei R B, Liu X B. Mechanical, dielectric, and thermal properties of polyarylene ether nitrile and boron nitride nanosheets composites. Polym Compos 39(S3): E1598–E1605 (2018).
113. Dai W, Yu J H, Wang Y, Song Y Z, Bai H, Nishimura K, Liao H W, Jiang N. Enhanced thermal and mechanical properties of polyimide/graphene composites. Macromol Res 22(9): 983–989 (2014).
114. Papageorgiou D G, Liu M F, Li Z L, Vallés C, Young R J, Kinloch I A. Hybrid poly(ether ether ketone) composites reinforced with a combination of carbon fibers and graphene nanoplatelets. Compos Sci Technol 175: 60–68 (2019).

115. Yan Z, Shi X L, Huang Y C, Deng X B, Liu X Y, Yang K. Tribological performance of Ni_3Al matrix self-lubricating composites containing multilayer graphene prepared by additive manufacturing. J Mater Eng Perform 27(1): 167–175 (2018).
116. Zhang L G, Qi H M, Li G T, Wang D A, Wang T M, Wang Q H, Zhang G. Significantly enhanced wear resistance of PEEK by simply filling with modified graphitic carbon nitride. Mater Design 129: 192–200 (2017).
117. Rodenas T, Luz I, Prieto G, Seoane B, Miro H, Corma A, Kapteijn F, Llabrés I X F X, Gascon J. Metal-organic framework nanosheets in polymer composite materials for gas separation. Nat Mater 14(1): 48–55 (2015).
118. Wu S, He F, Xie G X, Bian Z L, Luo J B, Wen S Z. Black phosphorus: Degradation favors lubrication. Nano Lett 18(9): 5618–5627 (2018).
119. Zhang Z Z, Yang M M, Yuan J Y, Guo F, Men X H. Friction and wear behaviors of MoS_2-multi-walled-carbonnanotube hybrid reinforced polyurethane composite coating. Friction 7(4): 316–326 (2019).
120. Zhang G L, Xie G X, Si L N, Wen S Z, Guo D. Ultralow friction self-lubricating nanocomposites with mesoporous metal–organic frameworks as smart nanocontainers for lubricants. ACS Appl Mater Interfaces 9(43): 38146–38152 (2017).
121. Peng S G, Guo Y, Xie G X, Luo J B. Tribological behavior of polytetrafluoroethylene coating reinforced with black phosphorus nanoparticles. Appl Surf Sci 441: 670–677 (2018).
122. Bhargava S, Koratkar N, Blanchet T A. Effect of platelet thickness on wear of graphene–polytetrafluoroethylene (PTFE) composites. Tribol Lett 59(1): 17 (2015).
123. Li S, Duan C J, Li X, Shao M C, Qu C H, Zhang D, Wang Q H, Wang T M, Zhang X R. The effect of different layered materials on the tribological properties of PTFE composites. Friction 8(3): 542–552 (2020).
124. Lv Y, Wang W, Xie G X, Luo J B. Self-lubricating PTFE-based composites with black phosphorus nanosheets. Tribol Lett 66(2): 61 (2018).
125. Chen J, Chen B, Li J Y, Tong X, Zhao H C, Wang L P. Enhancement of mechanical and wear resistance performance in hexagonal boron nitride-reinforced epoxy nanocomposites. Polym Int 66(5): 659–664 (2017).
126. Min Y J, Kang K H, Kim D E. Development of polyimide films reinforced with boron nitride and boron nitride nanosheets for transparent flexible device applications. Nano Res 11(5): 2366–2378 (2018).
127. Yang M M, Zhang Z Z, Zhu X T, Men X H, Ren G N. In situ reduction and functionalization of graphene oxide to improve the tribological behavior of a phenol formaldehyde composite coating. Friction 3(1): 72–81 (2015).
128. Qiu S L, Hu Y X, Shi Y Q, Hou Y B, Kan Y C, Chu F K, Sheng H B, Yuen R K K, Xing W Y. In situ growth of polyphosphazene particles on molybdenum disulfide nanosheets for flame retardant and friction application. Compos Part A: Appl Sci Manuf 114: 407–417 (2018).
129. Chen Z Y, Yan H X, Guo L L, Li L, Yang P F, Liu B. A novel polyamide-type cyclophosphazene functionalized rGO/WS_2 nanosheets for bismaleimide resin with enhanced mechanical and tribological properties. Compos Part A: Appl Sci Manuf 121: 18–27 (2019).
130. Zhang H, Wang L B, Chen Q, Li P, Zhou A G, Cao X X, Hu Q K. Preparation, mechanical and anti-friction performance of MXene/polymer composites. Mater Design 92: 682–689 (2016).
131. Kandanur S S, Rafiee M A, Yavari F, Schrameyer M, Yu Z Z, Blanchet T A, Koratkar N. Suppression of wear in graphene polymer composites. Carbon 50(9): 3178–3183 (2012).
132. Lahiri D, Hec F, Thiesse M, Durygin A, Zhang C, Agarwal A. Nanotribological behavior of graphene nanoplatelet reinforced ultra high molecular weight polyethylene composites. Tribol Int 70: 165–169 (2014).

133. Zhao J, Chen G Y, He Y Y, Li S X, Duan Z Q, Li Y R, Luo J B. A novel route to the synthesis of an Fe_3O_4/h-BN 2D nanocomposite as a lubricant additive. RSC Adv 9(12): 6583–6588 (2019).
134. Xiao Y L, Yao P P, Fan K Y, Zhou H B, Deng M W, Jin Z X. Powder metallurgy processed metal-matrix friction materials for space applications. Friction 6(2): 219–229 (2018).
135. Wang X J, Lu M Y, Qiu L, Huang H, Li D, Wang H T, Cheng Y B. Graphene/titanium carbide composites prepared by sol-gel infiltration and spark plasma sintering. Ceram Int 42(1): 122–131 (2016).

6 Working of Functional Components in Self-Healing Coatings for Anti-Corrosion Green Tribological Applications

An Overview

Tauseef Ahmed, H.H. Ya, Mohammad Azeem, Mohammad Azad Alam, and Hafiz Usman Khalid
Universiti Teknologi PETRONAS
Seri Iskandar, Malaysia

Abdul Munir Hidayat Syah Lubis
Universiti Teknikal Malaysia
Melaka, Malaysia

Mohammad Rehan Khan
National University of Sciences and Technology
Islamabad, Pakistan

Mian Imran
Institute of Space Technology
Islamabad, Pakistan

Adnan Ahmed
University of Engineering and Technology
Peshawar, Pakistan

CONTENTS

DOI: 10.1201/9781003139386-6

6.1 INTRODUCTION

Corrosion causes major economic risks which can contribute to serious health and environmental hazards [1]. Corrosion is a major concern in engineering applications including but not limited to offshore wind turbines, ships, the automobile industry, chromium replacement oil pipelines, etc. According to a recent estimate, China's annual corrosion expenditures total more than USD$310 billion, accounting for 3.34 percent of the country's GDP. If the same percentage is applied to the world economy, the annual cost of corrosion is USD$2.5 trillion. Chemical protective coatings are the most widely used "corrosion protection methods" as a solution, accounting for up to two-thirds of all anti-corrosion investment [2].

Due to the stringent economic effects, there has been a shear rise reported in the research of this domain. The percentage distribution of the key words searched in this domain is given in the study of Aisa et al. [3] as is shown in Figure 6.1. According to the data collected from "Engineering Village web-based information service" there is a ten-fold increase in the number of publications for the previous decade, 2000–2010 (Figure 6.1).

The data indicates that polymeric materials are the focus of current research and are being used in a variety of ways in modern engineering applications [4–7]. One can find that polymers are the essential part of anti-corrosion structures either in the form of the matrix of the coating or in the form of embedded self-healing mechanisms. These coatings are fundamentally organic composites with extreme complexity and variability of chemical formulations, containing macro-/micro-scale structures. In advance studies polymer coatings are considered as multi-layered engineering structures with the two main purposes of providing a mechanical shield and preserving the functionality of the coatings. In such multi-layer structure, the top layer is commonly known as the "topcoat" and the bottom layer is known as the "primer" as shown in Figure 6.2. The topcoat contains polymerizable healing agents while the primer layer is composed of corrosion inhibitors. UV rays, humidity, moisture, oxygen, and ions are examples of environmental causes, while local scratches/delamination and stress-related macro-cracks are examples of mechanical factors that endanger the barrier, operational, and aesthetics properties of coatings [8]. Self-healing coatings by definition refer to those coatings that can repair the coating damages and recover the coating performance with minimal or no external intervention,

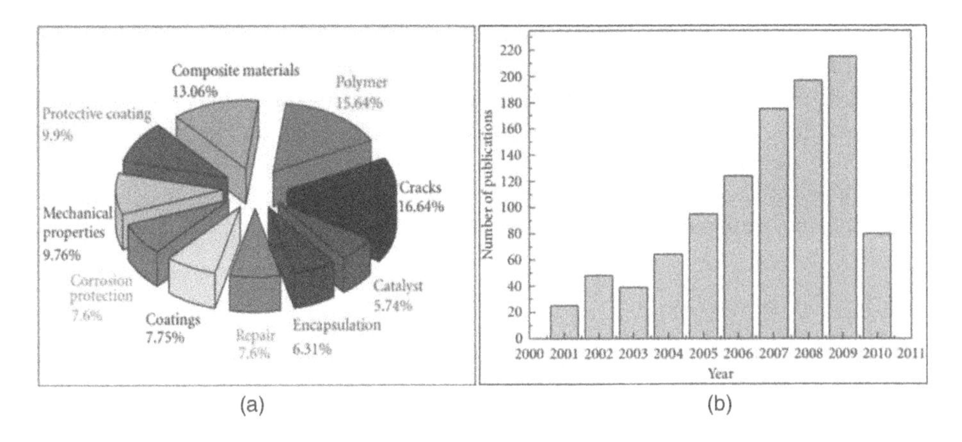

FIGURE 6.1 (a) Pie chart distribution of key words. (b) A rise in number of publications for a period of 10 years [3].

while the self-healing process refers to either the self-healing process by inhibiting the corrosion reaction or self-healing by closing the defect in the coatings.

In response to the rise in the number of publications in this field, there is need of an article reviewing previous work performed for the last 20 years. Literature comprises a range of good reviews on the topic of self-healing polymeric coatings. For instance, the study of P. Lan et al. [9] counts the self-healing coatings in connection with the super-hydrophobic ability of the coatings. Similarly, the subject of micro- and nano-encapsulation has been reviewed by a range of authors [8, 10, 11]. The review study of Iacono et al. [12] is dedicated to the subject of epoxy coatings. Sanka et al. [13] contributes to the subject of self-healing bulk polymer. On the other hand, the study of Aisa et al. [3] is related to the subject of thermosetting bulk polymers. Similarly, there is valuable work performed by Zhang et al. [8] which mainly focuses on common autonomous and non-autonomous self-healing mechanisms with some comparisons and limitations. The review work performed by Dmitry et al. [14] is dedicated to inorganic coatings. A good deal of literature review can be found in all of the aforementioned studies.

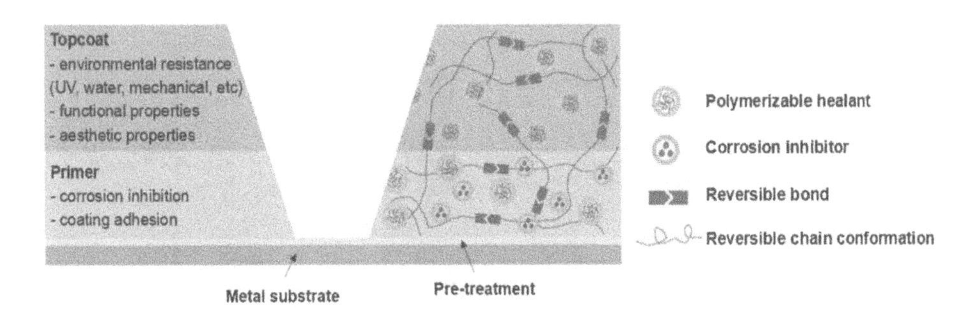

FIGURE 6.2 Schematic of a modern age self-healing polymer coating [8].

To the best of our knowledge there is no review in the literature on the subject of the self-healing models (working principles) of the functional components embedded in the coatings. Therefore, this article reviews the working principles and the associated variations followed in the research in recent times. We have classified the different models (techniques) of self-healing and reviewed the works of authors following these models. The contents of this review article are expected to aid researchers in their quest for new methods to complement previous ones.

6.1.1 Classification

The models of self-healing polymeric coatings (SHPC) can be classified based on (i) action of coatings which may be either intrinsic or extrinsic, and (ii) the technique of preventing corrosion which is performed either by corrosion inhibition or by crater filling. The overall classification of the SHPC based on the working principle of the functional component can be viewed as shown in Figure 6.3. The strategies for preventing corrosion are the main subject of this work.

The autonomous healing mechanisms are often enabled by embedding polymerizable healing agents or corrosion inhibitors in the coating matrices. For non-autonomous mechanisms, the healing effects are induced by external heat or light stimuli, which trigger the chemical reactions or physical transitions necessary for bond formation or molecular chain movement [8]. All examples in this group are non-autonomous healing mechanisms. Unlike coatings that are autonomously repaired by extrinsic curing agents, these coatings are cured by restoring the intrinsic chemical bonds and/or

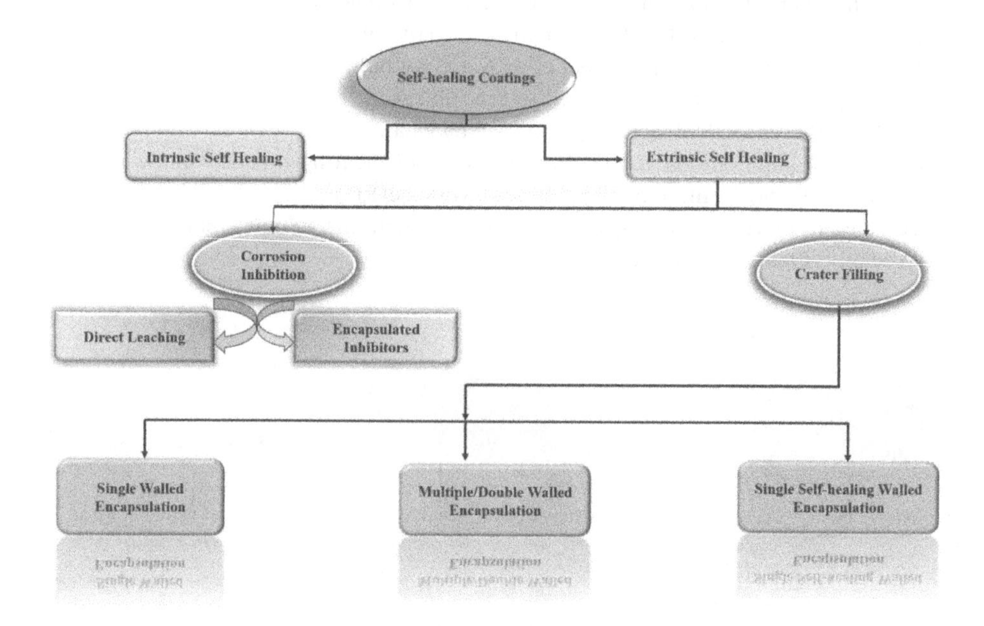

FIGURE 6.3 Classification of self-healing polymeric coatings based on the working of functional components of coatings.

physical conformations of the polymer networks in the coating matrices. Thus, all forms of intrinsic self-healing are non-autonomous while all forms of extrinsic self-healing are autonomous polymeric coatings as shown (Figure 6.3).

6.2 AUTONOMOUS SELF-HEALING COATINGS

Autonomous self-healing coatings without any actual physical interference have the potential to restore their bulk integrity or healing abilities. In some studies, embedding extrinsic polymerizable healing agents into the coating is the simplest strategy for achieving autonomous self-healing ability [8]. These curing agents are contained in micro-capsules. When the coating is affected by mechanical impacts, the capsules rupture, releasing the curing chemicals, which polymerize to form a protective film, preserving the coating's barrier properties. Some research, on the other hand, proposes that corrosion inhibitors be integrated directly or encapsulated to achieve self-healing. The corrosion inhibitor will drain into the coating defect and prevent electrochemical reactions on the metal surface substratum. This kind of self-coating mechanism would not fill the crater/defect in the coating that occurs during an operation.

Repairs of the corrosion resistance properties of coatings can be enabled or at least partly supported non-autonomously using external stimuli that are not produced by corrosion activities. Heat and light are the most common stimuli for this kind of coating so they can be easily applied in service environments. Many of the references in this field fall under the category of inherent or intrinsic healing processes [15]. Unlike those autonomously cured by extrinsic curing agents, these coatings are cured by restoring the innate chemical bonds and/or physical conformations of the polymer networks in the coating matrices. As a result, their curing efficiency is theoretically independent of the material surface. External stimulus is necessary since it provides the activation energy needed for bond breakage/reformation. Under a heat stimulus, the polymers' flow properties are improved, which enhances the reactions of the broken bonds by pushing them closer together [16]. This kind of healing mechanism is carried out by artificially adding the heat sources needed to cure the coatings (e.g., by a heat gun) or producing them from service conditions (e.g. sunshine, abrasion). The controlled release of the material of the capsules and the commonly flexible self-repair properties are a significant benefit of such non-autonomous coatings. There are non-autonomous self-healing mechanisms such as those based on superabsorbent polymers and conducting polymeric materials which are not widely studied and therefore out of the scope of this article. Autonomous self-healing is discussed below.

6.2.1 CRATER-FILLING TECHNIQUE

Where a hole or crater is formed in the coating material, this self-healing technique is used. The crater/defect is caused by mechanical disruption to the coatings during operation or shipping, which causes the contained polymerizable healing materials to spontaneously discharge. These materials can polymerize into coatings of a certain strength and thickness by interacting with catalysts incorporated in the coating

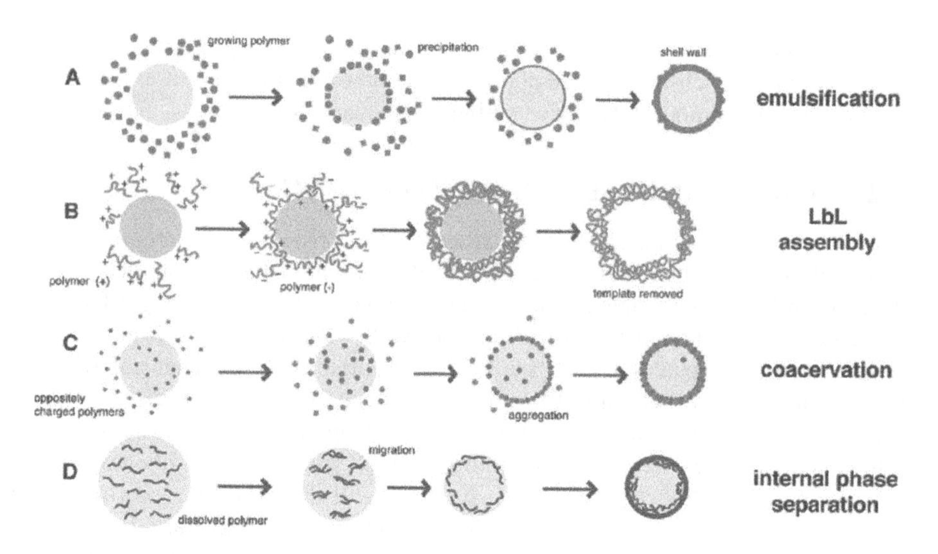

FIGURE 6.4 Fabrication methods of micro-capsules [16].

or with moisture in the surrounding atmosphere to fill coating defects and maintain the coating's barrier properties [8]. The curing agents, catalysts, or both are encapsulated during the crater-filling process. This method is linked to a set of parameters. According to the literature, a satisfactory self-healing efficiency is primarily determined by the micro-capsule's mechanical/chemical properties, size, and quantity. Since most of the reactive curing agents in this type of coating are liquid, the walls of an optimal micro-capsule must be sufficiently durable to preserve capsule integrity and coating strength. The capsules must, however, be robust enough to withstand externally applied stress [14].

There are four principal methods of obtaining micro-/nano-capsules: emulsification, layer-by-layer (LBL) assembly, coacervation, and internal phase separation [17]. The schematic of capsule fabrication is shown in Figure 6.4. Poly(methy methacrylate) (PMMA), poly(urea–formaldehyde) (PUF), methylene diphenyl diisocyanate (MDI), melamine–urea–formaldehyde (MUF), epoxy resin, polystyrene, poly(allylamine), polyvinyl alcohol (PVA), polyurethane, and phenol–formaldehyde are some of the materials widely reported in the literature used for the production of micro-capsules.

6.2.1.1 Single Wall Encapsulation

This kind of encapsulation refers to the case when the healing agents or the catalyst are encapsulated in a capsule having a protective wall with a single layer. It is a typical defect-filling technique when healants and catalysts are separately encapsulated; that is why it is sometimes called a two-part healing mechanism. The schematic of this kind of self-healing system is shown in Figure 6.5. A range of authors have performed research on the self-healing ability of two-part self-healing mechanisms [8, 18–20].

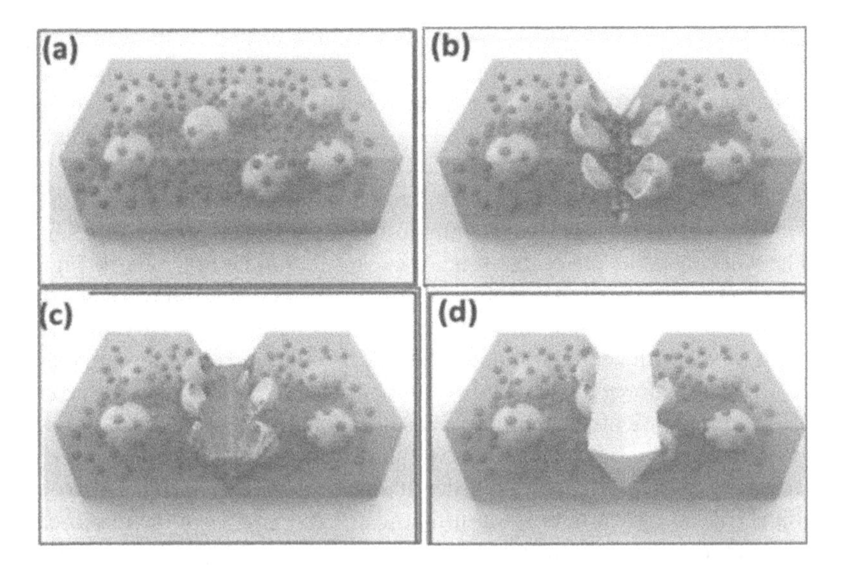

FIGURE 6.5 Schematic of self-healing by encapsulation method. (a) Embedded micro-capsules and catalysts, (b) rupture of capsules, (c) initiation of self-healing process, (d) healed defect in the coating [21].

In order to successfully seal the defect, the micro-capsules preferably should be embedded in the primer and positioned near the metal substrate. However, the inclusion of micro-capsules could minimize the adhesion of the primer to metals, and it was proposed that the micro-capsules be put in a separate layer on top of the primer [18].

One downside of using two-part healants is the unequal distribution of micro-capsules in the coating, which makes it impossible to chemically react and cure the two components in the necessary ratios [8]. Therefore, research is underway to introduce systems without catalysts to reduce the complexity. S.J. Garcia introduced what he calls a "new surface reactive system" [22]. According to the self-healing mechanism proposed in his work, the silyl ester's intended purpose was to create a hydrophobic coating on both the metal and polymer surfaces. The resulting coating should shield the metal substrate from corrosive compounds, slowing the rate of corrosion of the exposed metal. However, it is still required to target the initiation of the corrosion reaction, which is not counted in his work. Therefore, alternative techniques are adopted as included below.

6.2.1.2 Single Self-Healing Wall Encapsulation

This is a type of crater-filling model based on the capsule having a protective wall of a single layer. Additionally, the wall of capsules itself acts as a healing agent. Thus, there is no need to embed the healing chemicals separately in the coating, and the catalyst is trapped inside the wall (self-healing wall) of the micro-capsule. Therefore, this model is known as a one-part self-healing mechanism or single reactive healant. The concept of a single self-healing wall encapsulation is mainly intended to minimize the complexity of the self-healing mechanism.

Liquid isocyanates and silanes are the two candidates of single-part polymerizable healing agents successfully applied for this model. They have unique benefits for corrosion resistance applications which depend upon reaction with water. The inflow of water allows the exposure of micro-capsules, which swell and break and at the same time stimulates the polymerization of isocyanates to fill the defect. Using this model Wang et al. [23] fabricated micro-capsules containing isophorone diisocyanate, hexamethylene diisocyanate and less toxic hexamethylene diisocyanate trimer. The working principle of this kind of encapsulation is shown in the Figure 6.6. Seawater seeps into the crevice's mouth (Figure 6.6(c)). The inflow of self-healing materials from the micro-capsules is triggered by water (Figure 6.6(d)). Seawater mixes and reacts with healing products. Seawater will now reach through the healed crevice all the way down to the metal bottom because the scrape now has an exposed space. Curing ingredients from micro-capsules along the re-penetrated crevice may also be used to treat the itch. The first two phases should be performed three to four times to create a successful multi-self-healing process (Figure 6.6(f)). Micro-capsule swelling is caused by seawater infiltrating the crevice. Cracks in the micro-capsules are caused by the swelling phenomenon, which is a necessary step in initiating this process.

However, later on it was found by Wang et al. [23] that this method cannot be applied for craters with sizes wider than 30 micro-meter. Thus, the serviceability of this model is challenged. The restricted lifespan of self-healing abilities is a crucial concern for water-reactive healants. Moisture can likely penetrate into the intact coating in service conditions and react with the healing agents, thus reducing the ability of the coating to further repair as damage occurs.

D. Sun et al. [24] addresses the limitations associated with this technique. According to his study, a micro-capsule can be fabricated with the outer wall acting as a water-resistant layer and the inner wall plays the role of "curing/healing agent." The self-healing action of the inner wall is initiated after the rupture of the outer wall.

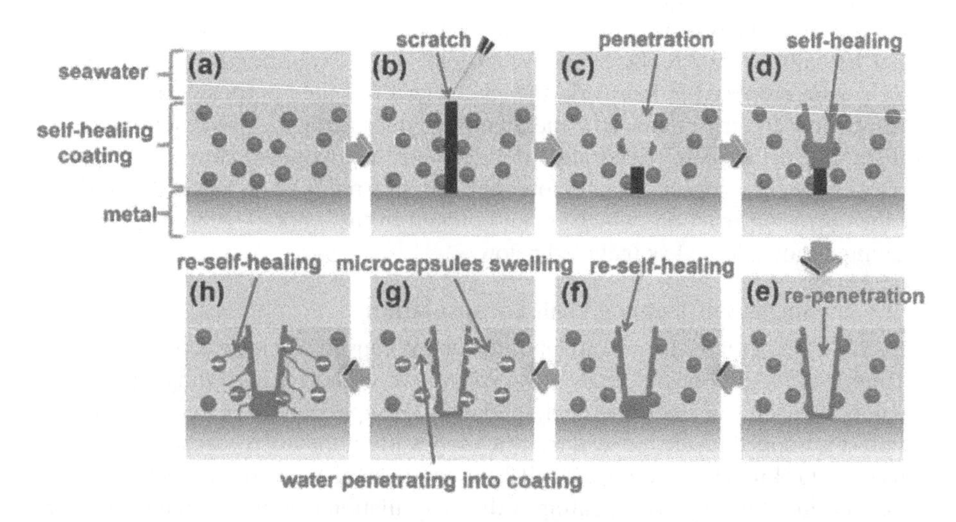

FIGURE 6.6 Self-healing mechanisms of water-triggered smart coating in seawater [23].

As a result, the exterior wall avoids excessive water/moisture penetration. It can be shown that he used a double-wall micro-capsule to enclose the 4,4′-methylenebis (cyclohexyl isocyanate) (HMDI) healing agent, with an inner wall formed by TEPA's reaction with isocyanates and an outer wall of PUF (polyurea-formaldehyde). Even if the total wall thickness was just 1.5 micro-meter, the double-walled micro-capsule demonstrated superior water resistance and better long-term anti-corrosion efficiency.

The use of a reactive silyl ester as an organic reactive healing agent encapsulated within polymer containers is another example of catalyst-free self-healing materials. Silyl esters have the ability to react with water/humidity and metallic substrates, creating a hydrophobic metal shield that prevents the metal at the scribe from further corrosion [22]. In a slightly modified form this method is adopted by Mahmoudian et al. [25]. Previously, it was believed that the healing reaction needed the involvement of a catalyst, but Mahmoudian et al. demonstrated that some air-drying natural oils can be used as self-healing agents that do not need the involvement of a catalyst and can shape the healing layer simply by making contact with atmospheric oxygen. Other natural drying oils that are commonly used as curing agents in the design and fabrication of self-healing coatings include linseed oil (LO), tung oil, and rapeseed oil.

6.2.1.3 Multiple Wall Encapsulation

This is a kind of encapsulation technique in which the wall of the capsule is fabricated having two or multiple layers. The healing agents and catalysts are the essential elements of this kind of encapsulation, entrapped between the alternating layers of the capsule. The profile image of a multi-layered capsule is shown in Figure 6.7. According to this system the outermost layer functions as a protective layer while the inner layers separate the residing chemical to retain their functionalities. Thus, not only the cost of separate encapsulation is avoided but also the limitation associated with moisture penetration is addressed.

Due to its significance a range of authors adapted to an embedded micro-capsule following multi-walled encapsulation [27–30]. In the study of Mookhoek et al. [27] it is reported that the liquid polyurea formaldehyde is stabilized with dibutyl phthalate. The outer cover is reported to be made up of a polyurethane layer. A novel fabrication process is reported in the study of Y. Yang et al. [31]. This method employs

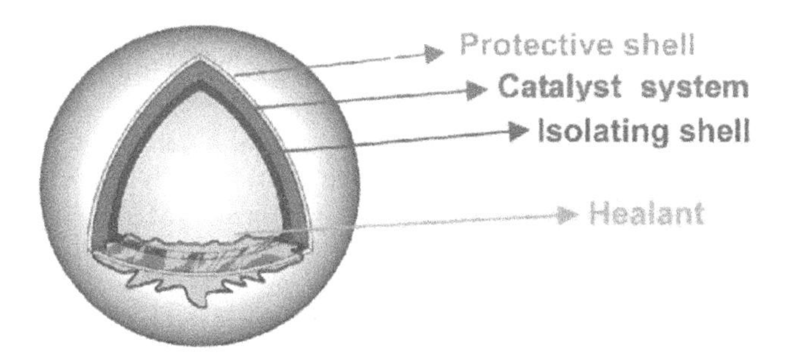

FIGURE 6.7 Multi-layered capsule for self-healing in autonomous coating [26].

Pickering stabilizers in the form of liquid-filled micro-capsules, as well as an interfacial polymerization technique to create dual micro-capsules with two distinct liquid components. The synthesis was split into two sections. First, in-situ urea and formaldehyde polymerization was used to make DBP micro-capsules (1.4 micro-meter). The DBP-filled PUF micro-capsules were then used as Pickering stabilizers in the fabrication of larger micro-capsules (140 micro-meter) containing dicyclopentadiene using an isocyanate–alcohol interfacial polymerization reaction (DCPD). However, it has been stated that when micro-capsules come into contact with a propagating fracture, they are at risk of breaking up simultaneously. The predicted healing reaction would be inefficiently carried out due to a shortage of healing agent supply if the inner small capsules were not fully opened after the outer capsule was demolished. The shape of a micro-capsule has also been discovered to be very significant in multi-layered encapsulation systems. Since circular capsules could not carry the healing agent over long distances to the cracks, elongated capsules were investigated to enhance healing efficiency. Therefore, Mookhoek et al. [27] fabricated elongated DCPD micro-capsules by combining DCPD and TDI in anisotropic droplets.

6.2.1.4 Major Implications

The micro-capsules should be blended in the primer and positioned near the metal substrate to efficiently seal the defect. Micro-capsules, on the other hand, can minimize the primer's adhesion to metals, so it is best to put them on top of the primer in a separate sheet [18]. Another way to enhance the coatings' self-healing potential is to eliminate the complexity in the coating matrix by eliminating isolated phases of healing agents and catalysts. Since oxidative curing agents, such as drying oils, do not require the use of a catalyst, they may be used [32]. Finally, problems such as the capsule's initial encounter with a fracture, its progression and coalescence, the capsules' eventual breakage, and the application of the curing agent to cracked portions are crucial for the coatings' smooth operation [26].

6.2.2 Corrosion Inhibition Technique

This technique (model) is based on the use of chemicals as corrosion inhibitors. The corrosion inhibitor's activity curtails corrosion reactions for a set period of time. However, it does not fully restore the coating's physical barrier. The anode or degradation of the metallic layer is delayed in this procedure due to inhibitors leached from the coating matrix. Alternatively, inhibitors that increase the passivity of the oxide layer to exposed metals trigger the inhibition, while inhibitors that induce oxide and hydroxide accumulation at cathodic sites often slow cathodic reactions [8]. Molybdates, tungstates, phosphates, nitrites, vanadates, borates, and rare earth salts and the organic corrosion inhibitors including benzotriazole (BTA), 8 hydroxyquinoline (8-HQ) and aliphatic amines, mercaptobenzothiazole (MBT), and imidazoline are the most common corrosion inhibitors used.

6.2.2.1 Inorganic Nano-Filler (Direct Leaching)

This is a type of self-healing technique which depends upon the dissolution of ions from the coating followed by the oxidation at the location of the defect. It is comparatively a traditional way to induce self-healing ability and is based on the concept of polymer-filler interactions in polymeric composite [6, 33]. In this procedure, corrosion inhibitors are directly combined with coating materials to cause a self-healing reaction. A schematic representation of the self-healing process by this mechanism is shown in the study of Yubiki et al. as shown Figure 6.8 [33]. A bare aluminum alloy substrate is exposed to a corrosive solution as a result of the defect in the coating, followed by an anodic dissolution. The anodic process led to the generation of aluminum cations:

$$Al \rightarrow Al^{3+} + 3e^-$$

Negatively charged OH^- ions were generated between the coating and metal substrate according to the following cathodic reaction:

$$O_2 + 2H_2O + 4e^- \rightarrow 4OH^-$$

The resulting OH^- ions would penetrate the coating. The increase in pH near the TiO_2 particles promotes the release of BPA from the surface of the TiO_2 particles as shown (Figure 6.8). The released BPA gradually penetrates the bare aluminum alloy to form a protective film, thereby repairing the defect.

The study was conducted by Mirzakhandeh [18], Carneiro [19], and Mardel [20]. All of these experiments were based on doping cerium nitrates and phosphates into organic coatings for zinc, galvanized steel, and aluminum alloy plate self-healing. Lithium salts, such as lithium carbonate, have recently been added to AA2024-T3 to improve the successful self-healing potential of polyurethane coatings. For active defense of the alloy AA2024-T3, lithium salts, such as lithium carbonate, were directly applied to polyurethane coatings, since leached lithium salts can create an alkaline condition in the coating break, causing a multilayer, defensive aluminum oxide/hydroxide film to form [21, 22].

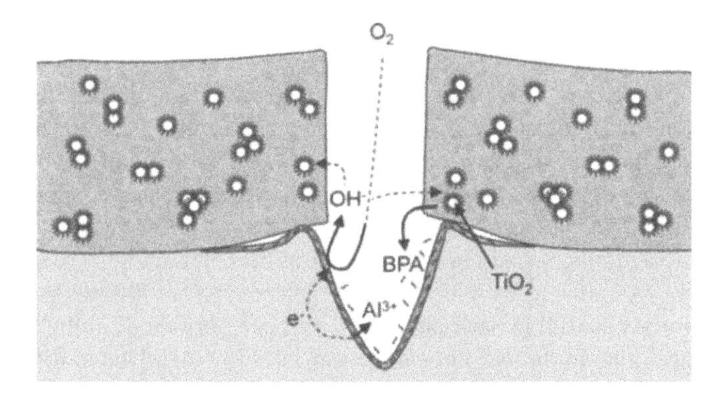

FIGURE 6.8 Schematic representation of direction leaching of corrosion-inhibiting inorganic nano-fillers [33].

For smooth functioning, the inhibitors must be properly leached from the matrix and reach the affected locations. Secondly the compatibility between the inhibitor and the coating material plays an influential role in the performance of the coatings developed to work according to this technique. Additionally, unwanted interactions between inhibitors and coatings must be reduced in order to sustain inhibitive ability over the service time. Inhibitor accumulation must be avoided since it creates ducts for faster water penetration [34]. Through covering the nano-particle's exterior surface with organic molecules, these problems may be overcome or at least mitigated. To improve compatibility with polyester and epoxy coatings, meso-porous silica nano-particles were treated with organo-silanes. The purpose of application of silanization of inorganic filler particles is solely to increase the inorganic filler and coating matrix interfacial interaction [35].

Literature reports two prominent adaptations/modifications to the application of direct leaching, firstly, the decomposition of nano-particles and secondly using a combination of inorganic nano-particles instead of a single nano-particle. The organic inhibitors emitted by the nano-particles form insoluble chelates on aluminum alloys as they decompose, avoiding pitting corrosion, which is intensified by the dealloying of the active copper-rich S-phase (Al2CuMg) intermetallic [36, 37]. The second modification is a promising approach. Multiple corrosion inhibitors, such as rare earth cations and organic inhibitors, are incorporated into the coating in this modification. As a result, a desired anti-corrosion synergy between various inhibitors can be achieved [38].

6.2.2.2 Encapsulation of Inorganic Nano-Fillers

According to this technique, inhibiting chemicals are encapsulated using micro- or nano-containers to preserve the inhibitors and to release them in a sustained and a more regulated way [21]. The encapsulation is in the form of specialized inorganic containers which are surface treated for a regulated release of corrosion inhibitors and a better interfacial interaction with coating matrix. The containers referred to here are different from polymeric containers discussed in Section 6.2.1. The containers in the form of inorganic particles do not take part actively in the self-healing process but rather act as storage for the inhibiting materials. Also, the encapsulation should not be confused with the silanization which is meant for polymer-filler interfacial interaction. The containers are evenly dispersed in the passive matrix; these nano-containers keep the active material in a "trapped" state, preventing unwanted contact between the active component and the matrix as well as random leakage. If there are variations in outside surroundings or the coating is disturbed by the external effects, the nano-containers react to these stimuli and release the encapsulated active material.

According to a number of authors, meso-porous SiO2 has a higher loading potential (roughly 80 wt%) due to its wide specific surface area (>1000 m2/g). Fibrous or particulated porous scaffolds have been used to store corrosion inhibitors for self-healing coatings in addition to micro-/nano-particles [39–41]. Due to differences in chemical nature inorganic nanocontainers/fillers are chemically incompatible with polymeric coating; therefore, inorganic containers are coated by surface-graft precipitation polymerization. For example Li et al. [42] coated silica nano-tubes with poly(methacrylic acid) (PMAA). According to his work surface-graft polymerization with double-bond modified nickel hydrazine/silica core–shell rod templates was used to make the silica/polymer hybrid nano-tubes, which were then etched to remove the nickel–hydrazine core. It is concluded that by fine-tuning the polymeric outer shells, it is possible to create nano-containers with a controllable release mechanism (pH, temperature, or redox reactions). The schematic and the TEM images of the micro-container obtained via this method are shown in Figure 6.9.

The selection of nano-particles for use as a container is based on their micro-structural properties. For example, halloysites are characterized by hollow or micro-particle structure, working similarly to polymeric nano-containers for encapsulating active materials. The ion exchange property of the clay particle plays an important role in storing the inhibitors. The interlayer galleries in the micro-structure of clay particles is thus of importance for creating the ion exchange ability and the resulting anti-corrosive substrate protection [43]. Similarly, anionic clays (hydrotalcite (HT)-like compounds), also known as layered double hydroxide (LDH) nano-containers, are anion-exchange materials made up of layers of positively charged mixed metal hydroxides in which anionic species and solvent molecules are intercalated [43].

The self-healing mechanisms reported earlier in this chapter are engineered to shield a substrate having a single material. However, the increased use of hybrid systems made up of joined dissimilar materials, such as aluminum alloy AA6061, in modern vehicles and airplanes necessitates advancements in self-healing safety. Hybrid materials, for example, have a high risk of galvanic corrosion due to their very different electrochemical potentials. Serdechnova et al. claimed that mixing anion-exchange LDH loaded with BTA- and cation-exchange bentonite loaded with Ce3+ in the same coating method shielded aluminum alloys and carbon fiber reinforced plastic (CFRP) from corrosion [44]. His innovative solution is based on a coating method that blends two types of nano-containers with two distinct inhibitors. When an aluminum alloy is galvanically combined with carbon fiber reinforced plastic, the nano-containers impart a stimulated release of all inhibitors, which work in a synergistic manner, giving inorganic containers new dimensions.

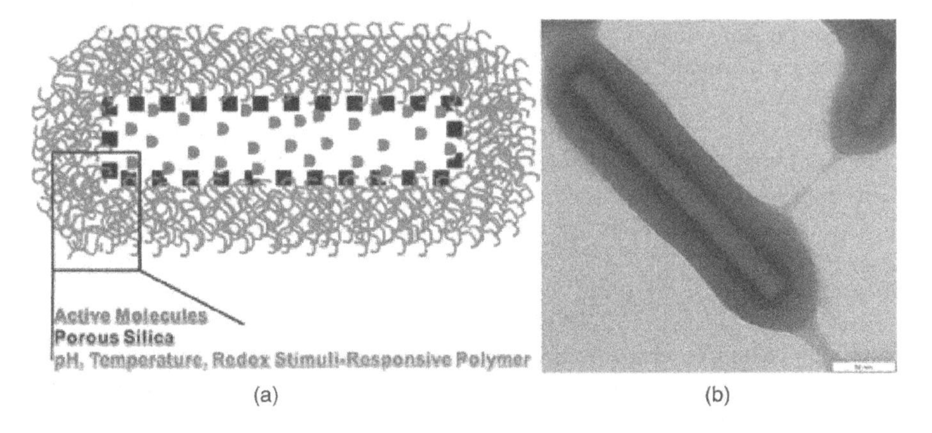

FIGURE 6.9 (a) Schematic representation of a coated inorganic nano-filler. (b) TEM image of coated nano-filler particles [42].

6.2.2.3 Major Implications

Achieving a fully optimized self-healing mechanism has always been a challenge. Therefore, experts make tradeoffs between the targeted functionality and the drawbacks of a particular self-healing system. The major drawbacks that experts need to incorporate in designing a healing system that works on the bases of inhibition is the loading capacity of the containers. It is reported that polymer micro-capsules usually provide 50–60 wt% healing agent loading, ceramic containers can have 20–30 wt% healing agent loading, and in extremely rare circumstances, 50–60 wt% healing agent loading can be obtained [14]. In the direct use of nano-particles, the barrier and adhesion properties of a coating are harmed by particle agglomeration and defects at the material/coating interface. This may be particularly troublesome if the particles are near metal substrates [45].

6.3 ANALYSIS TOOLS

The performance of self-healing anti-corrosion polymeric coatings is measured with the help of combining different microscopic techniques to either investigate the corrosion process or to investigate the action of coatings. Microscopic techniques such as electrochemical impedance spectroscopy (EIS), the scanning vibrating electrode technique (SVET) and the scanning ion-selective electrode technique (SIET) are reported in a wide range of literature [38, 46, 47]. When the duration of ingress of corrosive medium into the coating is short, EIS can be used to predict coating degradation and corrosion kinetics. SVET or localized impedance spectroscopy (LEIS) are commonly used to aid this technique [48, 49]. The SVET provides details on the initiation of corrosion in a small area of corrosion. Non-intrusive scanning and vibrating probe calculation are used in SVET. Furthermore, the electric field produced in a plane above the electrochemically active surface is mapped, allowing for real-time recording and quantification of the coating's local electrochemical behavior. This approach is particularly useful for researching the effects of corrosion inhibitors. Local variations in pH on the coating surface are mapped using the scanning ion-electrode technique. The use of the electrochemical techniques described above in combination creates a comprehensive image of the self-healing potential.

The self-healing actions of coating matrices is investigated with the help of techniques such as Xray photoelectron spectroscopy (XPS) and electron-probe micro-analysis (EPMA)[50]. XPS may be used to determine the chemical composition of a substance that has been mechanically affected. EPMA provides information on ion movement in the vicinity of a coating-damaged location. Coatings' self-healing abilities are often tested using microscopic techniques. The layer is sliced, immersed in a corrosive solvent, and the cut is analyzed using optical microscopy, scanning electron microscopy (SEM), and atomic force microscopy (AFM) and scanning Kelvin probe force microscopy (SKPFM) at fixed time intervals [51–53]. The scanning electrochemical microscope (SECM) is a tool that can be used to determine the self-healing properties of coatings. This technique can be used to collect microscopic data from corrosion-affected surfaces as well as quantify local variations in electrochemical reactivity [54]. It is also possible to monitor the release of metal ions at anodic sites and the absorption of oxygen at cathodic sites. The ability to control both insulating and conducting surfaces is a significant benefit of SECM.

6.4 CONCLUSION

The content of this review article discusses the working of a functional component of the coating and the evolutionary work performed by the researchers. There have been a variety of interesting methods introduced in the field of self-healing coatings, but it is important to remember that not only scientific but also other factors determine when a coating with a specific kind of function can be commercialized. It is important to consider socioeconomic criteria such as low cost and environmental friendliness. On the other hand, development in recent years has been so rapid and positive that one can be optimistic in finding alternatives for a variety of coatings.

REFERENCES

1. X. Li, D. Zhang, Z. Liu, Z. Li, C. Du, and C. Dong, "Materials science: Share corrosion data," *Nature News*, vol. 527, no. 7579, p. 441, 2015.
2. B. *Hou* et al., "The cost of corrosion in China," *NPJ Materials Degradation*, vol. 1, no. 1, pp. 1–10, 2017.
3. B. Aïssa, D. Therriault, E. Haddad, and W. Jamroz, "Self-healing materials systems: Overview of major approaches and recent developed technologies," *Advances in Materials Science and Engineering*, vol. 2012, pp. 1–17, 2012.
4. T. Ahmed, H. H. Ya, R. Khan, A. M. Hidayat Syah Lubis, and S. Mahadzir, "Pseudo-ductility, morphology and fractography resulting from the synergistic effect of $CaCO_3$ and bentonite in HDPE polymer nano composite," *Materials*, vol. 13, no. 15, p. 3333, 2020.
5. H. U. Khalid, M. C. Ismail, and N. Nosbi, "Permeation damage of polymer liner in oil and gas pipelines: A review," *Polymers*, vol. 12, no. 10, p. 2307, 2020.
6. T. Ahmed, H. Ya, S. Mahadzir, R. Khan, and M. A. Alam, "An overview: Mechanical and wear properties of HDPE polymer nanocomposites reinforced with treated/non-treated inorganic nanofillers," *Advances in Manufacturing Engineering*, pp. 231–241, 2020.
7. M. A. Alam, H. Ya, P. Hussain, M. Azeem, S. Sapuan, and T. Ahamad, "Experimental Investigations on the Surface Hardness of Synthesized Polystyrene/ZnO Nanocomposites," in *Advances in Manufacturing Engineering*: Springer, 2020, pp. 345–352.
8. F. Zhang et al., "Self-healing mechanisms in smart protective coatings: A review," *Corrosion Science*, vol. 144, pp. 74–88, 2018.
9. P. Lan, E. E. Nunez, and A. A. Polycarpou, "11466 Advanced Polymeric Coatings and Their Applications: Green Tribology," 2019.
10. M. Samadzadeh, S. H. Boura, M. Peikari, S. Kasiriha, and A. Ashrafi, "A review on self-healing coatings based on micro/nanocapsules," *Progress in Organic Coatings*, vol. 68, no. 3, pp. 159–164, 2010.
11. A. Stankiewicz, I. Szczygieł, and B. Szczygieł, "Self-healing coatings in anti-corrosion applications," *Journal of Materials Science*, vol. 48, no. 23, pp. 8041–8051, 2013.

12. S. Dello Iacono, A. Martone, and E. Amendola, "Corrosion-resistant self-healing coatings," in *AIP Conference Proceedings*, 2018, vol. 1990, no. 1, p. 020010: AIP Publishing LLC.
13. R. Sanka, B. Krishnakumar, Y. Leterrier, S. Pandey, S. Rana, and V. Michaud, "Soft self-healing nanocomposites," *Frontiers in Materials*, vol. 6, p. 137, 2019.
14. D. G. Shchukin, "Container-based multifunctional self-healing polymer coatings," *Polymer Chemistry*, vol. 4, no. 18, pp. 4871–4877, 2013.
15. Y. Yang and M. W. Urban, "Self-healing polymeric materials," *Chemical Society Reviews*, vol. 42, no. 17, pp. 7446–7467, 2013.
16. M. Zheludkevich and A. Hughes, "Delivery Systems for Self Healing Protective Coatings," in *Active Protective Coatings*: Springer, 2016, pp. 157–199.
17. M. Zheludkevich, J. Tedim, and M. Ferreira, ""Smart" coatings for active corrosion protection based on multi-functional micro and nanocontainers," *Electrochimica Acta*, vol. 82, pp. 314–323, 2012.
18. A. Kumar, L. Stephenson, and J. N. Murray, "Self-healing coatings for steel," *Progress in Organic Coatings*, vol. 55, no. 3, pp. 244–253, 2006.
19. M. W. Keller, S. R. White, and N. R. Sottos, "A self-healing poly (dimethyl siloxane) elastomer," *Advanced Functional Materials*, vol. 17, no. 14, pp. 2399–2404, 2007.
20. C. Mangun, A. Mader, N. Sottos, and S. White, "Self-healing of a high temperature cured epoxy using poly (dimethylsiloxane) chemistry," *Polymer*, vol. 51, no. 18, pp. 4063–4068, 2010.
21. H. Choi, K. Y. Kim, and J. M. Park, "Encapsulation of aliphatic amines into nanoparticles for self-healing corrosion protection of steel sheets," *Progress in Organic Coatings*, vol. 76, no. 10, pp. 1316–1324, 2013.
22. S. García et al., "Self-healing anticorrosive organic coating based on an encapsulated water reactive silyl ester: Synthesis and proof of concept," *Progress in Organic Coatings*, vol. 70, no. 2–3, pp. 142–149, 2011.
23. W. Wang, L. Xu, X. Li, Z. Lin, Y. Yang, and E. An, "Self-healing mechanisms of water triggered smart coating in seawater," *Journal of Materials Chemistry A*, vol. 2, no. 6, pp. 1914–1921, 2014.
24. D. Sun, H. Zhang, X.-Z. Tang, and J. Yang, "Water resistant reactive microcapsules for self-healing coatings in harsh environments," *Polymer*, vol. 91, pp. 33–40, 2016.
25. M. Mahmoudian, E. Nozad, M. G. Kochameshki, and M. Enayati, "Preparation and investigation of hybrid self-healing coatings containing linseed oil loaded nanocapsules, potassium ethyl xanthate and benzotriazole on copper surface," *Progress in Organic Coatings*, vol. 120, pp. 167–178, 2018.
26. D. Y. Zhu, M. Z. Rong, and M. Q. Zhang, "Self-healing polymeric materials based on microencapsulated healing agents: From design to preparation," *Progress in Polymer Science*, vol. 49, pp. 175–220, 2015.
27. S. D. Mookhoek, "Novel routes to liquid-based self-healing polymer systems," 2010.
28. H. Jin, C. L. Mangun, D. S. Stradley, J. S. Moore, N. R. Sottos, and S. R. White, "Self-healing thermoset using encapsulated epoxy-amine healing chemistry," *Polymer*, vol. 53, no. 2, pp. 581–587, 2012.
29. M. M. Caruso et al., "Robust, double-walled microcapsules for self-healing polymeric materials," *ACS Applied Materials & Interfaces*, vol. 2, no. 4, pp. 1195–1199, 2010.
30. M. T. Gokmen, B. G. De Geest, W. E. Hennink, and F. E. Du Prez, ""Giant" hollow multilayer capsules by microfluidic templating," *ACS Applied Materials & Interfaces*, vol. 1, no. 6, pp. 1196–1202, 2009.
31. Y. Yang, Z. Wei, C. Wang, and Z. Tong, "Versatile fabrication of nanocomposite microcapsules with controlled shell thickness and low permeability," *ACS Applied Materials & Interfaces*, vol. 5, no. 7, pp. 2495–2502, 2013.

32. C. D. Dieleman, P. J. Denissen, and S. J. Garcia, "Long-term active corrosion protection of damaged coated-AA2024-T3 by embedded electrospun inhibiting nanonetworks," *Advanced Materials Interfaces*, vol. 5, no. 12, p. 1800176, 2018.
33. A. Yabuki, W. Urushihara, J. Kinugasa, and K. Sugano, "Self-healing properties of TiO_2 particle–polymer composite coatings for protection of aluminum alloys against corrosion in seawater," *Materials and Corrosion*, vol. 62, no. 10, pp. 907–912, 2011.
34. A. Hughes, J. Laird, C. Ryan, P. Visser, H. Terryn, and A. Mol, "Particle characterisation and depletion of Li_2CO_3 inhibitor in a polyurethane coating," *Coatings*, vol. 7, no. 7, p. 106, 2017.
35. S. Kongparakul et al., "Self-healing hybrid nanocomposite anticorrosive coating from epoxy/modified nanosilica/perfluorooctyl triethoxysilane," *Progress in Organic Coatings*, vol. 104, pp. 173–179, 2017.
36. M. Zheludkevich, K. Yasakau, S. Poznyak, and M. Ferreira, "Triazole and thiazole derivatives as corrosion inhibitors for AA2024 aluminium alloy," *Corrosion Science*, vol. 47, no. 12, pp. 3368–3383, 2005.
37. D. Snihirova, M. Taryba, S. V. Lamaka, and M. F. Montemor, "Corrosion inhibition synergies on a model Al-Cu-Mg sample studied by localized scanning electrochemical techniques," *Corrosion Science*, vol. 112, pp. 408–417, 2016.
38. M. Montemor et al., "Evaluation of self-healing ability in protective coatings modified with combinations of layered double hydroxides and cerium molibdate nanocontainers filled with corrosion inhibitors," *Electrochimica Acta*, vol. 60, pp. 31–40, 2012.
39. E. Shchukina, D. Shchukin, and D. Grigoriev, "Effect of inhibitor-loaded halloysites and mesoporous silica nanocontainers on corrosion protection of powder coatings," *Progress in Organic Coatings*, vol. 102, pp. 60–65, 2017.
40. K. Zhang, L. Wang, and G. Liu, "Copper (II) 8-hydroxyquinolinate 3D network film with corrosion inhibitor embedded for self-healing corrosion protection," *Corrosion Science*, vol. 75, pp. 38–46, 2013.
41. Y.-H. Liu, J.-B. Xu, J.-T. Zhang, and J.-M. Hu, "Electrodeposited silica film interlayer for active corrosion protection," *Corrosion Science*, vol. 120, pp. 61–74, 2017.
42. G. L. Li, Z. Zheng, H. Möhwald, and D. G. Shchukin, "Silica/polymer double-walled hybrid nanotubes: synthesis and application as stimuli-responsive nanocontainers in self-healing coatings," *ACS Nano*, vol. 7, no. 3, pp. 2470–2478, 2013.
43. H. Wei et al., "Advanced micro/nanocapsules for self-healing smart anticorrosion coatings," *Journal of Materials Chemistry A*, vol. 3, no. 2, pp. 469–480, 2015.
44. M. Serdechnova, S. Kallip, M. G. Ferreira, and M. L. Zheludkevich, "Active self-healing coating for galvanically coupled multi-material assemblies," *Electrochemistry Communications*, vol. 41, pp. 51–54, 2014.
45. D. Borisova, H. Möhwald, and D. G. Shchukin, "Influence of embedded nanocontainers on the efficiency of active anticorrosive coatings for aluminum alloys. Part II: Influence of nanocontainer position," *ACS Applied Materials & Interfaces*, vol. 5, no. 1, pp. 80–87, 2013.
46. N. P. Tavandashti and S. Sanjabi, "Corrosion study of hybrid sol–gel coatings containing boehmite nanoparticles loaded with cerium nitrate corrosion inhibitor," *Progress in Organic Coatings*, vol. 69, no. 4, pp. 384–391, 2010.
47. W. Trabelsi, E. Triki, L. Dhouibi, M. Ferreira, M. Zheludkevich, and M. Montemor, "The use of pre-treatments based on doped silane solutions for improved corrosion resistance of galvanised steel substrates," *Surface and Coatings Technology*, vol. 200, no. 14–15, pp. 4240–4250, 2006.
48. D. Snihirova, L. Liphardt, G. Grundmeier, and F. Montemor, "Electrochemical study of the corrosion inhibition ability of "smart" coatings applied on AA2024," *Journal of Solid State Electrochemistry*, vol. 17, no. 8, pp. 2183–2192, 2013.

49. S. Neema, M. Selvaraj, J. Raguraman, and S. Ramu, "Investigating the self healing process on coated steel by SVET and EIS techniques," *Journal of Applied Polymer Science*, vol. 127, no. 1, pp. 740–747, 2013.
50. K. Aramaki, "XPS and EPMA studies on self-healing mechanism of a protective film composed of hydrated cerium (III) oxide and sodium phosphate on zinc," *Corrosion Science*, vol. 45, no. 1, pp. 199–210, 2003.
51. A. S. Hamdy and D. Butt, "Novel smart stannate based coatings of self-healing functionality for AZ91D magnesium alloy," *Electrochimica Acta*, vol. 97, pp. 296–303, 2013.
52. T. C. Mauldin and M. Kessler, "Self-healing polymers and composites," *International Materials Reviews*, vol. 55, no. 6, pp. 317–346, 2010.
53. S. Li and J. Fu, "Improvement in corrosion protection properties of TiO_2 coatings by chromium doping," *Corrosion Science*, vol. 68, pp. 101–110, 2013.
54. A. Pilbáth, T. Szabó, J. Telegdi, and L. Nyikos, "SECM study of steel corrosion under scratched microencapsulated epoxy resin," *Progress in Organic Coatings*, vol. 75, no. 4, pp. 480–485, 2012.

7 Nano-Indentation and Indentation Size Effect on Different Phases in Lamellar Structure High Entropy Alloy

Norhuda Hidayah Nordin and
Mohd Hafis Sulaiman
International Islamic University Malaysia
Selangor, Malaysia

Leong Zhaoyuan
The University of Sheffield
Sheffield, UK

CONTENTS

7.1 INTRODUCTION

Conventionally, typical alloys with major constituent elements with minor alloying (Fe-based, Ti-based or superalloys) have been used in many advanced applications. An alternative way in alloys development has emerged since an interest in exploring a central phase region of the phase diagram where there were alloys with different components in equal atomic proportion was investigated by Cantor [1] and Yeh [2]. Based on thermodynamic concept, a maximum value of configurational entropy is obtained when all elements involved are in equal atomic proportion ($\Delta S_{conf} = -R(X_A \ln X_A + X_B \ln X_B)$ and subsequently increase with an increasing number of elements in the system ($\Delta S_{conf} = R \ln N$, where N is the number of elements). Interestingly, instead of the predicted complex alloy, a simple solid solution alloy has been obtained with

DOI: 10.1201/9781003139386-7

improved properties compared to conventional alloys. Thus, high entropy alloys, known as a new class of materials, have shown promising potential for applications such as in structure, transportation, tool, and coating. Besides, this new design of alloys also can be classified as green materials because it undergoes simple processes and produces sustainable products due to its efficacy in service.

Most of the published literature on HEAs focuses on successful formation of single-phase solid solutions [3, 4]. HEAs with FCC structure mostly possess good ductility but lower strength, while BCC types exhibit higher strength with limited plasticity [5]. To that end, extensive studies have aimed to improve the balance of mechanical behaviours of HEAs. For example, Borkar et al. noted a balance of strength and ductility for HEAs with a composite structure consisting of a disordered solid solution phase and ordered precipitate phase [6]. In addition, previous studies have reported that eutectic HEA alloys with alternating soft FCC and hard BCC have an exceptional combination of high fracture strength and excellent tensile ductility at both room and elevated temperatures [7]. Moreover, the lamellar nature of these eutectic structures also requires low energy phase boundaries and implies high temperature creep resistance, better castability and controllable micro-structure [8].

To develop multiple phase HEAs for optimised mechanical behaviours, a detailed understanding of micro-structure formation and evolution is required. One approach that has been used for this is through computer-aided thermodynamic calculations. He et al. [7] used this method to design a eutectic high entropy alloy (EHEA), where the modification of a single-phase HEA composition into a eutectic alloy was performed via the careful selection of alloying elements. Binary phase HEAs can be designed by selecting an alloying element with a low enthalpy of mixing and large difference in atomic size, in order to favour precipitation of a secondary phase, resulting in alloy strengthening [9, 10]. It follows that manipulation of micro-structure requires a measured change in processing parameter or chemical composition.

Boron is one commonly used micro-alloying element in steels for improving hardenability [11]. It possesses a well-known grain-refinement effect when used as an alloying addition to several systems, while also potentially having a strengthening effect on the grain boundaries, which increases ductility and reduces intergranular fracture [12]. Additionally, boron addition is known to stabilise the lamellar micro-structure that has been shown to be desirable in HEAs [7, 8]. One strategy that may be considered is to thus include metalloid elements, such as boron, to tune the mechanical properties through micro-structural modification and refinement.

7.2 EXPERIMENTAL PROCEDURE

The series of lamellar structured HEA of $FeCoNi(B_x Al_{1-x})_{0.1}Si_{0.1}$ (where x = 0, 0.4, 0.7 and 1.0 are abbreviated as B0, B0.4, B0.7, and B1.0, respectively) has been produced using vacuum arc melting.

Nano-indentation was carried out using TI Premier nano-indentation tester (equipped with scanning probe microscope) fitted with Berkovich diamond tips to determine the hardness values for each phase in the alloys. The test method was performed on an array of 3 × 3 indents within a 50 μm × 50 μm sample area, each performed to a peak force of 4 mN. The indented samples were then cleaned by using acetone and were observed under an optical microscope for image analysis.

7.3 RESULT AND DISCUSSIONS

XRD patterns of $FeCoNi(B_0Al_{1.0})_{0.1}Si_{0.1}$ HEA in Figure 7.1 presented that as-cast samples possessed a single BCC/B2 phase. The introduction of boron into the HEA system induced the formation of secondary phases (presented in FCC and the Fe_2B intermetallic structure). With the increase of boron, the BCC/B2 diffraction peaks weakened, whilst the intensity of peaks corresponding to the FCC and Fe_2B structure increased. This phenomenon suggested that the substitution of aluminium with boron repressed the formation of BCC/B2 while stimulating the formation of the Fe_2B. The B1.0 sample displayed only FCC, and the Fe_2B phase presented with a very minimum of BCC/B2 phase remaining in the system.

Comparing the binary enthalpies of mixing of the alloying elements shows that the Fe-Si (-35 kJ/mol), Fe-Co (-38 kJ/mol), and Fe-Ni (-40 kJ/mol) are the most negative enthalpies of mixing. However, due to the low amount of Si addition, the Fe-B (-26 kJ/mol), Co-B (-24 kJ/mol), Ni-B (-24 kJ/mol), Ni-Al (-22 kJ/mol), and Co-Al (-19 kJ/mol) would dominate. A statistical analysis of the pairing suggests the segregation of B/Al rich phase Fe-B, Co-B, and Fe-Al, and the remaining Co, Ni, Fe, and Si.

Figure 7.2 shows the Scanning Electron Microscope (SEM) backscattered electron images for the as-cast sample of $FeCoNi(B_{0.7}Al_{0.3})Si_{0.1}$ and $FeCoNi(B_{1.0})Si_{0.1}$ high entropy alloys. The lighter areas in the micro-graph correspond to heavier elements and B/Al rich regions correspond to the darker areas. As CoNiFe is a known FCC former (CoCrFeNi is a known FCC composition, with Cr being a BCC stabiliser)

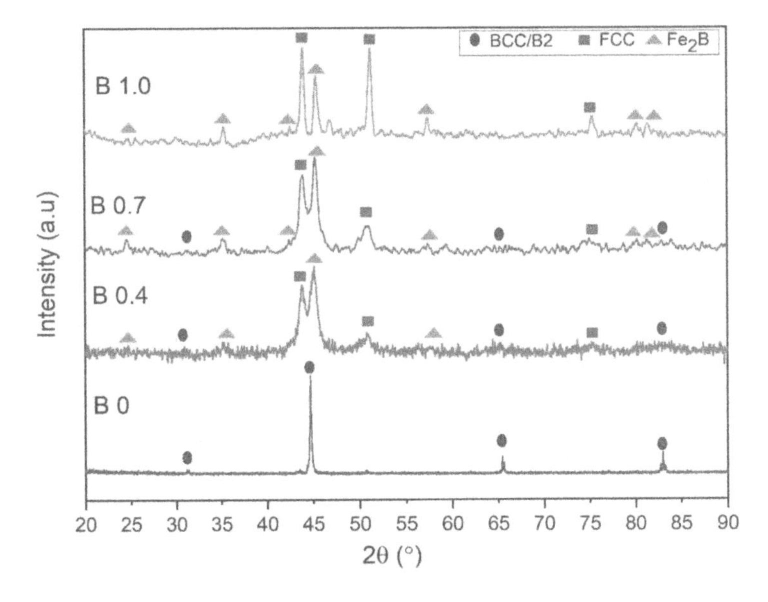

FIGURE 7.1 XRD patterns of $FeCoNi(B_xAl_{1-x})_{0.1}Si_{0.1}$ HEA with increasing boron content.

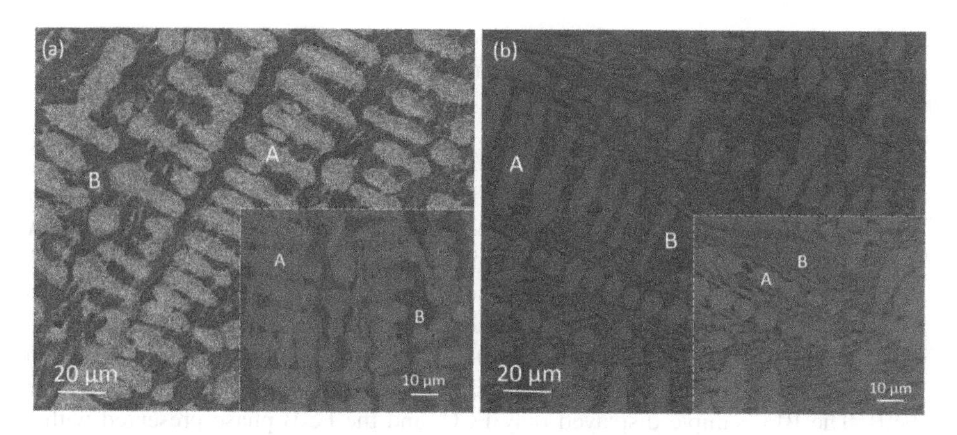

FIGURE 7.2 Scanning Electron Microscope backscattered images of (a) $FeCoNi(B_{0.7}Al_{0.3})Si_{0.1}$ and (b) $FeCoNi(B_{1.0})Si_{0.1}$.

we therefore consider region A in the SEM micro-graph to be representative of a represented FCC solid solution. Region B showed a lamellar structure consisting of FCC and Fe2B phase. There was also a dark region in the images that showed the BCC/B2 phase. However, this phase was not clearly seen in the images as the phase was diminished with increasing boron content.

In order to better reveal the relationship between boron content and mechanical properties of each phase, the nano-indentation method was used to distinguish the hardness of the primary phase (BCC/B2 or FCC) and the lamellar structured phases (FCC + Fe2B). It was noted that the lamellar spacing in the alloy was relatively low (0.1–4 µm). Therefore, the primary and lamellar phases were characterised by traditional nano-indentation method (by using an optical microscope as the observation technique). The B0.7 and B1.0 alloys were chosen for this investigation due to the existence of all phases (BCC/B2, FCC, and Fe2B phases) in B0.7 and only two phases (FCC and Fe2B) in the B1.0 alloy.

Figures 7.3(a) and 7.4(a) show the optical microscope image of the surface for the whole sample area (bound by 4 black rectangles) after the indentation test for B0.7 and B1.0, respectively. It can be seen that the dendritic FCC phase was lighter while the lamellar space containing FCC and Fe2B phases was much darker. However, the BCC/B2 phase in the B0.7 sample was hard to distinguish using the optical microscope due to low volume fraction and low image contrast. The image of the indents area (each rectangle) was then magnified to observe the indent marks that are located in each phase, which can be seen in Figures 7.3(b) and 7.4(b).

Within each area bounded by the dashed black rectangle, there are 9 indentations with different indent sizes. From that image, smaller indent marks were observed which are located at the lamellar structure while the bigger indent marks can be seen at the primary FCC phase. Therefore, it can be deduced that the FCC phase is softer than the lamellar phase. The indentation size can determine the hardness of a material which is subject to deformation, as presented in Equation 7.1:

$$H = P_{max} A_r \tag{7.1}$$

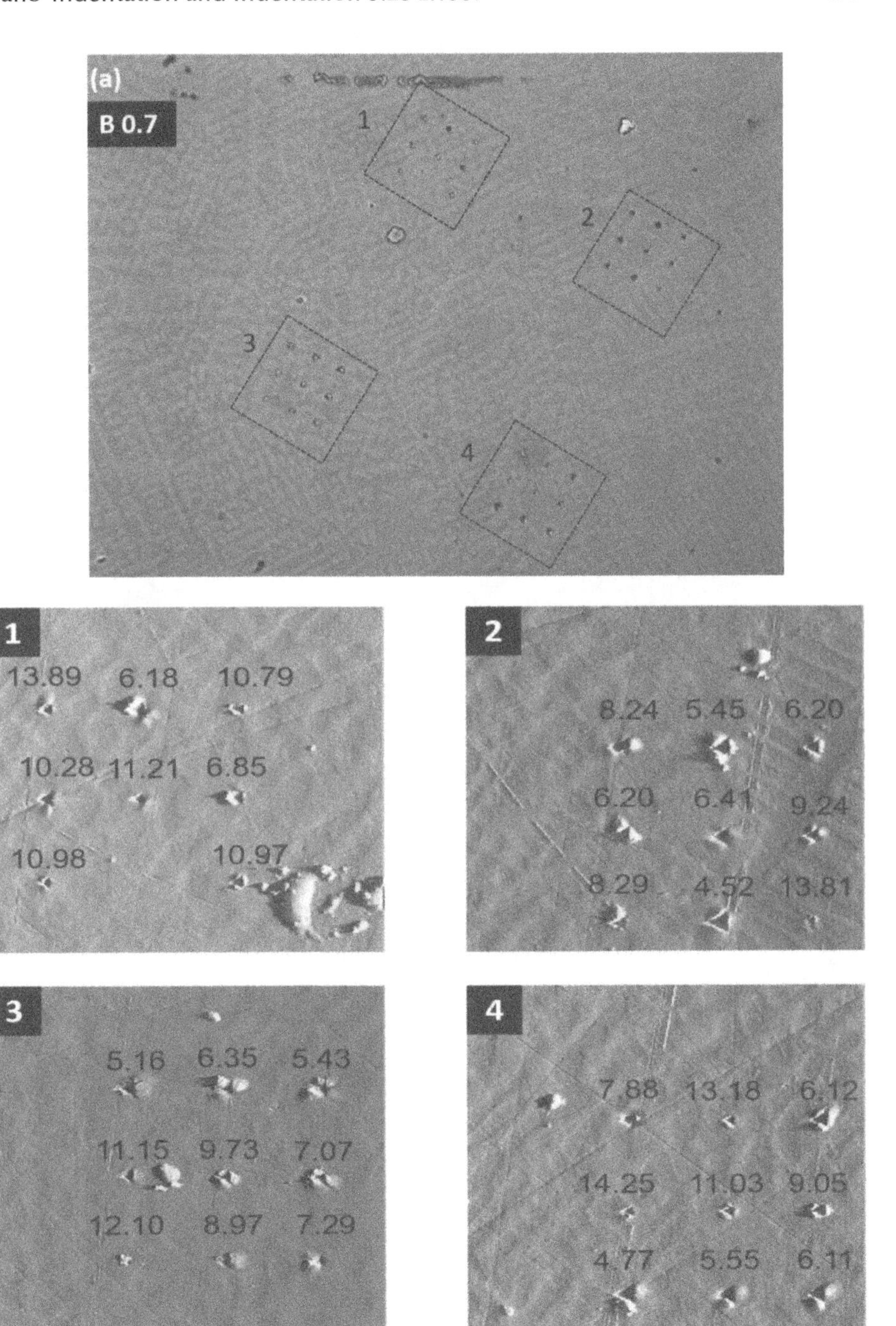

FIGURE 7.3 Optical microscope image of the sample surface after the indentation test for (a) the whole sample area, (b) magnified image of each indentation area with the hardness values (inside the dashed square region) for B0.7 alloy

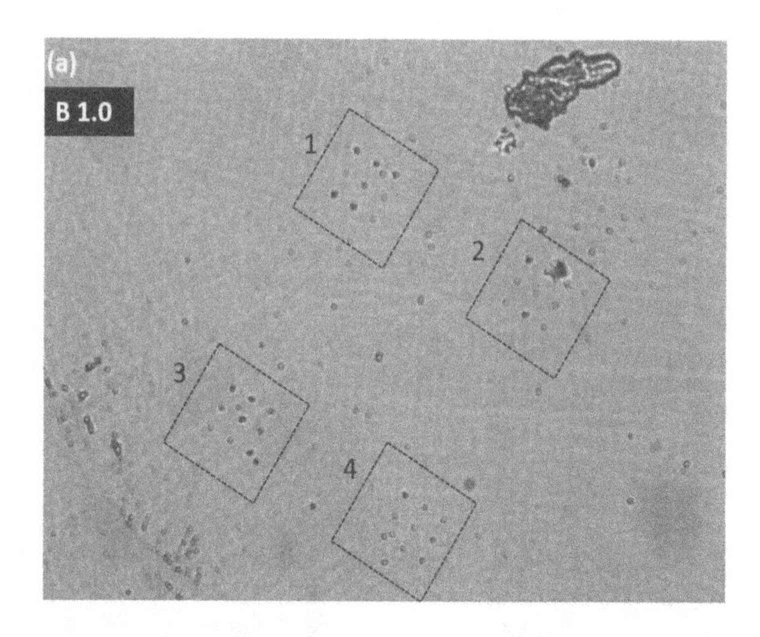

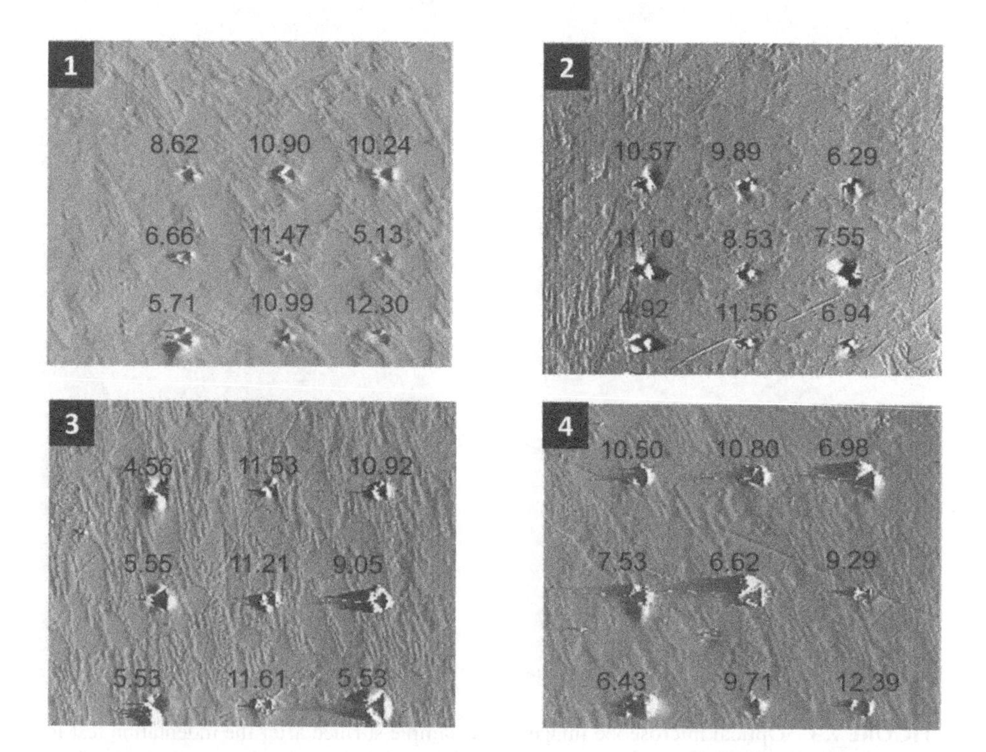

FIGURE 7.4 Optical microscope image from the sample surface after the indentation test for (a) the whole sample area, (b) magnified image of each indentation area with the hardness values (inside the dashed square region) for B1.0 alloy.

where P_{max} is the maximum load while A_r is the indentation area. Therefore, according to the equation, it can be interpreted that the smaller indents can be seen in some hard materials rather than in soft materials with a given load.

This was proved by the hardness values which were recorded during the test, given in the image (see Figure 7.3(b)). For B0.7, the lowest hardness value was in the range of 4–7 GPa, indicating the FCC phase, while the highest hardness value was recorded in the range of 10–15 GPa which is located exactly at the lamellar region. This is in accordance with the study by Castro et al. which reported that the hardness of Fe2B layers in tool-borided steel is approximately 14 GPa, tested using the Berkovich nano-indentation technique at 250 mN applied load [13, 14]. However, lower hardness value was expected in this study due to the low volume fraction of boron (< 0.1 at%) in the composition. Meanwhile the range of FCC phase was determined according to Almasri and Voyiadjis [15] who performed experimental studies on nano-indentation in FCC metal. The image of the indents also revealed that there were some intermediate indents located between FCC and lamellar structure. The intermediate indents were expected to be BCC/B2 phase, which gave the hardness value in the range of 7–10 GPa, which was out of the range between FCC and Fe2B phase. The distribution of hardness on each phase for the B0.7 sample was illustrated in Figure 7.5.

The same scenario can be seen in the B1.0 sample as in Figure 7.4(b). However, as the sample was only composed of FCC and Fe2B phases, the hardness can be determined directly, either FCC or Fe2B phase. It was noted that the highest hardness value was recorded up to 12.39 GPa. This is due to the absence of BCC/B2 phase in the alloy, resulting in lower hardness compared to B0.7 alloy. The hardness distribution graphs for B1.0, as shown in Figure 7.6, showed that the distribution of the lamellar phase was higher than the B0.7 sample. This was in accordance with

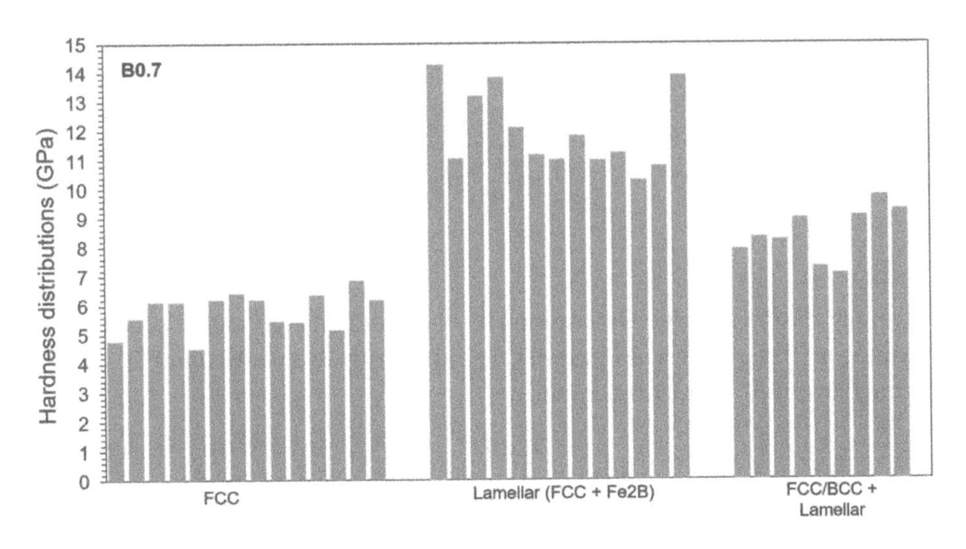

FIGURE 7.5 Distribution of hardness values of FCC, Fe2B and BCC/B2 phases in B0.7 sample.

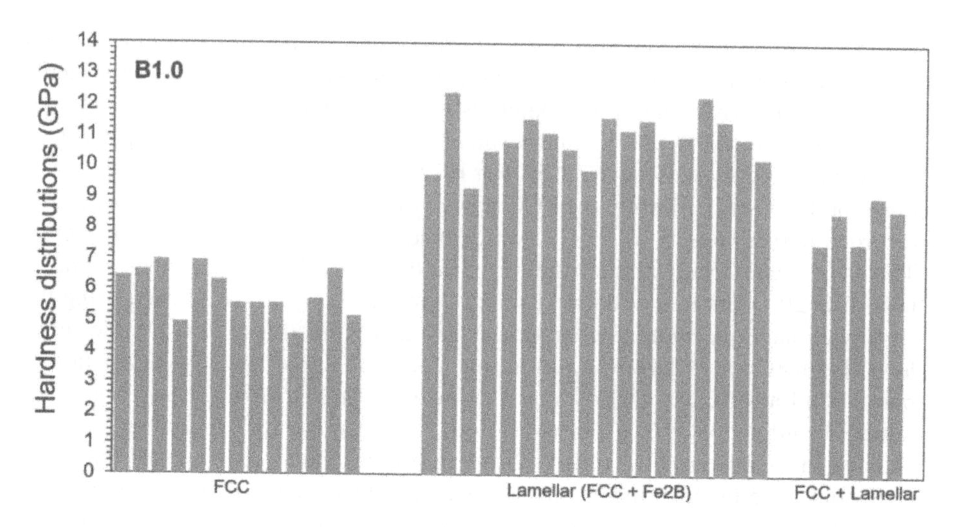

FIGURE 7.6 Distribution of hardness values of FCC, Fe2B, and BCC/B2 phases in B1.0 sample.

the SEM images in Figure 7.2. Besides, hardness values from the range of 7–10 GPa indicated the intermediate indents between FCC and lamellar phase. Table 7.1 shows the average hardness of three different phases in this alloy (FCC, lamellar (FCC+Fe2B), and FCC/BCC + lamellar). It was shown that the average FCC and lamellar phase for B0.7 and B1.0 was almost similar which is 5.78 and 5.70 and 12.14 and 11.89 GPa respectively. However, the significant differences of average hardness values can be seen for the intermediate phases (FCC/BCC + lamellar). This might due to the existence of hard and brittle BCC/B2 phases in the B0.7 alloy.

Although nano-indentation can be used to qualitatively distinguish (by using optical microscope) the hardness of different phases in this alloy, the detailed and quantitative results need to be done such as elastic modulus to determine the elastic moduli of this alloy, which are underway, and are not covered herein.

The mechanical properties in HEAs are known to rely on the mechanical and crystal structure of the alloy [16, 17]. With regards to the bulk hardness of the

TABLE 7.1
Average Hardness Value Determined by Nano-Indentation and Microhardness Testing of Three Different Phases FCC, Lamellar (FCC + Fe2B) and FCC/BCC + Lamellar in B07 and B1.0 Sample

Sample	FCC (GPa)	Lamellar (GPa)	FCC/BCC + Lamellar (GPa)	FCC (Hv)	Lamellar (Hv)	FCC/BCC + Lamellar (Hv)
B0.7	5.78	12.14	8.90	589	1238	907
B1.0	5.70	11.89	8.16	581	1212	832

composition, this can be understood to depend on the volume fraction of the phases represented and its micro-structure [18]. Typical bulk hardness values are FCC phase and its derivatives: 100–300 HV; BCC and its derivatives: 300–300; intermetallic phases: 650–1,300; covalent compounds (silicide, carbides, borides, etc.): 1,000–4,000 [18]. In comparison to micro-hardness values, nano-indentation values are larger than micro-hardness values due to a depth effect [19]. There are also differences in the surface of the material (oxide formation, structural changes due to polishing, etc.). This is especially true for HEAs which are metastable alloys which may lead to silicide precipitation from solid solution during the polishing process. The earlier enthalpy of mixing analysis suggests Si enrichment of the Fe, Co, and Ni elemental additions which may play a role in the strengthening of the FCC structure to 600 HV. The various silicide is approximately 5.5 Mohs hardness (equivalent to approx. 600 Vickers hardness) which brings that in line with the hardness measured here. The various covalent type bonds (Fe-B, Co-B, etc.) are typically between 1,000 and 2,000 Vickers hardness [20], which is in agreement with the results shown here. A reduction in the maximum hardness measured may be due to the metastable and high-entropy nature of the alloy, which means that its semi-ordered structure may include various other additions that reduce the covalency of the phase.

7.4 CONCLUSION

High entropy alloys with lamellar structure have been designed to obtain a balance in strength and ductility for $FeCoNi(B_x\ Al_{1-x})_{0.1}Si_{0.1}$ high entropy alloys where it can be tailored by modification of boron content. A nano-indentation test along with the indent morphologies has distinguished the hardness of each phase in the alloy. The mechanical properties of these alloys are closely related to their structures where the micro-hardness varies with the change in the phase present. Changes in BCC/B2 structure and the presence of lamellar structure would affect the hardness value in the alloys.

ACKNOWLEDGEMENTS

The authors thank the International Islamic University Malaysia (IIUM) and Ministry of Education grant no. (FRGS/1/2019/TK05/UIAM/03/3) for the support of this research. The authors also would like to acknowledge Prof. Iain Todd, who initially funded this project, and for his invaluable contributions in the form of discussions pertaining to this work.

REFERENCES

1. Cantor, B., et al., *Microstructural development in equiatomic multicomponent alloys.* Materials Science and Engineering: A, 2004. **375**: pp. 213–218.
2. Yeh, et al., *Nanostructured high-entropy alloys with multiple principal elements: Novel alloy design concepts and outcomes.* Advanced Engineering Materials, 2004. **6**(5): pp. 299–303.
3. Zhang, Y., et al., *Solid-solution phase formation rules for multi-component alloys.* Advanced Engineering Materials, 2008. **10**(6): pp. 534–538.

4. Gao, M.C. and D.E. Alman, *Searching for next single-phase high-entropy alloy compositions.* Entropy, 2013. **15**(10): pp. 4504–4519.
5. Zhang, Y., et al., *Microstructures and properties of high-entropy alloys.* Progress in Materials Science, 2014. **61**: pp. 1–93.
6. Borkar, T., et al., *A combinatorial assessment of Al x CrCuFeNi 2 (0< x< 1.5) complex concentrated alloys: Microstructure, microhardness, and magnetic properties.* Acta Materialia, 2016. **116**: pp. 63–76.
7. He, F., et al., Stability of lamellar structures in CoCrFeNiNb x eutectic high entropy alloys at elevated temperatures. Materials & Design, 2016. **104**: pp. 259–264.
8. Lu, Y., et al., *A promising new class of high-temperature alloys: eutectic high-entropy alloys.* Scientific Reports, 2014. **4**: pp. 1–5.
9. Ma, S. and Y. Zhang, *Effect of Nb addition on the microstructure and properties of AlCoCrFeNi high-entropy alloy.* Materials Science and Engineering: A, 2012. **532**: pp. 480–486.
10. Rogal, Ł., et al., *Microstructure and mechanical properties of the new Nb25Sc25Ti25Zr25 eutectic high entropy alloy.* Materials Science and Engineering: A, 2016. **651**: pp. 590–597.
11. Liu, W.-J., et al., *Effect of micro-alloying element boron on the strengthening of high-strength steel Q690D.* Metallography, Microstructure, and Analysis, 2015. **4**(2): pp. 102–108.
12. Chaki, T., *Mechanism of boron-induced strengthening of grain boundaries in Ni3Al.* Philosophical Magazine Letters, 1991. **63**(3): pp. 123–126.
13. Rodríguez-Castro, G., et al., *Mechanical properties of FeB and Fe2B layers estimated by Berkovich nanoindentation on tool borided steel.* Surface and Coatings Technology, 2013. **215**: pp. 291–299.
14. Campos-Silva, I., et al., *A study of indentation for mechanical characterization of the Fe2B layer.* Surface and Coatings Technology, 2013. **232**: pp. 173–181.
15. Almasri, A.H. and G.Z. Voyiadjis, *Nano-indentation in FCC metals: Experimental study.* Acta Mechanica, 2009. **209**(1): p. 1.
16. Guo, S., C. Ng, J. Lu and C.T. Liu, *Effect of valence electron concentration on stability of FCC or BCC phase in high entropy alloys*, Journal of Applied Physics, 2011. **109**(10): pp. 1–5.
17. Leong, Z.Y., *A new semi-empirical method based on a distorted tetragonal scheme for the structure prediction and alloy design of multiple-component alloys*, Thesis, University of Sheffield, 2017. http://etheses.whiterose.ac.uk/id/eprint/17022 (accessed February 1, 2018).
18. Tsai, M.-H. and J.W. Yeh, *High-entropy alloys: A critical review*, Materials Research Letters, 2014. **2**: pp. 107–123.
19. Qian, L., M. Li, Z. Zhou, H. Yang, X. Shi, *Comparison of nano-indentation hardness to microhardness*, Surface and Coatings Technology, 2005. **195**: pp. 264–271.
20. Lentz, J., A. Röttger, W. Theisen, *Hardness and modulus of Fe2B, Fe3(C,B), and Fe23(C,B)6 borides and carboborides in the Fe-C-B system*, Materials Characterization, 2018. **135**: pp. 192–202.

8 Improving Tribological Performance of Meso Scale Air Journal Bearing Using Surface Texturing

An Approach of Green Tribology

Nilesh D. Hingawe and Skylab P. Bhore
Rotor Dynamics and Vibration Diagnostics Lab, Motilal Nehru National Institute of Technology Allahabad
Prayagraj, India

CONTENTS

8.1 INTRODUCTION

Technological advancement in miniaturized rotary systems urges more research on the tribological study of the meso scale journal bearing. The meso scale journal bearing is an important component of miniature rotary systems which should accomplish superior tribological performance in terms of higher load capacity and lower frictional, wear, and energy losses. To achieve this, lubricant between the journal and bearing plays a vital role. The use of oil as a lubricant in the bearing, however, fails to achieve these tribological requirements. This is because for higher power generation the bearing needs to be operating at a higher rotational speed which causes a

DOI: 10.1201/9781003139386-8

higher lubricant shear rate. The increase in shear rate increases the lubricant (oil) temperature, and thus a significant reduction in viscosity takes place. This reduces the load-carrying capacity (LCC) and increases the friction coefficient (COF) of the bearing. Also, lubricant consumption, leakage, and pollution due to lubricant are serious concerns. In contrast, use of air as a lubricant eliminates these issues. An air bearing offers simple design, higher rotational speed, low friction, lesser noise, and maintenance-free operation (no leakage issue) [1]. Also, the lubricant (air) is freely available and environmentally friendly which fulfills the demand of green tribology. That is why air-lubricated meso scale bearings are widely used in miniature rotary systems like hard disk drives, laser scanners, micro motors, dentist drills, miniature precision machines, meso scale gas turbines, etc. [2]. Despite these benefits, the low viscosity of air generates weak hydrodynamic pressure in the bearing [3, 4]. This affects the LCC. Although upon increasing speed the LCC can be increased, shear loss and instability issues dominate [5]. It is also noted that the appearance of friction and wear in frequent start-stop applications reduces the bearing life [6]. To address these issues the feasible and possible approach is to use different gas lubricants, and/or modify the mating face(s). Although few attempts have been made by researchers on improving hydrodynamic performance (and hence LCC) by employing different gas lubricants viz. H_2, CH_4, O_2, and H_2O [7], and modifying the smooth surface into a random rough surface for friction reduction [8], the improvement was observed to be trivial. Nonetheless, surface texturing is a well-known, widely accepted, and proven most suitable technique to address these tribological issues (LCC and COF) of different machine components. Also, in comparison to other techniques, the improvement in tribological performance using surface texturing is an environmentally friendly approach that contributes to the attainment of the objectives of green tribology [9].

Surface texturing is a promising technique which significantly improved the tribological performance of machine components such as oil-lubricated bearings, mechanical and gas sealings, piston rings, cutting tools, bio-medical implants, etc. [10, 11]. However, due to a large number of texture parameters, it is essential to select the appropriate design [12]. Otherwise, it may negatively influence the results [13]. For macro-scale oil-lubricated journal bearings the texture parameters: shape, dimensions, and zones (position) are the most important parameters. The commonly used texture shapes are square, cylindrical, and spherical; but their size is not consistent over the cited results [14]. Their size usually varies in the range of 25–100 μm. Other than this, the research is focused on full and partial texturing. For both short and long bearings, Brizmer and Kligerman [15] found that partial texturing is preferable over full texturing. Kango et al. [16] examined the influence of textured and grooved patterns. For both patterns, the convergent zone of the bearing was found effective. Also, Meng et al. [17] analyzed the hydrodynamic characteristics of compound-shaped textures in the active pressure region 90°–180°. Further, Shinde and Pawar [18] numerically investigated the optimum groove position for LCC and frictional torque. The optimum grooves positioned in the 90°–175° zone of the bearing showed the best performance. With reference to these studies, Zhang et al. [19] optimized the texture arrangement in the convergent zone of the bearing. The optimum semi-elliptical arrangement obtained higher LCC and lesser COF than both half-textured and smooth bearing. In contrast to these results, Tala-Ighil et al. [20, 21] found improvement in when textures were employed at the outlet of the active pressure zone of the bearing.

In the case of meso scale air journal bearing, as far as authors know, except the research work carried out by Hingawe and Bhore [22–24] no study focuses on employing textures in the meso scale air journal bearing. They recently examined that among the common texture shapes (square, diamond, cylindrical, spheroid, and ellipsoid), a square-shaped profile outperformed for both the maximization of LCC and minimization of COF. They also observed that partial texturing in the convergent zone gives results superior to the divergence zone of the bearing. However, they do not emphasize on the effect of texture size and its appropriate position in the bearing operating at variable bearing parameters. Other studies on meso scale air bearings are limited to grooved patterns [5]. To maximize the LCC, Zhang et al. [25] numerically observed that the geometrical characteristics of herringbone groove play a crucial role in both journal and thrust bearings. Also, Guenat and Schiffmann [26] performed an optimization study to maximize the machining tolerance of grooved parameters. Similarly, for herringbone grooved meso scale oil-lubricated bearings, Chen et al. [27] optimized the geometrical characteristics using the Taguchi L_{27} approach. The output responses viz. LCC and fluid leakage showed significant improvement.

The above research review reveals that surface texturing is a well-established and promising technique of improving the tribological performance of oil- and gas-lubricated machine components. But the selection of appropriate texture design parameters such as shape, size, and position is very essential. Furthermore, it is observed that the meso scale air bearing has numerous applications in miniature systems, and it is essential to enhance its tribological performance. Despite this, limited research on surface-textured meso scale air journal bearings is carried out so far. With this research gap, in the present numerical study, a detailed tribological investigation on the effect of texture size and its appropriate position in meso scale air bearings is carried out. A square-shaped texture whose size varies from 25 μm to 100 μm in the convergent zone (0°–90°, 90°–180°, and 0°–180°) of the bearing, is considered. This analysis is done at variable bearing parameters viz. eccentricity ratio, rotational speed, and fluid film clearance.

8.2 NUMERICAL MODEL

8.2.1 Compressible Reynolds Equation

The present study is based on numerical simulation. Detailed numerical investigation is carried out in a thin-film model of the COMSOL 5.2 package. In a thin-film model pressure change along the film thickness is neglected. Flow is considered laminar and the lubricant (air) is assumed to be Newtonian. Considering these assumptions, the steady-state compressible Reynolds equation is given by following equation [23]:

$$\frac{1}{r^2}\frac{\partial}{\partial\theta}\left(ph^3\frac{\partial p}{\partial\theta}\right)+\frac{\partial}{\partial z}\left(ph^3\frac{\partial p}{\partial z}\right)=6\mu\,\omega\frac{\partial(ph)}{\partial\theta} \tag{8.1}$$

In the above equation, film thickness (h) relation for smooth and textured surfaces as shown in Figure 8.1 is given below:

$$\mathrm{h}(\theta,\mathrm{z})_{\mathrm{Smooth}}=\mathrm{c}(1+\varepsilon\cos\theta) \tag{8.2}$$

$$\mathrm{h}(\theta,\mathrm{z})_{\mathrm{Texture}}=\mathrm{h}(\theta,\mathrm{z})_{\mathrm{Smooth}}+\Delta\mathrm{h}(\theta,\mathrm{z}) \tag{8.3}$$

where, $\Delta h(\theta, z)$ is the texture depth.

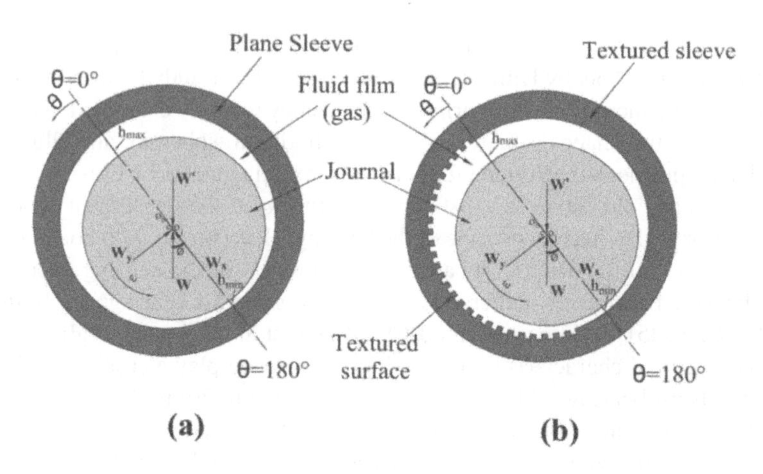

FIGURE 8.1 (a) Smooth air journal bearing; (b) Textured air journal bearing.

At both ends of the bearing the atmospheric pressure is considered and the no-slip condition is assumed at the surface. With these boundary conditions, compressible Reynolds equation (8.1) is discretized by the finite element method. To get an accurate result the convergence criteria is set to be

$$\sum\sum\left|\frac{\left(p_{i,j}\right)_{n+1}-\left(p_{i,j}\right)_{n}}{\left(p_{i,j}\right)_{n+1}}\right|\leq 1\times10^{-5} \tag{8.4}$$

Further, load carrying capacity (LCC) and frictional force (F) are evaluated using the following equations [23]:

$$\text{LCC}=\sqrt{\left(\int_0^L\int_0^{2\pi} p\,r\cos\theta d\theta dz\right)^2+\left(\int_0^L\int_0^{2\pi} p\,r\sin\theta d\theta dz\right)^2} \tag{8.5}$$

$$\text{F}=\int_0^L\int_0^{2\pi}\left(\mu\frac{U}{h}+\frac{h}{2}\frac{\partial p}{\partial x}\right)rd\theta dz \tag{8.6}$$

Based on this, the friction coefficient (COF) is calculated as

$$\text{COF}=\frac{\text{F}}{\text{LCC}} \tag{8.7}$$

The above procedure created within COMSOL 5.2 software is described using a flow chart in Figure 8.2.

8.2.2 Mesh Independence and Numerical Validation

In the developed model, a physics-controlled meshing structure is adopted. It offers extra-fine mesh in the textured region and coarse mesh in the non-textured region.

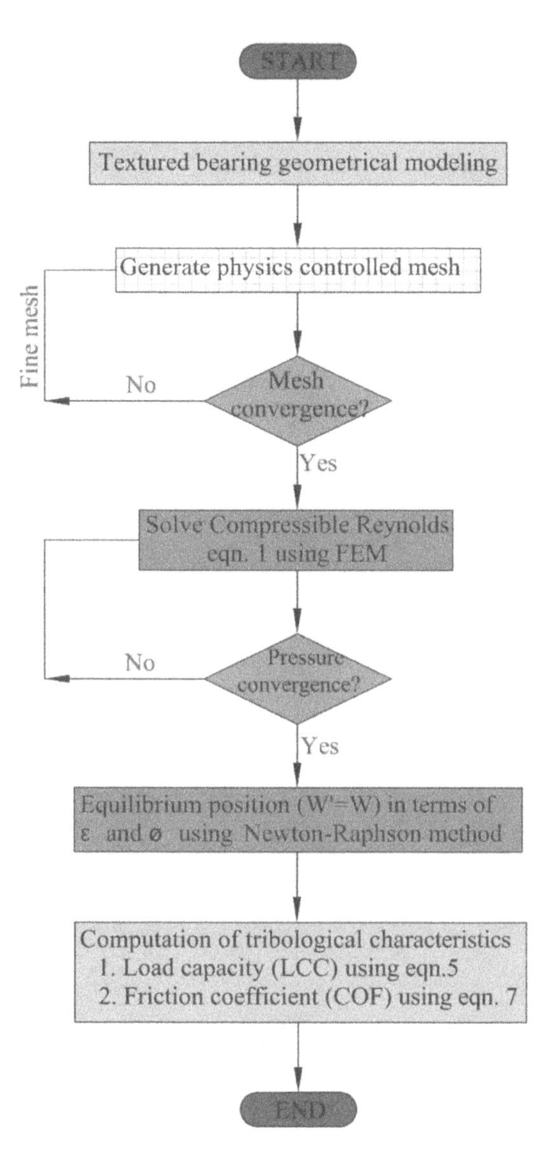

FIGURE 8.2 Solution scheme using COMSOL 5.2 software.

To get an accurate result with the least computation time, a mesh convergence test was carried out, as shown in Table 8.1. Based on this, a normal mesh (combination of extra-fine mesh in the textured region and coarse mesh in the smooth region) having 378,043 elements is adopted for the present analysis.

For the developed numerical model, hydrodynamic pressure distribution obtained for the smooth air journal bearing is validated with the result obtained by Orr [28], as shown in Figure 8.3. Good agreement between the results was obtained.

TABLE 8.1
Mesh Independence Test

Load Capacity	No. of Elements	Type of Mesh
1.3007 N	58,350	Extremely coarse
1.3013 N	138,329	Extra coarse
1.3017 N	219,453	Coarser
1.3019 N	276,333	Coarse
1.3020 N	**378,043**	**Normal**
1.3020 N	458,799	Fine
1.3020 N	575,688	Finer
1.3020 N	879,274	Extra fine

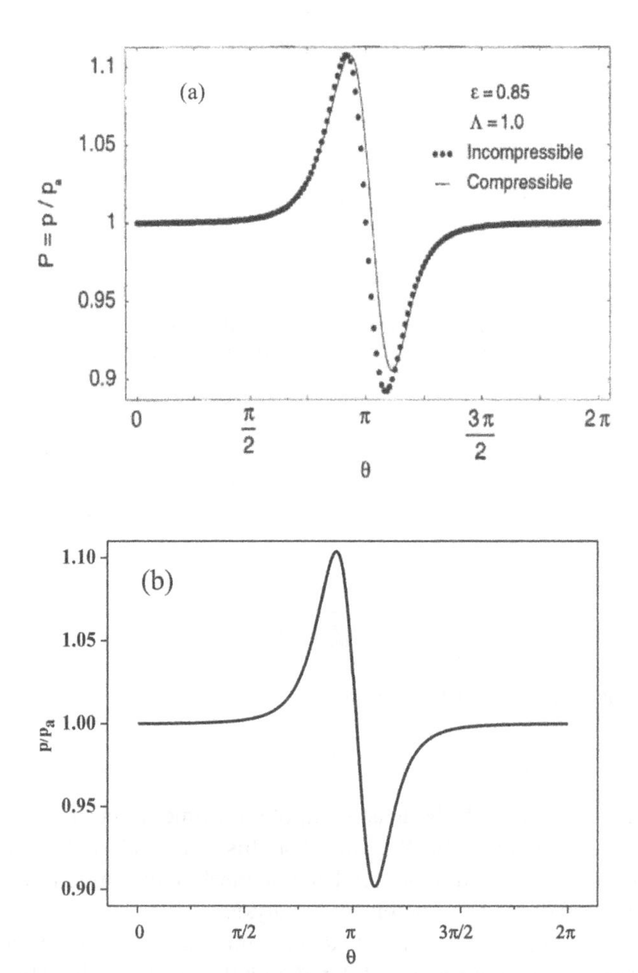

FIGURE 8.3 Pressure distribution for meso scale air bearing: (a) Orr [28], (b) present model.

TABLE 8.2
Specification of Meso Scale Air Journal Bearing System

Parameters	Baseline Values	Variable Range
Bearing diameter, d (mm)	2	-
Bearing l/d ratio	1	-
Clearance, c (μm)	4	2–7
Eccentricity ratio, ε	0.6	0.3–0.8
Rotational speed (rpm)	8000	5,000–50,000
Air density, **r** (kg/m^3)	1.225	-

8.2.3 Texture Design

Bearing geometrical characteristics are set to be fixed parameters, and the operating parameters such as eccentricity ratio, rotational speed, and fluid film clearance are varied in a given range (see Table 8.2) [2, 27].

Based on the earlier research carried out on textured meso scale air journal bearings by Hingawe and Bhore [23], a square-shaped texture with area ratio of 30% and depth of 2 μm is considered for the present analysis. For a given area ratio the texture size is varied in the range of 25–100 μm as Ts1 (25 × 25), Ts2 (50 × 50), Ts3 (75 × 75), and Ts4 (100 × 100) as given in Table 8.3. The change in texture size also changes the number of textures, spacing between textures along the length, and circumferential direction of bearing which is presented in Figure 8.4. Effect of these parameters is examined in partially textured zones such as 0°–90°, 90°–180°, and 0°–180° which was proposed in Section 8.1 (see Figure 8.5).

8.3 RESULTS AND DISCUSSION

Effect of texture size and its appropriate position in the convergent zone of the meso scale air journal bearing is analyzed for the tribological characteristics viz. LCC and COF. The study is carried out at variable bearing parameters such as eccentricity ratio, rotational speed, and fluid film clearance as follows.

TABLE 8.3
Texture Specifications in Bearing

Texture Specifications	Texture Ts1	Texture Ts2	Texture Ts3	Texture Ts4
Texture dimensions (μm)	25 × 25	50 × 50	75 × 75	100 × 100
Texture spacing along θ_L°	2.5°	5°	7.5°	10°
Texture spacing along z_L (μm)	50	100	150	200
Textures no. along θ_L° x z_L				
(a) For 0°–90° and 90°–180°	37 × 40	19 × 20	13 × 13	10 × 10
(b) For 0°–180°	73 × 40	37 × 20	25 × 13	19 × 10

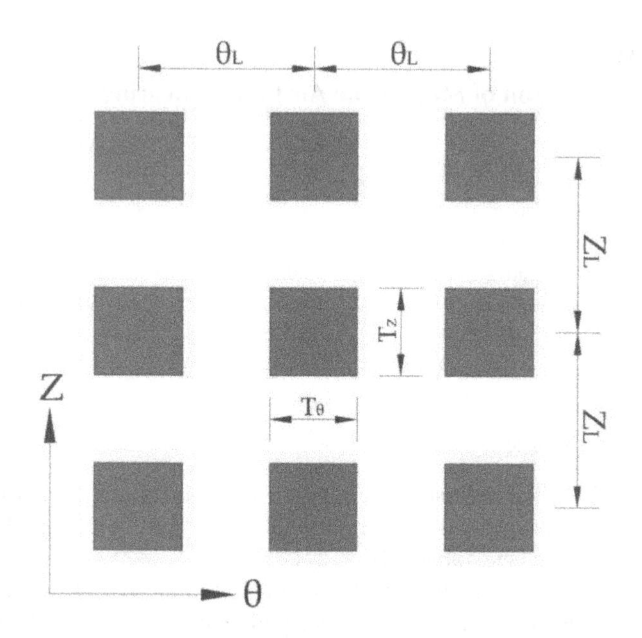

FIGURE 8.4 Texture geometrical arrangement.

8.3.1 Eccentricity Ratio

At variable eccentricity ratio, the LCC of smooth and differently sized textures (Ts1, Ts2, Ts3, and Ts4) positioned in 0°–90°, 90°–180°, and 0°–180° of the bearings is shown in parts (a), (b), and (c), respectively, of Figure 8.6. Compared to a smooth bearing the noticeable improvement in LCC is obtained by the textured bearings regardless of eccentricity ratio. This is due to the fact that textures on the bearing surface act as a lubricant (air) reservoir which contributes to the generation of an additional hydrodynamic effect. It causes further separation of mating faces that leads to an enhancement in pressure and thus LCC of the bearing [10]. It is revealed that among differently sized textures, Ts1 outperformed, followed by Ts2. However,

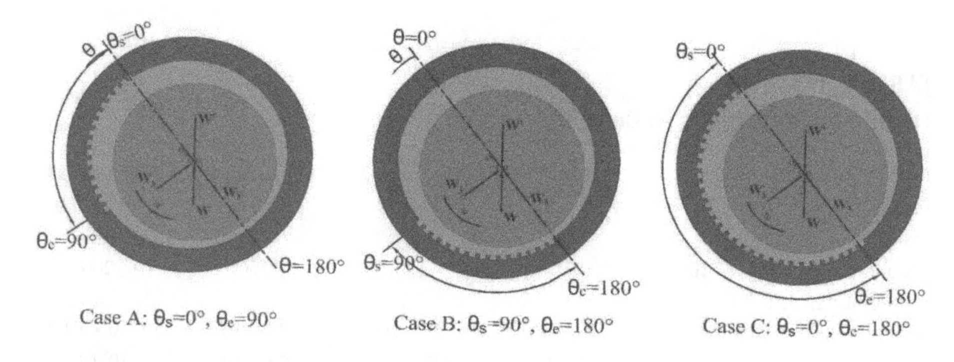

FIGURE 8.5 Partially textured bearings.

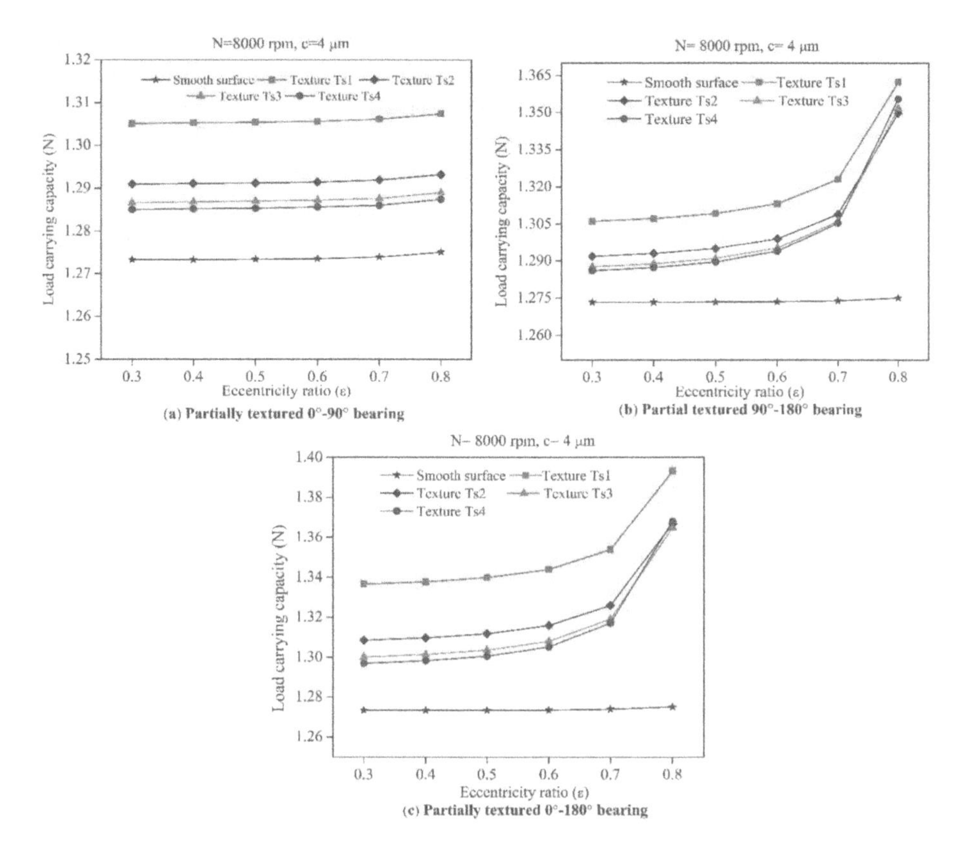

FIGURE 8.6 Load carrying capacity of partially textured (a) 0°–90° (b) 90°–180°, and (c) 0°–180° bearings at variable eccentricity ratio.

a further increase in texture size (Ts3, and Ts4) is found to be insignificant. Each texture has the same area ratio, and therefore the volume of lubricant accumulation in each textured bearing is the same. But texture Ts1 (smaller size textures with large numbers) gained superior compressibility effect. Also, the distance between textures both along the circumference and length is the minimum, which offers better guidance of lubricant. However, increase in texture size (Ts2, Ts3, and Ts4), i.e. increase in texture area, reduces the compressibility effect. In addition, with the increase in distance between textures both along the circumference and length, the fluid flow retards. These factors affect the formation of additional hydrodynamic pressure and hence LCC of the bearing [29, 30]. Therefore, texture Ts1 has better capability of hydrodynamic improvement than others, while it is important to note that this improvement is least effective when textures are positioned at the inlet of convergent zone 0°–90° of the bearing. But employing the same number of textures in 90°–180° gives better result. Whereas, texturing in the entire convergent zone (0°–180°) is most beneficial. For Ts1 (specifically at ϵ=0.8), compared to smooth bearing the partially textured 0°–90°, 90°–180°, and 0°–180° bearings obtained 2.53%, 6.84%, and 9.26% improvement, respectively. Based on this, it is stated that the amount of lift generated by partially textured convergent zone 0°–180° is considerably larger. Since the compressibility effect is less significant at lower eccentricity ratios, LCC is reasonably lesser. In contrast, a sharp increase in LCC is obtained at higher eccentricity ratios because of the greater compressibility effect [24]. Therefore, texturing is more beneficial in the meso scale air bearing operating at higher eccentricity ratios than at lower eccentricity ratios.

Friction coefficient obtained by smooth and differently sized textures positioned in convergent zones of the bearing at variable eccentricity ratio is shown in Figure 8.7. The significant reduction in COF is observed by the textured bearings rather than the smooth bearing. Employing textures on the bearing surface further increases the fluid film thickness between journal and bearing operating under hydrodynamic lubrication. Development of an additional pressure due to increase in film thickness causes greater separation of faces. As a result an increase in LCC, a decrease in local shear stress, and thus a reduction in COF is obtained [12]. For all the texture sizes, constant depth is considered. For a given eccentricity ratio, therefore, texture size has a minor effect on decreasing the shear rate of fluid film. This depicts that bearing load primarily helps in friction reduction. Considering this, it is found that the texture Ts1 is outperformed. However, upon increase in texture size

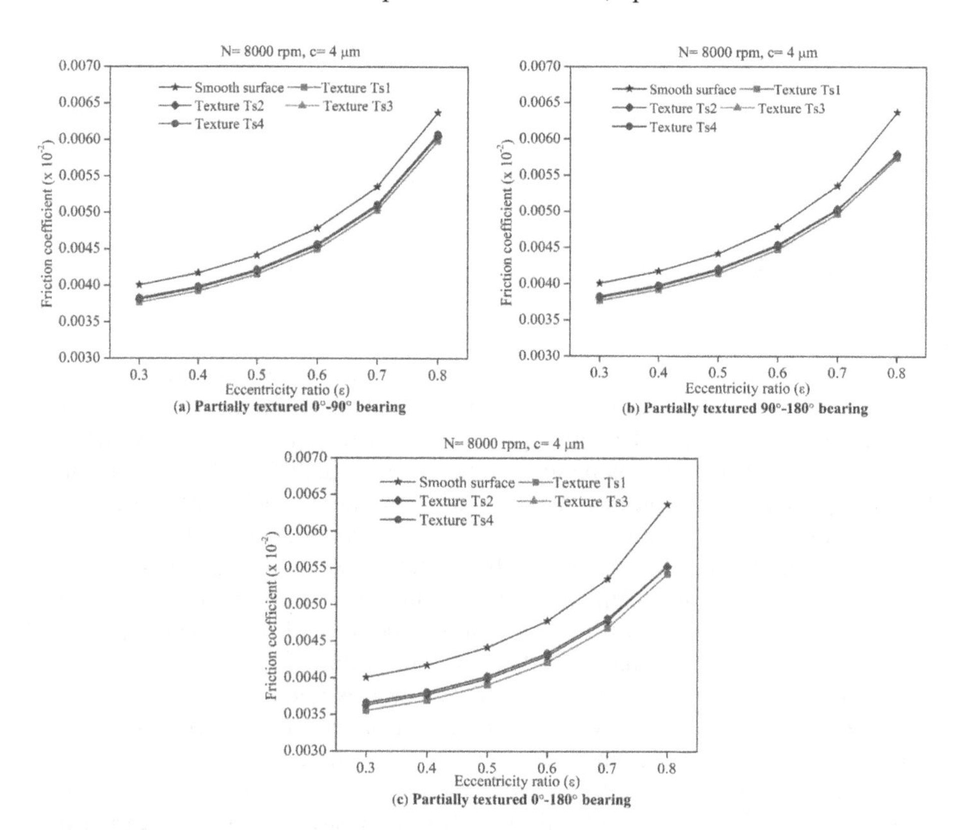

FIGURE 8.7 Friction coefficient of partially textured (a) 0°–90° (b) 90°–180°, and (c) 0°–180° bearings at variable eccentricity ratio.

viz. Ts2, Ts3, and Ts4 a minor change in the result is observed. The deviation in friction reduction between smooth and textured bearings in convergent zone 0°–90° is almost the same over the given range of eccentricity ratios. But for other cases, the deviation in improvement is more at higher eccentricity ratios (0.7–0.8) due to sharp improvement in LCC (see Figure 8.6). For texture Ts1, the friction reduction at an eccentricity ratio of 0.3 is 5.99%, 5.99%, and 11.47% for 0°–90°, 90°–180°, and 0°–180° textured bearing, respectively. But, for an eccentricity ratio of 0.8, it is, respectively, 6.12%, 9.89%, and 14.91%. This implies that small-sized textures positioned in the entire convergent zone give better friction reduction in a higher eccentricity ratio than at lower.

8.3.2 Rotational Speed

LCC obtained by the smooth and differently sized textures positioned in different convergent zones of bearing operating in the range of 5,000–50,000 rpm is shown in Figure 8.8. In comparison with a smooth bearing all the textured bearings (Ts1, Ts2, Ts3, and Ts4) improved the LCC. But it strongly depends upon the position in

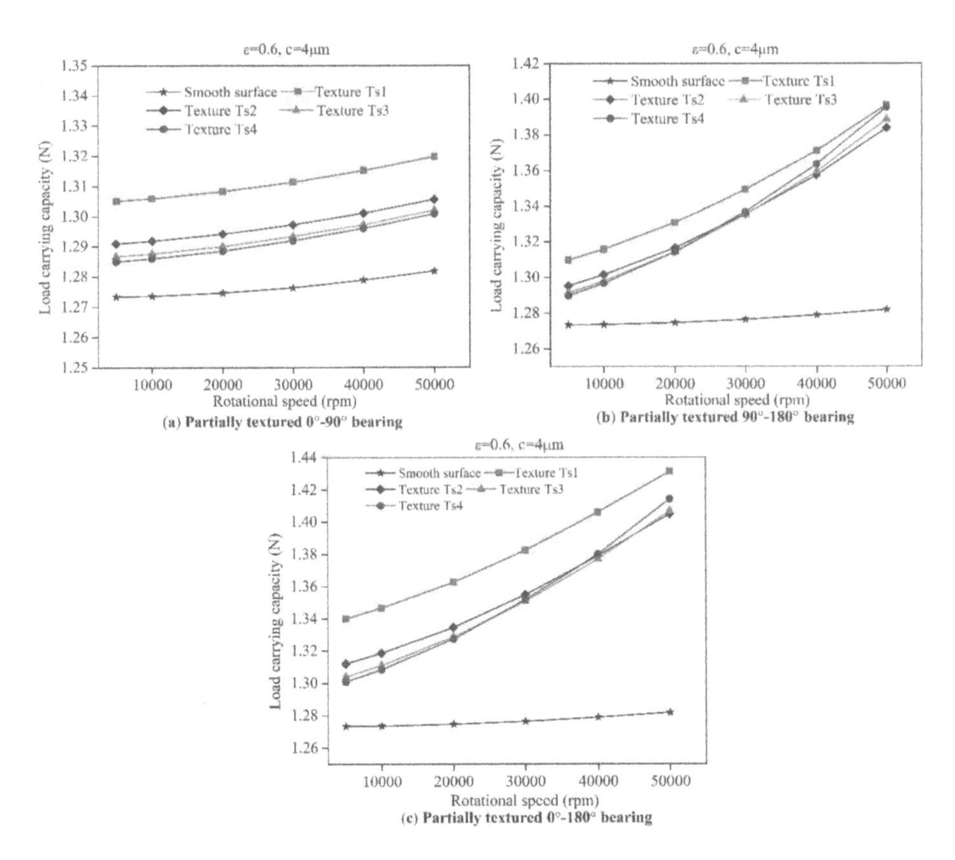

FIGURE 8.8 Load carrying capacity of partially textured (a) 0°–90°, (b) 90°–180°, and (c) 0°–180° bearings at variable rotational speed.

the bearing. For 0°–90°, texture Ts1 performed best, followed by Ts2. In contrast, a further increase in size (Ts3 and Ts4) showed a negligible change in a given operating range. But, for 90°–180° it is found that Ts2, Ts3, and Ts4 have obtained almost the same LCC up to 30,000 rpm. Upon further increasing rotational speed, interestingly it is observed that texture Ts4 gives better improvement than Ts2 and Ts3, and at the highest rotational speed (50,000 rpm) the load capacity obtained by Ts1 and Ts4 is almost same. This depicts that surface texture significantly increases the LCC but its size is almost insignificant at a higher rotational speed. Also for a partially textured 0°–180° bearing, Ts1 is most effective, and increasing texture size (Ts2, Ts3, and Ts4) is insignificant.

The development in hydrodynamic pressure by each partially textured bearing for texture Ts1 at 50,000 rpm is shown in Figure 8.9. It is revealed that texturing in the minimum film thickness region (90°–180°) develops better hydrodynamic pressure than at the inlet of bearing (0°–90°). But, combining these two, i.e. texturing in the entire convergent zone (0°–180°), offers superior results. The obtained results also supported by the earlier results obtained in oil and air lubricated bearings discussed in Section 8.1.

Increase in rotational speed obviously increases the shear rate and hence COF of both smooth and textured bearings [31]. But the rate of increase is lesser for textured bearings (see Figure 8.10). This is because textures increase the film thickness which further helps to improve bearing load and decreases the shear stress (and hence frictional force) which ultimately reduces the COF [32]. Although texture size is influence the friction reduction, it is observed that textures position play a vital role. Texturing in 0°–90° zone gives minor improvement up to 30,000 rpm. But for 90°–180° and 0°–180° it is up to 20,000 rpm, and 15,000 rpm, respectively. Upon further increasing speed, each partially textured bearing significantly improved the results. Specifically, texturing in 0°–180° is most effective, and the maximum improvement was achieved at 50,000 rpm. For texture Ts1, compared to the smooth bearing, the COF reduced by 6.17%, 11.34%, and 16.57% for 0°–90°, 90°–180°, and 0°–180° partially textured bearings, respectively.

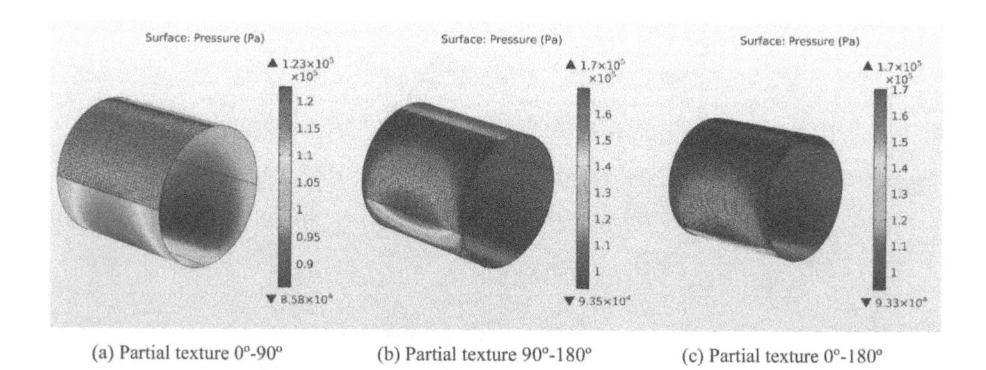

(a) Partial texture 0°-90° (b) Partial texture 90°-180° (c) Partial texture 0°-180°

FIGURE 8.9 Hydrodynamic pressure of partially textured bearings.

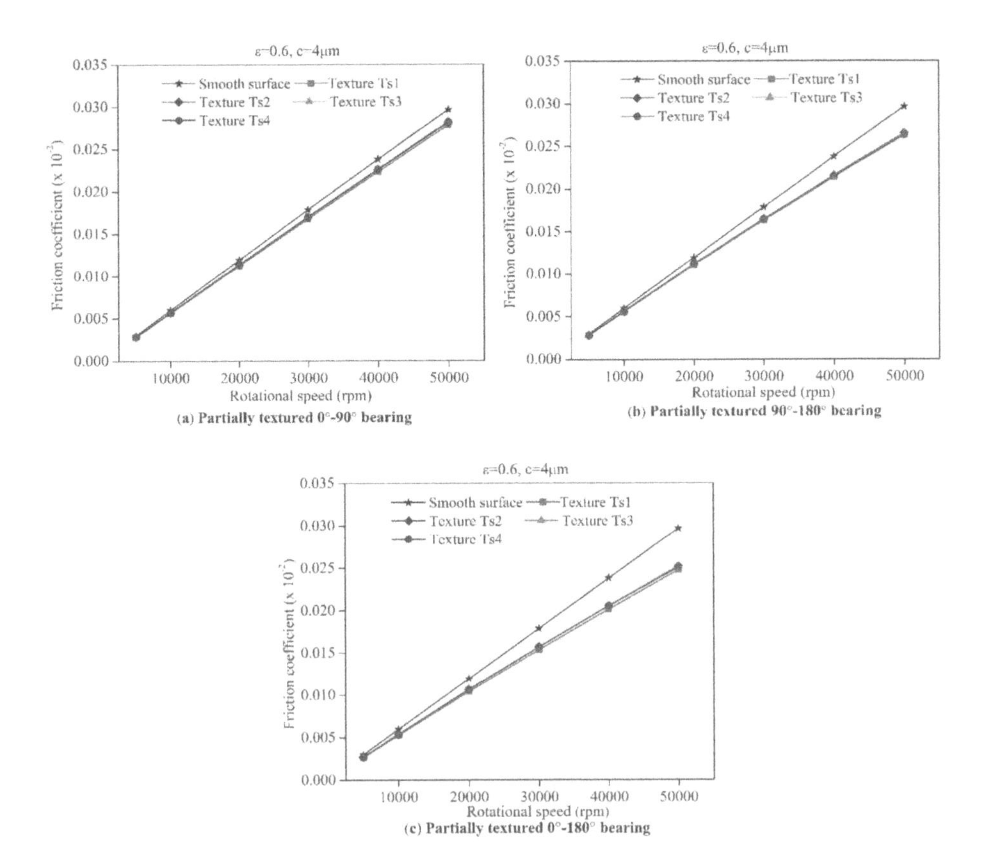

FIGURE 8.10 Friction coefficient of partially textured (a) 0°–90° (b) 90°–180°, and (c) 0°–180° bearings at variable rotational speed.

8.3.3 Fluid Film Clearance

Effect of texture size and its appropriate position in improvement in LCC of the bearing at variable fluid film clearance of 2–7 µm is shown in Figure 8.11. Similar to eccentricity ratio and rotational speed, for fluid film clearance the texture Ts1 outperformed, followed by Ts2, while with further increase in texture size (Ts3 and Ts4) the negligible change in LCC is observed for all the partially textured cases. Here also, textures positioned in the 0°–180° zone performed better than 0°–90° and 90°–180°. The observed phenomenon of texturing in an entire convergent zone (0°–180°) of the bearing creates similar virtual steps found in the film thickness discontinuities of Rayleigh step bearings [10, 33]. The improvement is significant at lesser fluid film clearance, but texture size is insensitive. In contrast, upon increasing clearance (4–7 µm) almost no change in the result is obtained but texture size plays a significant role.

Although surface texturing is beneficial for the reduction of COF in a given range of fluid film clearance, the texture size has minor significance. As shown

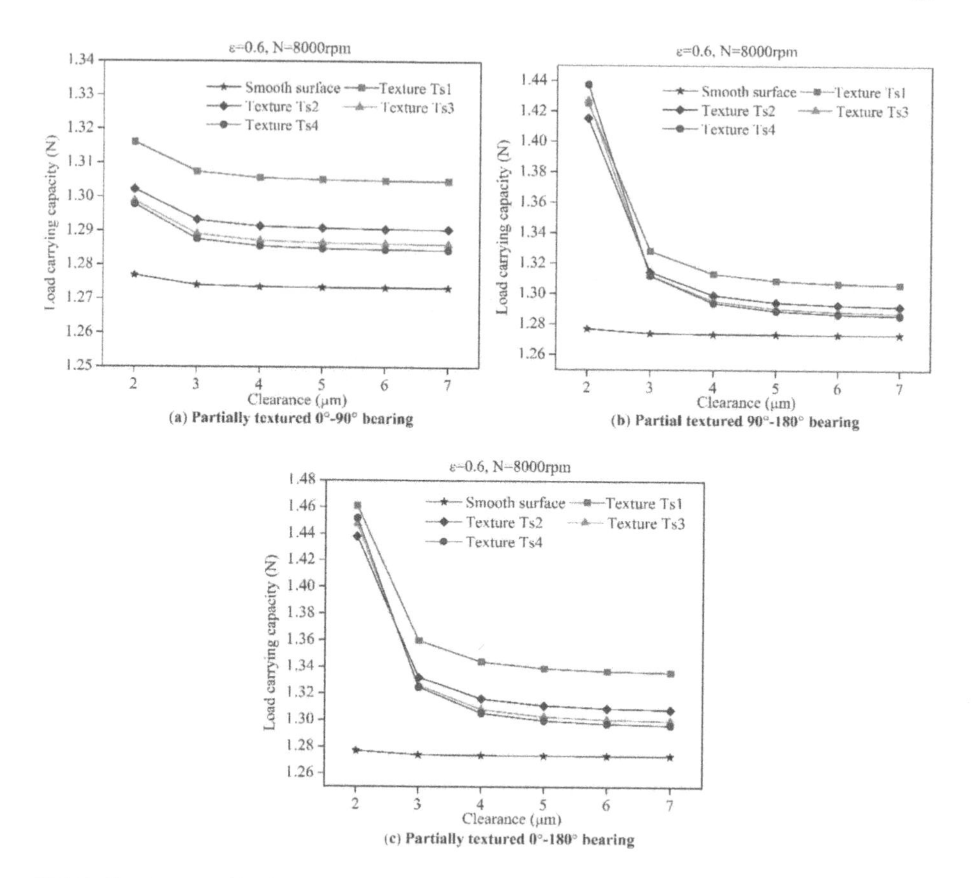

FIGURE 8.11 Load-carrying capacity of partially textured (a) 0°–90°, (b) 90°–180°, and (c) 0°–180° bearings at variable fluid film clearance.

in Figure 8.12, all sized textures have obtained almost the same friction reduction. However, texture position plays a vital role in friction reduction. For texture Ts1 at 2 μm clearance, in comparison with a smooth bearing the reduction in COF by partially textured 0°–90°, 90°–180°, and 0°–180° bearings is about 10.05%, 16.96%, and 23.98%, respectively. Partially textured 0°–180° has maximum extent in the convergent zone, and thus it has capability to develop significant hydrodynamic pressure. As a result, employing textures in an entire convergent zone of the bearing is advantageous from the viewpoint of both LCC and COF of the bearing.

8.4 CONCLUSION

In this study, the surface texturing approach of green tribology has been employed for improving the tribological performance of the meso scale air journal bearing. The potential use of important texture parameters: size (Ts1, Ts2, Ts3, and Ts4) and its position (0°–90°, 90°–180°, and 0°–180°), were investigated numerically to maximize the load-carrying capacity and minimize the friction coefficient. The analysis

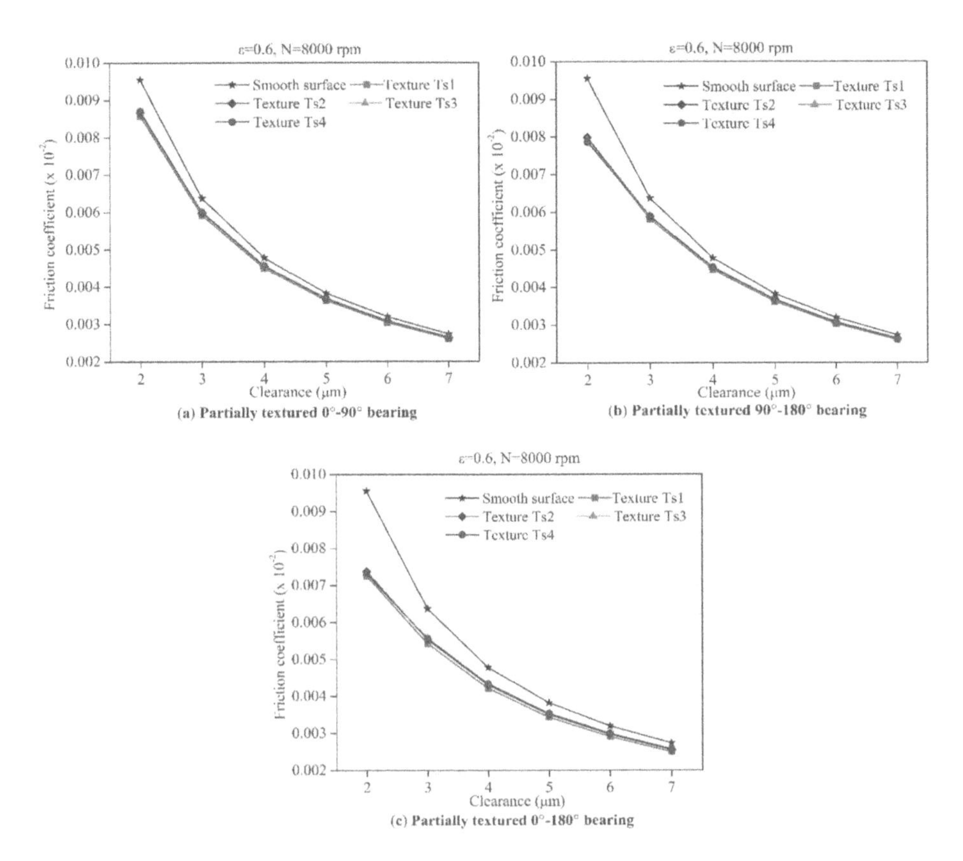

FIGURE 8.12 Friction coefficient of partially textured (a) 0°–90°, (b) 90°–180°, and (c) 0°–180° bearings at variable fluid film clearance.

was carried out at variable bearing parameters such as eccentricity ratio, rotational speed, and fluid film clearance. It is concluded that:

- Among differently sized textured bearings, textured Ts1 (smallest in size with largest in numbers) bearing has obtained the highest load-carrying capacity and lowest friction coefficient than the smooth bearing, irrespective of eccentricity ratio, rotational speed, and fluid film clearance.
- Texture Ts1 significantly improved the tribological performance, mainly due to gaining superior compressibility effect inside the texture, and also due to minimum distance provided between textures both along the circumference and length, which offered better guidance of the lubricant.
- The highest tribological improvement was observed with textures positioned in the entire convergent zone (0°–180°), followed by active pressure convergent zone 90°–180° of the bearing, whereas, texturing at the inlet of the convergent zone, i.e. 0°–90°, was least effective.
- The improved results obtained due to appropriate texture design can be significantly enhanced further when the bearing operates at higher eccentricity ratio, higher rotational speed, and at lower fluid film thickness.

REFERENCES

1. Bhore, S. P. and Darpe, A. K. 2014. Rotordynamics of micro and mesoscopic turbomachinery – a review. *Journal of Vibration Engineering & Technologies* 2(1):1–9.
2. Kim, D. and Bryant, M. D. 2004. Hydrodynamic performance of meso scale gas journal bearings. *ASME International Mechanical Engineering Congress and Exposition* 47136:1089–1098.
3. Hwang, T. and Ono, K. 2003. Analysis and design of hydrodynamic journal air bearings for high performance HDD spindle. *Microsystem Technologies* 9:386–394.
4. Murthy, A. N., Etsion, I., and Talke, F. E. 2007. Analysis of surface textured air bearing sliders with rarefaction effects. *Tribology Letters* 28:251–261.
5. Gu, L., Guenat, E., and Schiffmann, J. 2020. A review of grooved dynamic gas bearings. *Applied Mechanics Reviews* 72(1):010802.
6. Kim, H. J., Park, I. K., Seo, Y. H., Kim, B. H., and Hong, N. P. 2014. Wire tension method for coefficient of friction measurement of micro bearing. *International Journal of Precision Engineering and Manufacturing* 15(2):267–273.
7. Zhang, X., Chen, Q., and Liu, J. 2017. Steady characteristics of high-speed micro-gas journal bearings with different gaseous lubricants and extreme temperature difference. *Journal of Tribology* 139(2):021703.
8. Zhang, W. M., Meng, G., and Peng, Z. K. 2013. Gaseous slip flow in micro-bearings with random rough surface. *International Journal of Mechanical Sciences* 68:105–113.
9. Matele, S. and Pandey, K. N. 2018. Effect of surface texturing on the dynamic characteristics of hydrodynamic journal bearing comprising concepts of green tribology. *Proceedings of the Institution of Mechanical Engineers, Part J: Journal of Engineering Tribology* 232(11):1365–1376.
10. Gachot, C., Rosenkranz, A., Hsu, S. M., and Costa, H. L. 2017. A critical assessment of surface texturing for friction and wear improvement. *Wear* 372:21–41.
11. Rosenkranz, A., Grützmacher, P. G., Gachot, C., and Costa, H. L. 2019. Surface texturing in machine elements–a critical discussion for rolling and sliding contacts. *Advanced Engineering Materials* 21(8):1900194.
12. Gropper, D., Wang, L., and Harvey, T. J. 2016. Hydrodynamic lubrication of textured surfaces: a review of modeling techniques and key findings. *Tribology International* 94:509–529.
13. Hingawe, N. D. and Bhore, S. P. 2019. Tribological performance of surface textured automotive components: A review. *Automotive Tribology* 287–306. https://doi.org/10.1007/978-981-15-0434-1_15
14. Sudeep, U., Tandon, N., and Pandey, R. K. 2015. Performance of lubricated rolling/sliding concentrated contacts with surface textures: A review. *Journal of Tribology* 137(3):031501.
15. Brizmer, V. and Kligerman, Y. 2012. A laser surface textured journal bearing. *Journal of Tribology* 134(3):031702.
16. Kango, S., Sharma, R. K., and Pandey, R. K. 2014. Comparative analysis of textured and grooved hydrodynamic journal bearing. *Proceedings of the Institution of Mechanical Engineers, Part J: Journal of Engineering Tribology* 228(1):82–95.
17. Meng, F. M., Zhang, L., Liu, Y., and Li, T. T. 2015. Effect of compound dimple on tribological performances of journal bearing. *Tribology International* 91:99–110.
18. Shinde, A. B. and Pawar, P. M. 2017. Multi-objective optimization of surface textured journal bearing by Taguchi based Grey relational analysis. *Tribology International* 114:349–357.
19. Zhang, H., Hafezi, M., Dong, G., and Liu, Y. 2018. A design of coverage area for textured surface of sliding journal bearing based on genetic algorithm. *Journal of Tribology* 140(6):061702.

20. Tala-Ighil, N. and Fillon, M. 2015. Surface texturing effect comparative analysis in the hydrodynamic journal bearings. *Mechanics & Industry* 16(3):302.
21. Tala-Ighil, N. and Fillon, M. 2015. A numerical investigation of both thermal and texturing surface effects on the journal bearings static characteristics. *Tribology International* 90:228–239.
22. Hingawe, N. D. and Bhore, S. P. 2019. Effect of partial texture on the hydrodynamic performance of meso scale air bearings for mesoscale turbo-machines. *Proceedings of the ASME 2019 Gas Turbine India Conference* https://doi.org/10.1115/GTINDIA2019-2517.
23. Hingawe, N. D. and Bhore, S. P. 2019. Tribological performance of a surface textured meso scale air bearing. *Industrial Lubrication and Tribology* 72(5):599–609.
24. Hingawe, N. D. and Bhore, S. P. 2021. Investigation on Hydrodynamic Characteristics of Textured Meso Scale Gas Bearing, in *Mechanisms and Machine Science*. Springer, *Singapore*, 525–535.
25. Zhang, Q., Guo, G., and Bi, C. 2005. Air bearing spindle motor for hard disk drives. *Tribology Transactions* 48(4):468–473.
26. Guenat, E. and Schiffmann, J. 2019. Multi-objective optimization of grooved gas journal bearings for robustness in manufacturing tolerances. *Tribology Transactions* 62(6):1041–1050.
27. Chen, C. Y., Liu, C. S., Li, Y. C., and Mou, S. 2015. Geometry optimization for asymmetrical herringbone grooves of miniature hydrodynamic journal bearings by using Taguchi technique. *Proceedings of the Institution of Mechanical Engineers, Part J: Journal of Engineering Tribology* 229(2):196–206.
28. Orr, D. J. 2000. Macro-scale investigation of high speed gas bearings for MEMS devices (*Doctoral dissertation), Massachusetts Institute of Technology*, Cambridge, MA.
29. Qiu, M., Delic, A., and Raeymaekers, B. 2012. The effect of texture shape on the load-carrying capacity of gas-lubricated parallel slider bearings. *Tribology Letters* 48(3):315–327.
30. Grützmacher, P. G., Rosenkranz, A., Szurdak, A., Grüber, M., Gachot, C., Hirt, G., and Mücklich, F. 2019. Multi-scale surface patterning–an approach to control friction and lubricant migration in lubricated systems. *Industrial Lubrication and Tribology*. https://doi.org/10.1108/ILT-07-2018-0273.
31. Shinde, A. B. and Pawar, P. M. 2017. Effect of partial grooving on the performance of hydrodynamic journal bearing. *Industrial Lubrication and Tribology* 69(4):574–584.
32. Qiu, M., Minson, B. R., and Raeymaekers, B. 2013. The effect of texture shape on the friction coefficient and stiffness of gas-lubricated parallel slider bearings. *Tribology International* 67:278–288.
33. Fowell, M. T., Medina, S., Olver, A. V., Spikes, H. A., and Pegg, I. G. 2012. Parametric study of texturing in convergent bearings. *Tribology International* 52:7–16.

9 Textured Tool Surfaces for Improved Lubrication and Friction in Sheet Metal Forming

Mohd Hafis Sulaiman, Norhuda Hidayah Nordin, and N.A. Sukindar
International Islamic University Malaysia
Selangor, Malaysia

M.J.M. Ridzuan
International Islamic University Malaysia Perlis
Perlis, Malaysia

CONTENTS

9.1 INTRODUCTION

Galling is undesirable, and it occurs in situations where lubricant film breaks down; scratching of a mating material on the sliding surface occurs, Figure 9.1(a), and thereby subsequent scoring of the softer mating material due to pick-up on the harder sliding surface, Figure 9.1(b), occurs. Galling is generally a gradual process, but can also occur abruptly and spread rapidly as the pick-up and cold-welding of the softer material to the hard surface induce more galling. Tribology knowledge plays

DOI: 10.1201/9781003139386-9

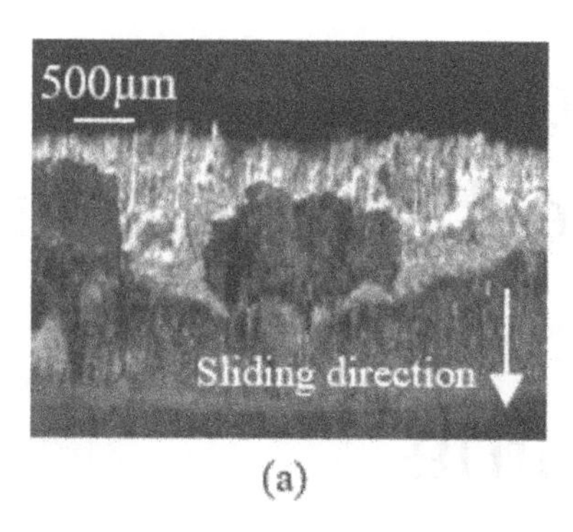

(a)

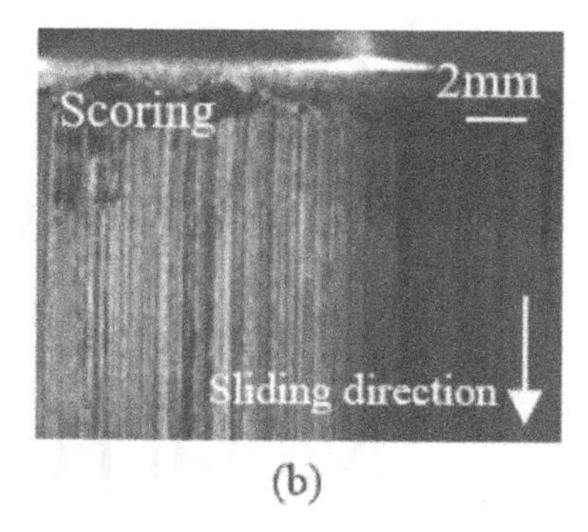

(b)

FIGURE 9.1 (a) Pick-up of workpiece material on the tool surface, and (b) scoring on the workpiece surface [3].

an important role to avoid metal surfaces from galling, eventually preventing material failure or loss of functionality. Thus, tribology knowledge has large economic relevance in metal shaping industries [1] and heavy-duty transportation [2]. As outlined in Figure 9.2, frictional losses in the engine, transmission, and other components comprise a major portion of heavy-duty vehicles' energy consumption profile. Only 34% of the total energy supplied by a fuel source is utilized to physically move the vehicle, while 66% of that energy is lost to the surroundings, either as thermal losses or through the aforementioned components. Consequently, the application of tribological knowledge in heavy-duty vehicle manufacturing can have profound fuel efficiency and economic benefits.

Lubricant is a material used to minimize the friction and wear between interacting surfaces. In addition to reducing friction and wear, lubricant also serves other functions such as removal of heat, corrosion prevention, transfer of power and providing a liquid seal at moving contacts. Over the last hundred years, petroleum-based lubricants have become established in most industrial and machinery applications. The complexity of lubricant formulations has been increased steadily over the last few decades, as the operating and performance requirements of the lubricants became more and more demanding. This complexity, however, poses a challenge as some components of lubricants completely degrade in a relatively short time, while others may not degrade at all. It is estimated that more than 50% of all lubricants used worldwide enter the environment due to spills, improper disposal, accidents, volatility, and total loss applications. Since over 95% of the lubricants that are lost in the environment are derived from petroleum, they contaminate the air, soil, and water and affect human and plant life to a great extent [4, 5]. The growing environmental concern surrounding the potential pollution and contamination of ecosystems due to the spillage or leakage of lubricants has necessitated the creation of a new class of renewable and biodegradable lubricants. Recent developments discovered that plant oil-based lubricants and derivatives are far superior to traditional mineral oils in terms of biodegradability [6–8].

Stringent environmental regulations regarding the use of conventional mineral oil-based lubricants have developed a wide interest to create and invent tailored

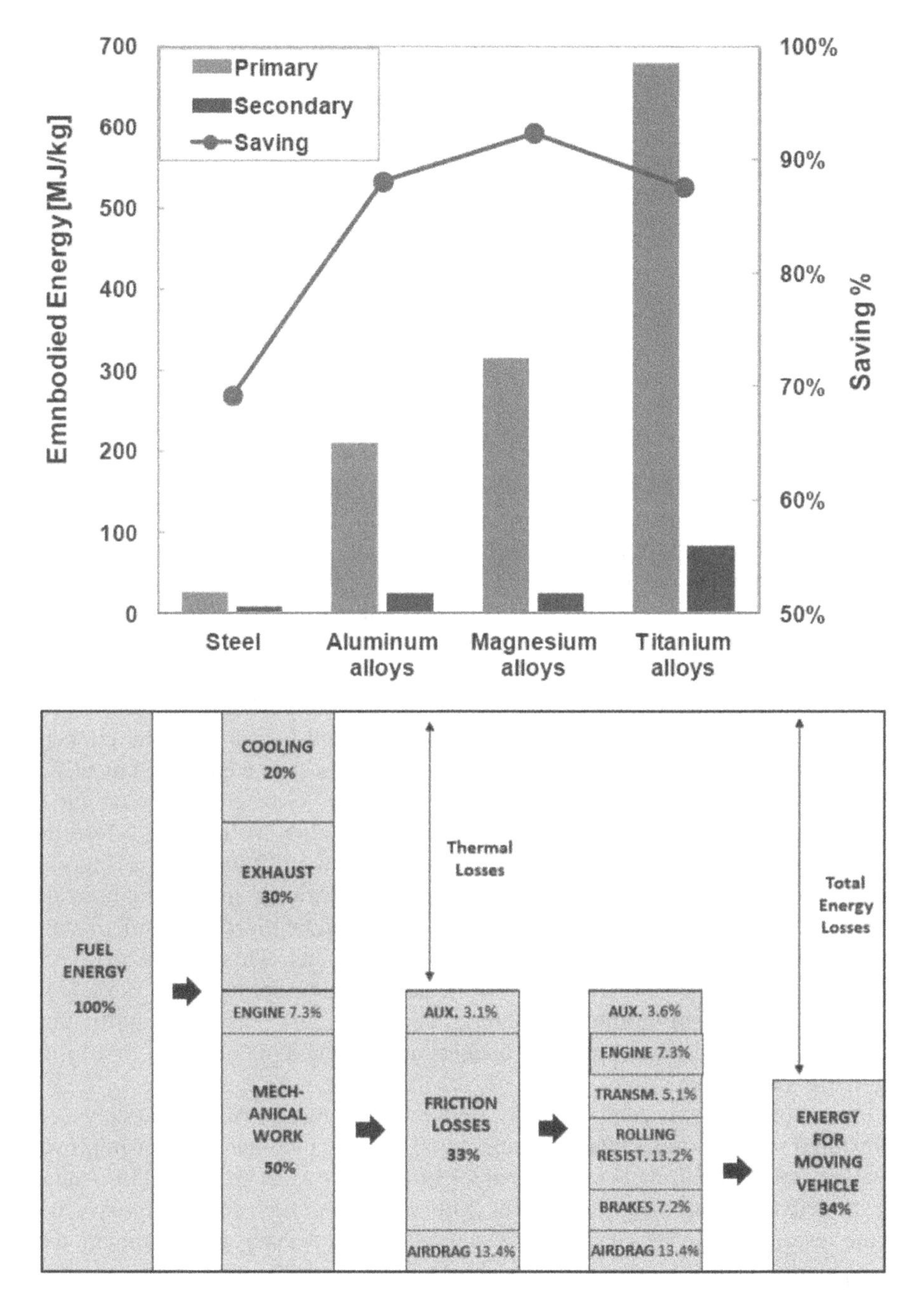

FIGURE 9.2 Energy consumption percentage in metal shaping industries [1] and heavy-duty vehicles [2].

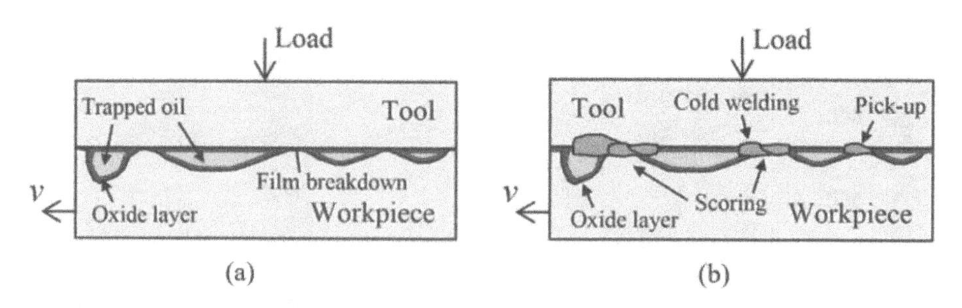

FIGURE 9.3 Schematic of (a) lubricant film breakdown and lubricant entrapment by pressurization, and (b) pick-up of soft workpiece material on the hard tool material surface [3].

surface topographies as one sustainable solution. Surface texturing has proved to be one of the promising techniques which helped in the improvement of tribological properties of tailored surface topographies. It has been applied in many engineering applications like engine [9], bearings [10], biomedical [3, 11], metal cutting [12], and metal forming [3]. Texturing the surface topography promotes lubricant entrainment, entrapment and pressurization in lubricant pockets, and possible subsequent lubricant escape to the neighboring sliding surfaces due to contact pressure difference, Figure 9.3 [13–15]. A small pocket angle towards the softer sliding surface facilitates escape of the trapped lubricant in the pockets, which reduces the friction and wear [16]. These effects can be enhanced by utilizing transverse roughness profiles and oblong pockets oriented perpendicular to the sliding direction [17, 18]. The lowest friction is found in sheet metal-forming tribological tests when the ratio between pocket area and total area is approximately 20% [19] and it is learned that increasing drawing speed enhances the effect [20]. Texturing of hard material surfaces would be more economical since a textured hard surface can be utilized in the long run. This is backed by the industrial experience of texturing the soft material surface, in which the texturing technique applied on the soft material surface is less feasible in multi-stage metal-forming operations, since the pockets are flattened out after the first forming operation [21].

The chapter presents comprehensive studies regarding the tribological behaviors of textured surface topographies manufactured on hard tool materials for improved resistivity towards galling in sheet metal forming. The studies are central to evaluating the tribo-mechanical performances with medium-to-high lubricant viscosities on the severity of the wear and product surface quality. A theoretical formulation of micro-hydrodynamic mechanism is established and hence the discussions are made separately for different contact area ratio of textured surface topographies, categorized by the high tribological severity of simulative tests for sheet metal forming; strip reduction testing. The broader impact of the chapter ends with a concluding remark about the present tribological textured surface topography work in manufacturing processes which are needed for exploring the tribological behavior of new and greatly classes of surface topography.

9.2 DEVELOPMENT OF TRIBOLOGICAL EXPERIMENTAL SYSTEM IN SHEET METAL FORMING

In order to investigate potential measures and the possibility of changing the most important process parameters influencing tribology systems, i.e. normal pressure, sliding length and tool temperature, under close control, full-scale testing of tribological improvements is time-consuming and costly. It is, however, imperative to simplify the testing of tribology systems utilized for sheet metal forming by pilot simulative testing, enabling cost-effective screening of candidate tribological concepts. Figure 9.4 shows laboratory simulation tests, where a blank slides against a deep drawing die. It functions to identify some unknown process parameters promoting a poor tribological system, and proposes suggestions to prevent the identified problem afterwards. The parameters include lubricant types, coatings, surface topography, process speed, tool material, sheet thickness, tool geometry, etc. Typical simulative tests emulating various sheet metal-forming processes are

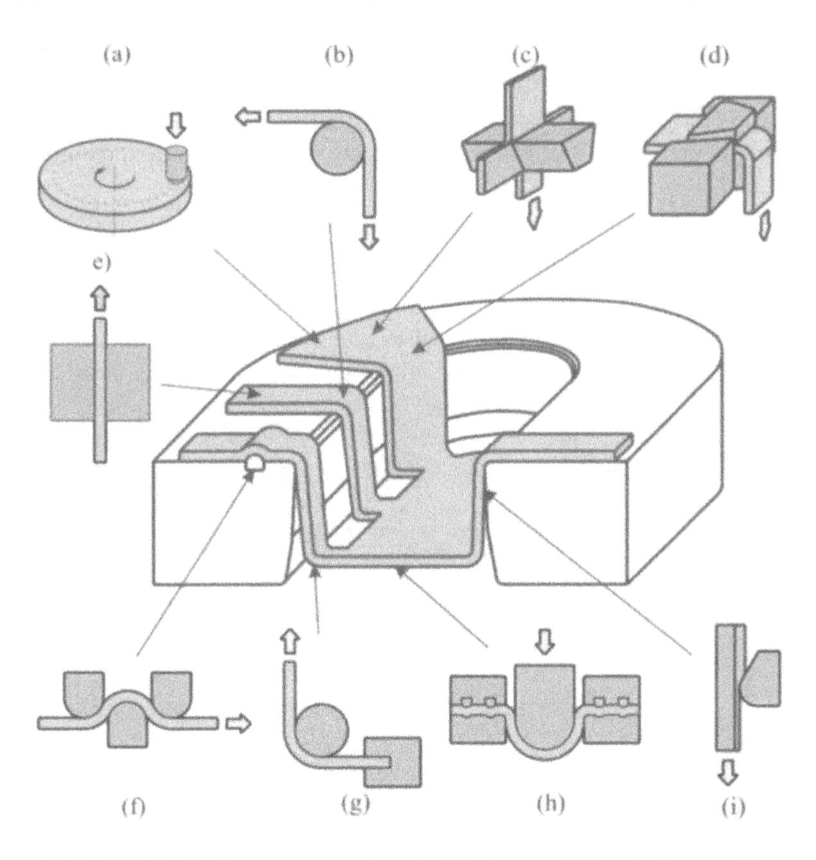

FIGURE 9.4 Tribological tests representing the friction conditions in the specific areas of the sheet metal forming: (a) pin-on-disc tribometer, (b) bending under tension, (c) drawing with tangential compression, (d) bending with tangential compression, (e) strip-drawing test, (f) draw-bead test, (g) strip-tension test, (h) hemispherical stretching, (i) strip-reduction testing; prepared based on [24].

TABLE 9.1
Sheet Tribo-Tests Characteristics [25]

Test Types	Normal Pressure	Surface Expansion	Tool Temperature	Tribological Severity
Bending-Under-Tension (BUT)	Low	0	Low	Low
Draw-Bead-Test (DBT)	Medium	0	Medium	Medium
Strip-Reduction-Test (SRT)	High	Medium	High	High
PUnching-Test (PUT)	Medium-high	Infinite	Very high	Very high

Strip-Bending-Under-Tension (BUT), Draw-Bead-Test (DBT), Strip-Reduction-Test (SRT), and PUnching-Test (PUT) [22, 23]. The above-mentioned simulative tests can be classified into ranks according to the severity of the tribological conditions, i.e. normal pressure, surface expansion and tool temperature, see Table 9.1. The BUT test possesses the lowest risk of galling, followed by DBT, SRT and lastly, PUT, considered as a high risk of galling. Two tribological simulation tests, BUT and SRT tests, are used in the present PhD project in order to investigate varying tribo-systems.

9.2.1 Strip-Reduction Test Setup between Flat Dies for Sheet Metal Forming

A Strip-Reduction Test (SRT) replicating an ironing process, as shown in Figure 9.5 (center) [26] was selected and performed to identify the most promising textured tool surface for reducing the friction and mitigating the wear in sheet metal forming.

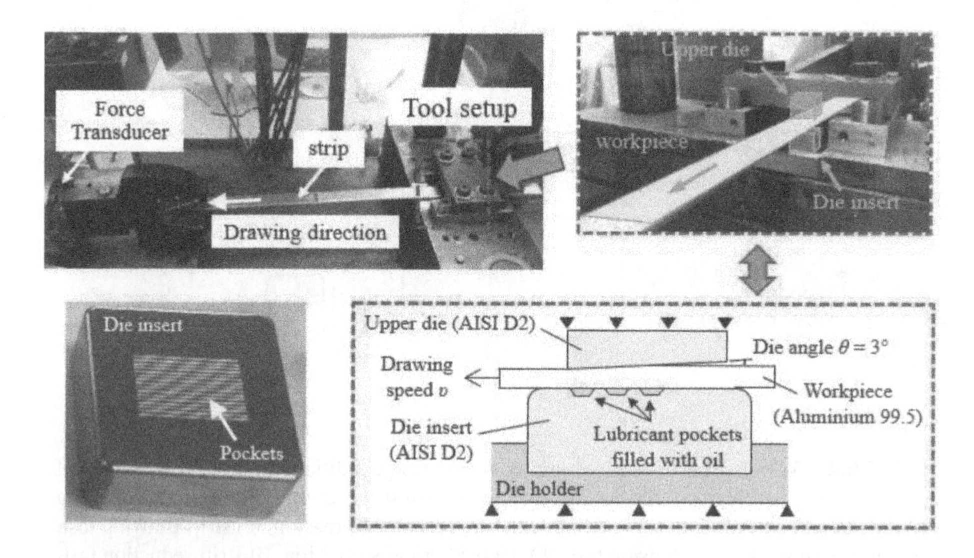

FIGURE 9.5 Schematics of the simulative laboratory tests emulating sheet metal forming.

The process conditions used in aluminum-forming production were adopted as a reference case to accurately study the tool-workpiece contact surface phenomena. The tool material was made of Sverker 21 corresponding to AISI D2 cold work tool steel, a high-carbon, high-chromium tool steel alloyed with molybdenum and vanadium. The tools were through-hardened and tempered to 60 HRC before the surface texturing procedure. The surface finish of the upper die and the die insert were mirror-polished to a low roughness, $Ra = 0.01 - 0.04$ μm.

The workpiece material was a commercially pure Al 99.5%, H111 with dimensions 480 mm × 20 mm × 4 mm. The 4 mm sheet thickness ensures a sufficient deformation region (tool/workpiece contact length) for a fairly large number of pockets to be within the deformation zone. This will reduce the experimental scatter due to the results being less sensitive to the exact number of pockets within the deformation zone. The sheet width was chosen to be large enough to ensure approximately plane strain deformation conditions resembling ironing. The as-received workpiece surface roughness was Ra = 0.21 μm. The stress-strain curve of the workpiece material was determined by plane strain compression testing. The material work hardening turned out to follow Voce's model quite well, and this gives the determined material constants according to the Voce flow curve expression: $\sigma_f = \sigma_o + (\sigma_\infty - \sigma_o)\left[1 - exp(-n\varepsilon_{eff})\right] = 55 + (149 - 55)\left[1 - exp(-1.52\varepsilon_{eff})\right]$.

9.2.2 Textured Tool Surface Topographies

Three surface texture features are important parameters to promote the micro-hydrodynamic lubrication mechanism [27]; these are (1) small pocket angle γ, (2) shallow pocket depth d, and (3) plateau distance between pockets x. Figure 9.6 shows a schematic of the SRT die insert consisting of a deformation region ($X \times Y$ = 11.5 mm × 20 mm) and the transverse pocket length y = 16 mm. The contact area ratio α on a smooth tool surface has been studied by creating an oblong and a small gap y in between the oblong pockets. Therefore, the number of pocket columns n_{column} = 1 for *Tool A*, and n_{column} = 4 was manufactured onto

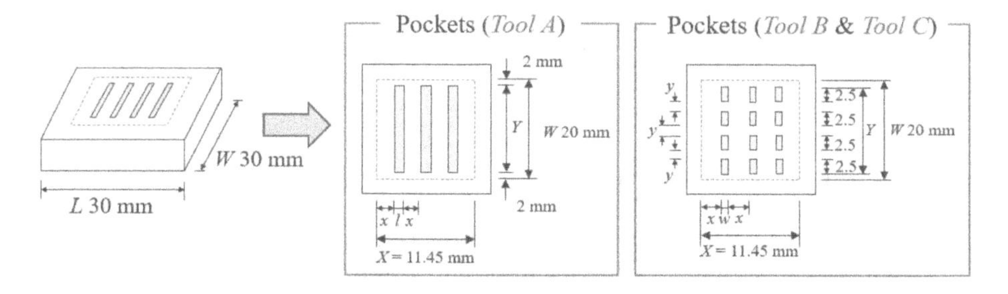

FIGURE 9.6 Different textured insert surfaces with varying distance between pockets longitudinal to the sliding direction for Tool A (oblong pocket geometries), Tool B and Tool C (longitudinal pocket length of 4 mm and 2.5 mm, respectively). Sliding occurs in vertical direction.

TABLE 9.2
Target Surface Texture Parameters

Tool Types	Tool A	Tool B	Tool C
Distance between pockets x (mm)	1 × w (x = 0.23 mm)	2 × w (x = 0.46 mm)	4 × w (x = 0.92 mm)
Number of pocket rows nrow	25	16	10
Number of pocket columns ncolumn	1	4	4
Distance between pocket columns y	-	0.8 mm	2 mm

Tool B and *Tool C* surfaces. Detailed information about the pocket geometries are described in Table 9.2. The pocket length y for *Tool B* is 4 mm and *Tool C* is 2.5 mm. This gives the plateau distance in longitudinal direction for *Tool B* as 0.8 mm and *Tool C* as 2 mm. The α–value for *Tool A*, *Tool B*, and *Tool C* are described in Table 9.3.

Three different inserts were manufactured, the only difference being the distance between the pockets x. A TiAl70-coated milling tool having a two-flute solid carbide ball-nose and a radius R of 1.25 mm was used for machining the transverse flat-bottomed lubricant pockets in the surface of the hardened tool. After machining and polishing, the resulting textured patterns were examined. The textured die insert pocket geometries and the surface were measured by a tactile roughness profilometer, and the images and roughness were then analyzed by an analytical software for microscopy, SPIP. Figures 9.7 and 9.8 represent the resulting, measured pockets with dimensions: length $y = 16$ mm, angle $\gamma = 5° \pm 0.5°$, width $w = 0.23 \pm 0.01$ mm, depth $d = 7 \pm 1$ µm, and distance between pockets of $x = 0.23$, 0.46, and 0.92 mm, respectively. Subsequent polishing of the tool surfaces was done in three steps with water-based polycrystalline diamonds of grain sizes 3, 1, and 0.25 µm, resulting in a final roughness, $Ra = 0.01 - 0.04$ µm. The pocket depths were reached within the tolerance gap, whereas the pocket angles turned out to be somewhat smaller than the target value. However, this only promotes the micro-hydrodynamic lubrication mechanism and prevents mechanical interlocking.

TABLE 9.3
Notation of Tool Types with Varying Contact Area Ratio α

Tool Types / Plateau Distance x	Notation / Contact Area Ratio α (%)		
Tool A ($x = 0.23$ mm)	A1 ($\alpha = 60\%$)	A2 ($\alpha = 74\%$)	A3 ($\alpha = 84\%$)
Tool B ($x = 0.46$ mm)	B1 ($\alpha = 60\%$)	B2 ($\alpha = 74\%$)	B3 ($\alpha = 84\%$)
Tool C ($x = 0.92$ mm)	C1 ($\alpha = 75\%$)	C2 ($\alpha = 84\%$)	C3 ($\alpha = 90\%$)

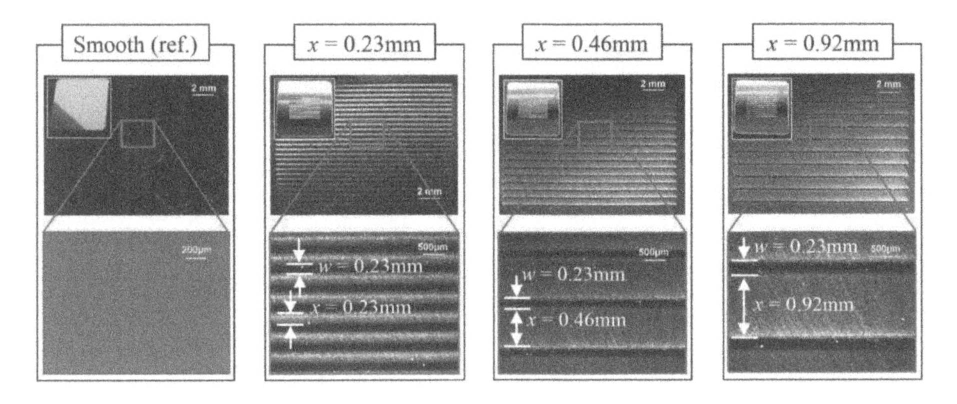

FIGURE 9.7 Die inserts with (a) smooth surface as reference, and with varying pocket interspacing x of (b) 0.23 mm, (c) 0.46 mm, and (d) 0.92 mm. The flat plateaus were polished down to a roughness Ra 0.01 – 0.04 μm.

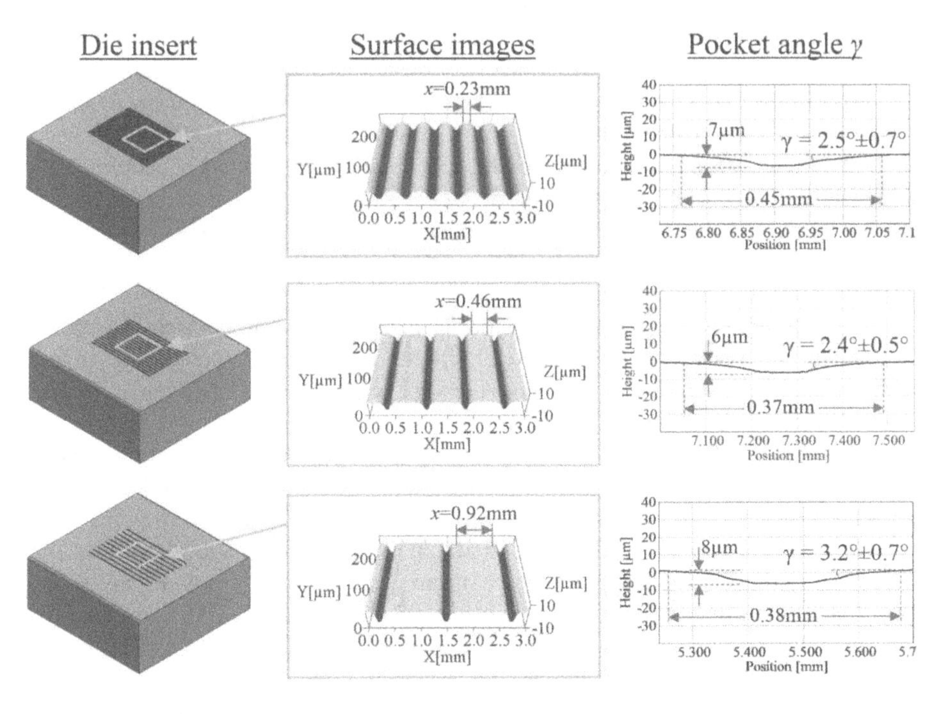

FIGURE 9.8 Textured die insert pocket geometries with varying pocket plateau distance.

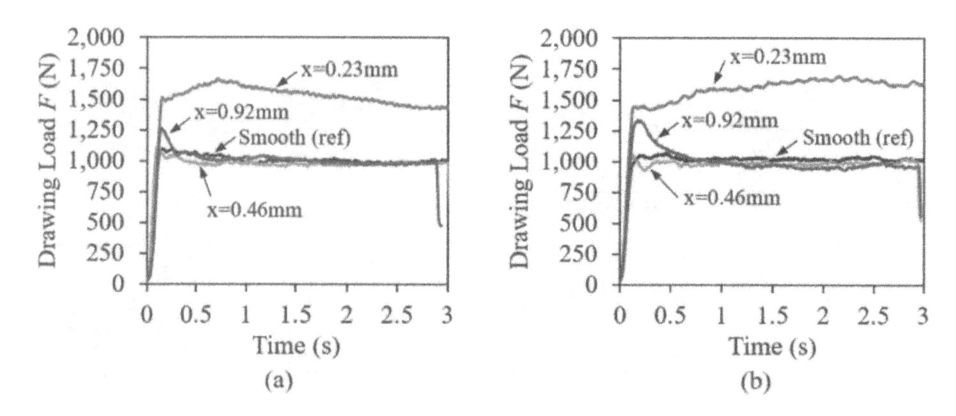

FIGURE 9.9 Forming load for *Tool A* at speed $\upsilon = 65$ mm/s for (a) Rhenus oil $\eta = 800$ cSt and (b) Rhenus oil $\eta = 300$ cSt.

9.3 TEXTURED TOOL SURFACE IN SHEET METAL FORMING

9.3.1 Effects of Plateau Distance between Pockets

The drawing load reaches steady-state condition after a short time, as seen in Figure 9.9, which shows the results for the two different lubricants at lower drawing speed $\upsilon = 65$ mm/s. No load difference is observed at lower speed except that the transverse pocket with $x = 0.23$ mm leads to a larger forming load. Figure 9.10 shows similar forming load patterns for the four different lubricants at larger drawing speed $\upsilon = 240$ mm/s in which the drawing load reaches steady-state after a short time. The influence of tool texture is significant at higher speed, regardless of the lubricant applied. The small distance between the pockets ($x = 0.23$ mm) leaves only a small, flat plateau between the pockets, see Figure 9.8. This promotes metal flow into and out of the pockets, which will provide a mechanical gripping effect of the workpiece. The indent depth of the workpiece onto the surface pocket is smaller than the pocket width; hence the pockets are not completely emptied when the workpiece goes in and out of the pockets. Marks of the die insert texture on the strip can be seen at the end of the reduction zone, Figure 9.13 (left).

The positive influence of high drawing speed is explained by micro-plasto-hydrodynamic lubrication, which is promoted by high sliding speed and high lubricant viscosity [21]. Since no improvements were noted on the drawing load when testing tool textures at the lower speed (65 mm/s; Figure 9.9), the rest of the discussion is focused on the tool texture at larger speed (240 mm/s; Figure 9.10). It is noticed here that the tool texture with $x = 0.46$ mm and $x = 0.92$ mm (two to four times the pocket width w) has reduced the drawing load as compared to the smooth tool surface when testing with the larger viscosity lubricants, while testing with the low-viscosity pure mineral oil CR5-Sun 60 had the opposite effect. This could be due to the relationship between viscosity and micro-plasto-hydrodynamic lubrication which caused pocket indent and pressurization of trapped lubricant in a pocket as illustrated in Figure 9.11.

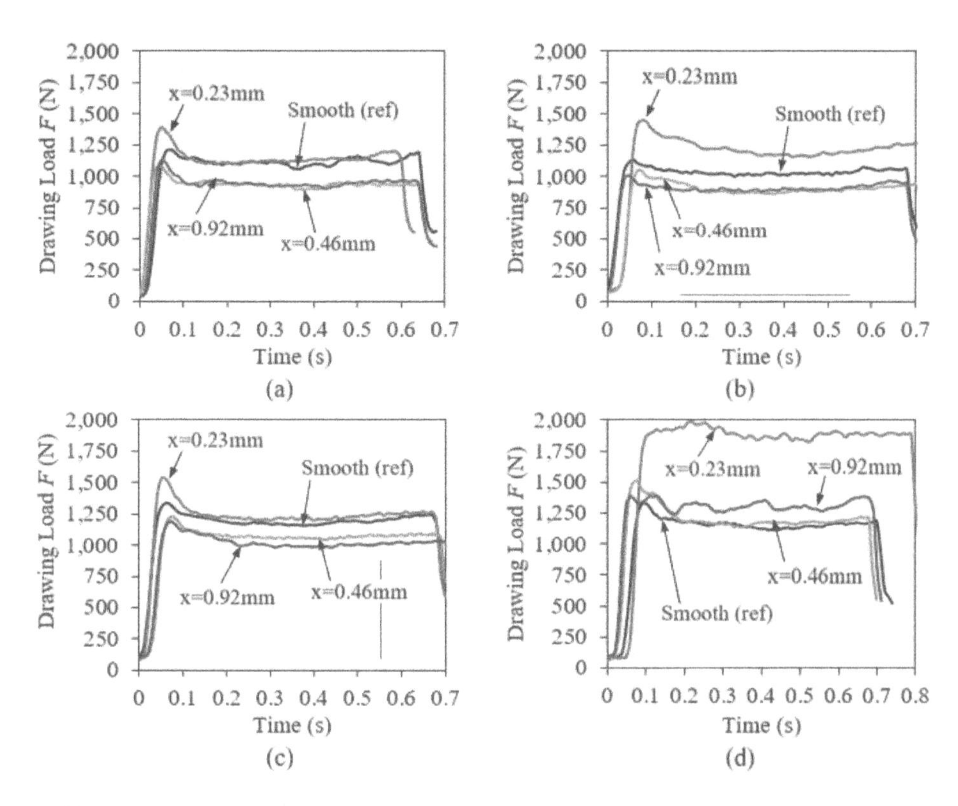

FIGURE 9.10 Drawing load F at speed υ = 240 mm/s for (a) Rhenus oil η = 800 cSt; (b) Rhenus oil η = 300 cSt; (c) mineral oil CR5 η = 660 cSt; and (d) mineral oil mixtures CR5-Sun 60 η = 60 cSt.

Figure 9.12 shows that the tool texture increases the sheet roughness as compared to the smooth die surface, regardless of the test lubricants investigated. The tool texture with pocket distance x = 0.23 mm gave smallest sheet roughness among the textured inserts. It is furthermore noticed that increasing viscosity leads to increasing roughness of the strip plateau. This may be explained by improved micro-plasto-hydrodynamic lubrication at higher viscosities, leading to an effective separation between the tool and the workpiece on the plateaus of the tool table mountain [22].

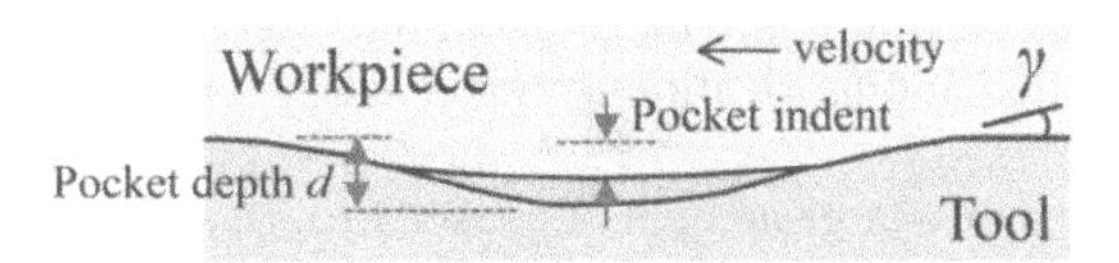

FIGURE 9.11 Illustration of pocket indent and the pressurization of the trapped lubricant in a pocket.

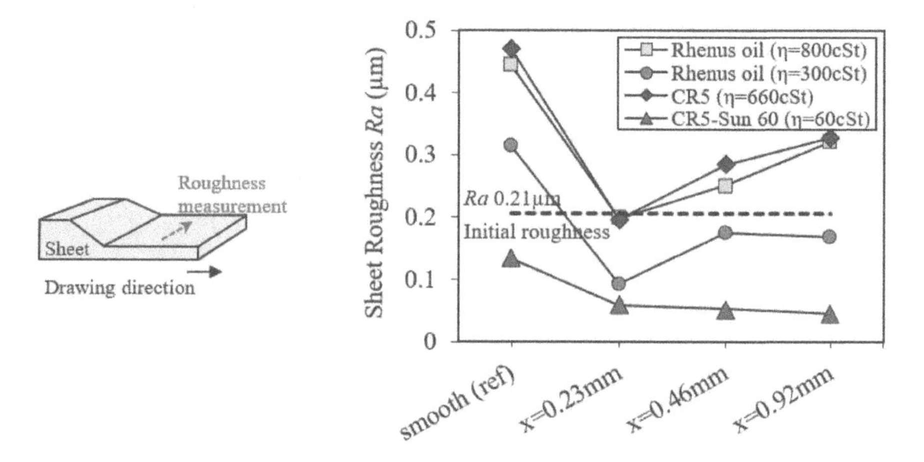

FIGURE 9.12 Sheet roughness on the plateaus at $\upsilon = 240$ mm/s drawing speed [15].

The sheet roughness profiles shown in Figure 9.13 confirm this. The *Ra* values on the plateaus are measured by a tactile roughness profilometer. They are based on an average of six measurements. As explained in both Figures 9.15 and 9.16, the viscosity η and the speed υ play an important role to improve lubrication. The Rhenus oil contains additives providing a protective boundary film, which can carry the load and prevent metal-to-metal contact. This contributes to lower friction and prevents lubricant film breakdown. The additives in the Rhenus oils furthermore prevent these oils from decomposition and vaporization [23].

9.3.2 Effects of Contact Area Ratio *A*

The advantages of larger drawing speed with increasing viscosity are studied for the influence of contact area ratio α [28]. Figure 9.14 shows measurements of average drawing loads in the steady-state condition as a function of contact area ratio at a drawing speed of 240 mm/s. Tool textures with too large amounts of pocket area, i.e., with low α-value, were found to increase the drawing load. Too small amounts of pocket area, on the other hand, may also lead to increased drawing load since the lubricant escape by micro-plasto-hydrodynamic lubrication may not be sufficient to cover the entire flat plateau. This implies an increase in drawing load due to increased metal-to metal contact. Otherwise, the drawing load decreases with larger α-value. It is noticed that an optimum α-value exists in which the contribution to mechanical interlocking of the workpiece into the pockets is limited and lubrication is enhanced by the micro-plasto-hydrodynamic lubrication.

The tool texture with a large amount of pocket area, i.e., low α-value, gave smallest sheet roughness in comparison to the smooth tool surface. This may be explained by improved micro-plasto-hydrodynamic lubrication at larger viscosities leading to

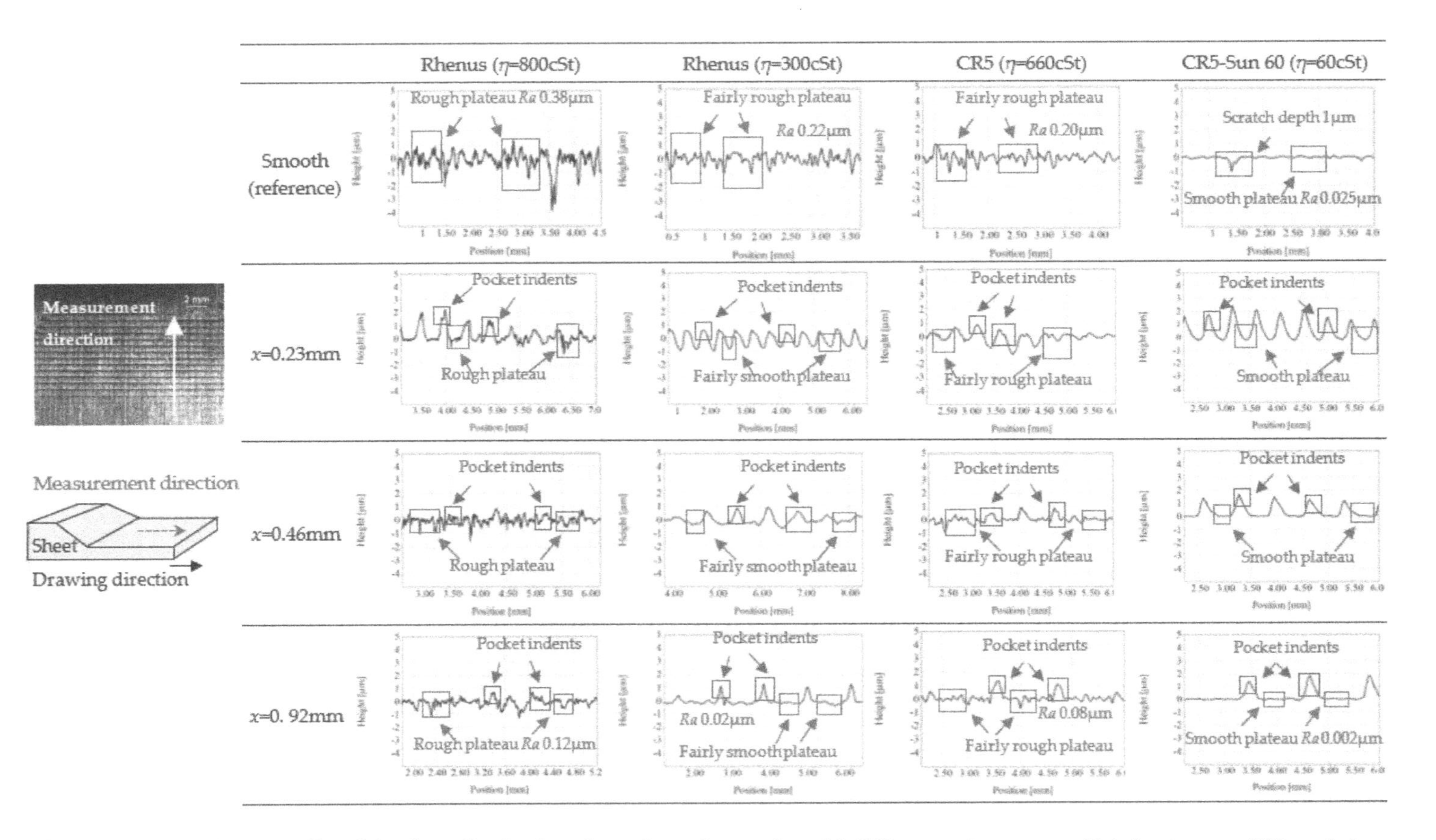

FIGURE 9.13 Roughness profiles of the sheets flowing into the pockets when testing with different tool textures and lubricants at $\upsilon = 240$ mm/s drawing speed [15].

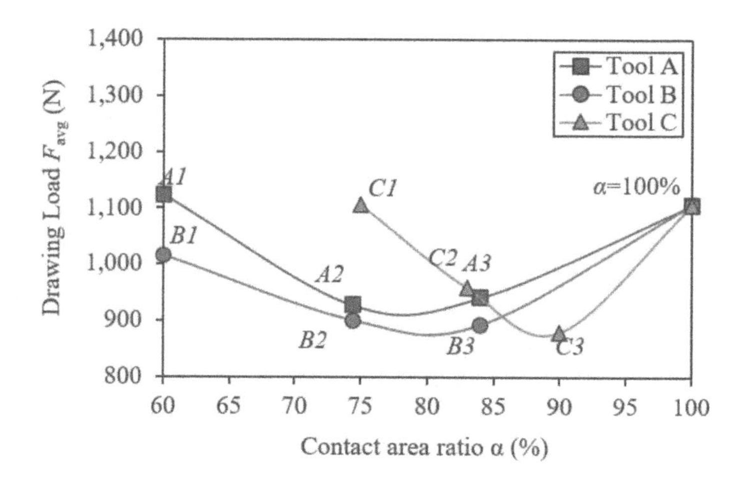

FIGURE 9.14 Influence of contact area ratio α on the drawing load.

an effective separation between the tool and the workpiece on the plateaus of the tool table mountain [13]. In addition, application of the Rhenus oil generates a thin, protective film to separate the tool/workpiece interface. LOM images presented in Figure 9.15 confirmed this, where almost no pick-up of aluminum is observed on the plateaus of the table mountain structure.

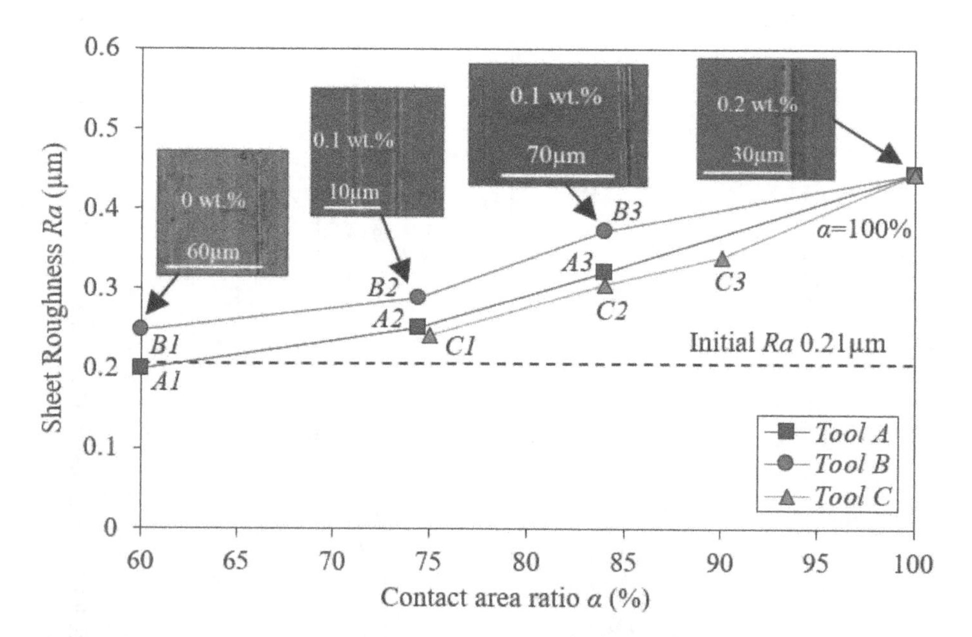

FIGURE 9.15 Influence of contact area ratio α on the sheet roughness in SRT. SEM pictures with the amount of aluminum pick-up are also seen.

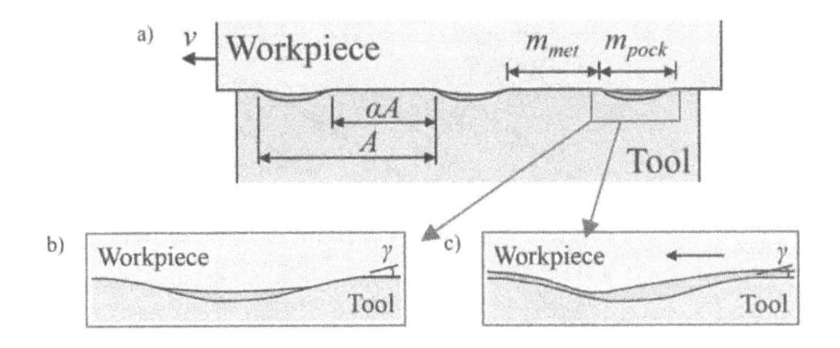

FIGURE 9.16 (a) Notation, (b) compression of lubricant trapped in pocket, (c) lubrication of plateaus by lubricant escaped from the pocket.

9.4 FORMULATION OF MICRO-HYDRODYNAMIC MECHANISM ON TEXTURED TOOL SURFACE

Contributions to friction in the lower, textured tool/workpiece interface play an important role in effective lubrication. As illustrated in Figure 9.16, the contributions to the friction include a contribution from the metal-to-metal contact area with relative area α and a contribution from the contact between the workpiece and the lubricant-filled pocket. Partial penetration of the workpiece material into the pocket and pressurization of the lubricant appear as shown in Figure 9.16(b), when loading is applied. When sliding is superimposed, the workpiece material forms a wave motion moving in and out of the pocket when passing it. At the same time the lubricant is dragged out of the pocket by viscous forces and thereby provides lubrication to the flat tool part by the micro-hydrodynamic mechanism as illustrated in Figure 9.16(c).

The overall friction factor m_{tex} representing the combined friction factor of the lower, textured tool surface is determined by Equation 9.1:

$$m_{tex} = \alpha m_{met} + (1-\alpha) m_{pock} \tag{9.1}$$

where $\alpha = A_{met}/A$ is the metal-to-metal contact area ratio between the flat plateau A_{met} and the total contact area A in the deformation zone. The relative area of contact between the pocket and the strip is then $(1-\alpha)$. m_{met} is the local friction factor between the strip and the flat plateau. The local pocket friction factor m_{pock} is given by Equation 9.2:

$$m_{pock} = m_{lub} + m_{wave} \tag{9.2}$$

where m_{lub} is the friction factor due to viscous drag forces between the strip and the trapped lubricant in the pockets and m_{wave} is the apparent friction factor caused by the material wave movement into and out of the pockets. The viscous drag effect of the lubricant in the pocket is assumed minimal, i.e. $m_{lub} = 0$. Thus, the local friction factor m_{pock} becomes:

$$m_{pock} = m_{wave} \tag{9.3}$$

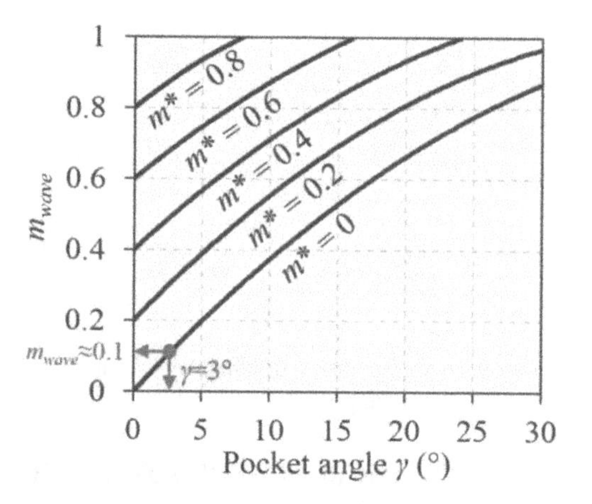

FIGURE 9.17 Apparent friction factor m_{wave} as a result of pocket angle γ and m^* [29].

The value of m_{wave} can be interpreted from Figure 9.17 [29], in which the metallic friction mechanism is based on a plastic wave formed by the workpiece surface moving into and out of a long groove with triangular cross-section in the tool. Implementing the value of m_{wave} to the present, textured tool illustrated in Figure 9.16, the plastic wave moves into and out of the pockets experiencing an apparent friction stress $\tau = m_{wave}k$, which is plotted as a function of the tool asperity slope γ in Figure 9.16. m^* is the local friction factor between the pocket surface and the workpiece. For $\gamma = 0°$ thus $m^* = m_{wave}$. Due to the wave motion in and out of the pocket, an extra contribution to the apparent friction factor m_{wave} appears, whereby it becomes larger than m^*. In the present case $\gamma = 3°$ and assuming $m_{lub} = m^* = 0$ due to the entrapped lubricant, Figure 9.17 shows $m_{wave} \cong 0.1$. Accordingly the overall friction factor m_{tex} of the textured tool in Equation 9.1 becomes:

$$m_{tex} = \alpha m_{met} + (1-\alpha) m_{pock}$$

$$m_{tex} = \alpha m_{met} + (1-\alpha) m_{wave} \tag{9.4}$$

Inserting the value from Figure 9.21, Equation 9.4 becomes:

$$m_{tex} = \alpha m_{met} + (1-\alpha) \times 0.1 \tag{9.5}$$

Based on a plane strain slab analysis [30], the normalized drawing stress σ_d in strip reduction through an inclined upper die and a flat lower die can be written as

$$\frac{\sigma_d}{2k} = \left[1 + \left(m_{low} + m_{up}\right)\frac{1}{2\tan\theta}\right]\ln\frac{h_1}{h_0} \tag{9.6}$$

where k is the mean shear flow stress in the deformation zone ($k = \sigma_f / \sqrt{3}$), m_{low} and m_{up} are the friction factors on the lower and upper tool surfaces, respectively, θ is the die angle, h_o is the initial sheet thickness and h_l is the final sheet thickness.

Noting that the stress–strain curve of the workpiece material Al 99.5% – H111 that follow Voce's model is described in Equation 9.7:

$$\sigma_f = \sigma_o + (\sigma_\infty - \sigma_o)\left[1 - exp(-n\varepsilon_{eff})\right] \tag{9.7}$$

Assuming zero prestrain and setting ε_l = the effective strain of the material after drawing, the following average flow stress in the deformation zone is determined as written in Equation 9.8:

$$\bar{\sigma}_f = \frac{1}{\varepsilon_1}\int_0^{\varepsilon_1}\sigma_f(\varepsilon_{eff})\,d\varepsilon_{eff} = \frac{1}{\varepsilon_1}\left\{\sigma_\infty\varepsilon_1 + \frac{\sigma_\infty - \sigma_0}{n}\left[exp(-n\varepsilon_1) - 1\right]\right\} \tag{9.8}$$

The smooth tool surface with no textures on the tool surface has an apparent contact area ratio $\alpha = 1$. For the textured tools, the plateau distances x= 0.23, 0.46 and 0.92 mm result in α = 0.60, 0.74 and 0.84, respectively. From the experimentally measured drawing force, it is possible to determine the overall friction factor on the lower tool by applying Equations 9.5 and 9.6 in the following way. The friction factors m_{low} and m_{up} are considered equal for the smooth, non-textured tool. The obtained friction factor $m_{low} = m_{up}$ is then applied to the untextured upper tool (m_{up}) for the experiments with the textured, lower tool surfaces. This leaves only the value of $m_{low} = m_{tex}$ as unknown, which is then determined, so that experimental and theoretical drawing loads are matching, Figure 9.18. Figure 9.19 shows the corresponding values of the overall friction factor m_{tex} on the textured tool surface. It is noticed that minimum drawing force and m_{tex} appear when $\alpha \cong 0.8$ in good accordance with experimental findings in literature for the plane strip drawing test [31].

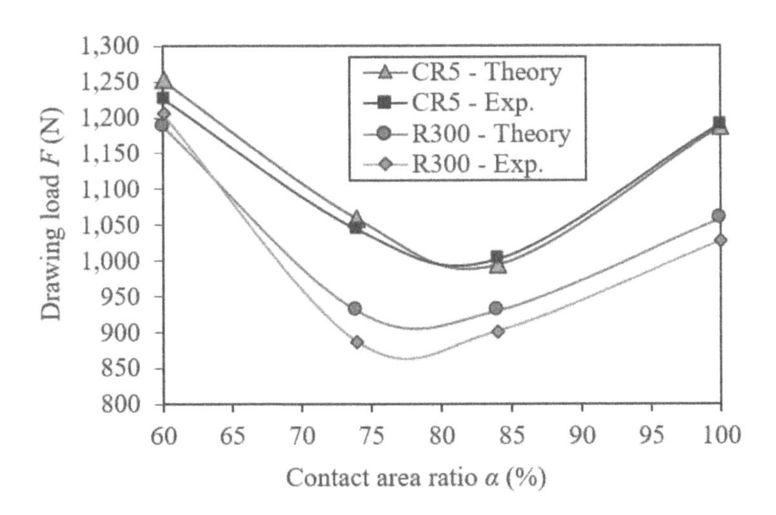

FIGURE 9.18 Theoretical and experimental drawing load as a function of contact area ratio.

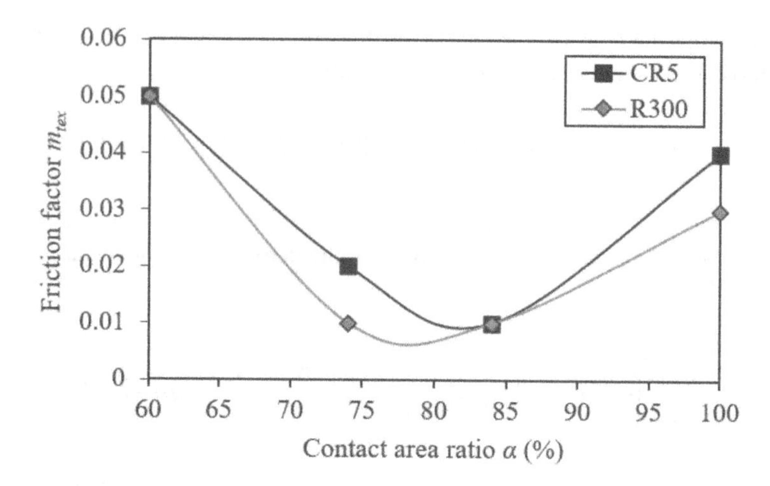

FIGURE 9.19 Friction factor m_{tex} as a function of contact area ratio for two different lubricants.

9.5 CONCLUSION

A technique to improve resistivity towards galling by applying textured tool surface topographies was studied. Oblong shallow pockets with small pocket angles, oriented perpendicular to the sliding direction with a distance of 2–4 times the pocket width, were tested. A strip reduction test, which emulates the tribological conditions in an ironing process, was used for experimental measurements of friction and determination of possible pick-up and galling. The study included testing of four different lubricants—two plain mineral oils with a low and a high viscosity, and two mineral-based oils with boundary lubrication additives having medium and high viscosity. The results confirmed that tool texture can lower friction and improve lubrication performance in comparison to that of a fine polished tool surface when the pocket distance is 2–4 times larger than the pocket width, which ensures a table mountain structure of the tool topography. The tool textures were advantageous at greater sliding speeds, when using higher viscosity oils, which facilitate the escape of trapped lubricant by micro-plasto-hydrodynamic lubrication.

A friction model for a soft workpiece deforming against a textured tool surface was proposed. The model takes into account the plastic wave motion appearing when the workpiece material flows into and out of local pockets between the flat plateaus of a table mountain-like tool surface topography. The model was evaluated by strip reduction tests, which emulate the tribological conditions in an ironing process. The study included testing of two different lubricants, a plain mineral oil with a high viscosity and mineral-based oil with boundary lubrication additives having a medium viscosity. It was found that an optimum amount of tool texture exists which reduces friction and thus draws load for the table mountain-like tool surface topography. The overall friction factor in the interface between workpiece and textured tool surface can be satisfactory as predicted by the model.

ACKNOWLEDGMENT

The authors would like to thank the Malaysia Technical University Network (MTUN) Matching Grant, Ministry of Higher Education, Malaysia (Ref:UniMAP/PPP/GRN IRPA/MTUN/9002-00099/9028-00020(1)) and Research Management Centre Grant 2020 (RMCG) (Grant No: RMCG20-036-0036). The authors are grateful for the significant contributions from Abda Jamil Shaharan, Tanjung Advance Sdn. Bhd. (TASB).

REFERENCES

1. Ingarao, G. 2017. "Manufacturing Strategies for Efficiency in Energy and Resources Use: The Role of Metal Shaping Processes." *Journal of Cleaner Production* 142 (January): 2872–86. https://doi.org/10.1016/j.jclepro.2016.10.182.
2. Shah, R., M. Woydt, and S. Zhang. 2021. "The Economic and Environmental Significance of Sustainable Lubricants." *Lubricants* 9 (2): 21. https://doi.org/10.3390/lubricants9020021.
3. Sulaiman, M.H. 2017. "Development and Testing of Tailored Tool Surfaces for Sheet Metal Forming." Technical University of Denmark (DTU), Kongens Lyngby, Denmark.
4. Singh, Y., A. Farooq, A. Raza, M.A. Mahmood, and S. Jain. 2017. "Sustainability of a Non-Edible Vegetable Oil Based Bio-Lubricant for Automotive Applications: A Review." *Process Safety and Environmental Protection* 111 (October): 701–13. https://doi.org/10.1016/j.psep.2017.08.041.
5. Syahir, A.Z., N.W.M. Zulkifli, H.H. Masjuki, M.A. Kalam, A. Alabdulkarem, M. Gulzar, L.S. Khuong, and M.H. Harith. 2017. "A Review on Bio-Based Lubricants and Their Applications." *Journal of Cleaner Production* 168 (December): 997–1016. https://doi.org/10.1016/j.jclepro.2017.09.106.
6. Mannekote, J.K., S.V. Kailas, K. Venkatesh, and N. Kathyayini. 2018. "Environmentally Friendly Functional Fluids from Renewable and Sustainable Sources – A Review." *Renewable and Sustainable Energy Reviews* 81 (June 2016): 1787–1801. https://doi.org/10.1016/j.rser.2017.05.274.
7. Amiril, S.A.S., E.A. Rahim, and S. Syahrullail. 2017. "A Review on Ionic Liquids as Sustainable Lubricants in Manufacturing and Engineering: Recent Research, Performance, and Applications." *Journal of Cleaner Production* 168: 1571–89. https://doi.org/10.1016/j.jclepro.2017.03.197.
8. Zainal, N.A., N.W.M. Zulkifli, M. Gulzar, and H.H. Masjuki. 2018. "A Review on the Chemistry, Production, and Technological Potential of Bio-Based Lubricants." *Renewable and Sustainable Energy Reviews* 82 (February): 80–102. https://doi.org/10.1016/j.rser.2017.09.004.
9. Kang, Z., Y. Fu, D. Zhou, Q. Wu, T. Chen, Y. He, and X. Su. 2021. "Reducing Engine Oil and Fuel Consumptions by Multidimensional Laser Surface Texturing on Cylinder Surface." *Journal of Manufacturing Processes* 64 (April): 684–93. https://doi.org/10.1016/j.jmapro.2021.01.052.
10. Yang, L., Y. Ding, B. Cheng, J. He, G. Wang, and Y. Wang. 2018. "Investigations on Femtosecond Laser Modified Micro-Textured Surface with Anti-Friction Property on Bearing Steel GCr15." *Applied Surface Science* 434 (March): 831–42. https://doi.org/10.1016/j.apsusc.2017.10.234.
11. Shivakoti, I., G. Kibria, R. Cep, B.B. Pradhan, and A. Sharma. 2021. "Laser Surface Texturing for Biomedical Applications: A Review." *Coatings* 11 (2): 124. https://doi.org/10.3390/coatings11020124.

12. Ranjan, P. and S.S. Hiremath. 2019. "Role of Textured Tool in Improving Machining Performance: A Review." *Journal of Manufacturing Processes* 43 (July): 47–73. https://doi.org/10.1016/j.jmapro.2019.04.011.
13. Bech, J., N. Bay, and M. Eriksen. 1999. "Entrapment and Escape of Liquid Lubricant in Metal Forming." *Wear* 232 (2): 134–39. https://doi.org/10.1016/S0043-1648(99)00136-2.
14. Dubar, L., C. Hubert, P. Christiansen, N. Bay, and A. Dubois. 2012. "Analysis of Fluid Lubrication Mechanisms in Metal Forming at Mesoscopic Scale." *CIRP Annals – Manufacturing Technology* 61 (1): 271–74. https://doi.org/10.1016/j.cirp.2012.03.126.
15. Sulaiman, M.H., P. Christiansen, and N. Bay. 2017. "The Influence of Tool Texture on Friction and Lubrication in Strip Reduction Testing." *Lubricants* 5 (1). https://doi.org/10.3390/lubricants5010003.
16. Popp, U., and U. Engel. 2006. "Microtexturing of Cold-Forging Tools – Influence on Tool Life." *Proceedings of the Institution of Mechanical Engineers, Part B: Journal of Engineering Manufacture* 220 (1): 27–33. https://doi.org/10.1243/095440505X32968.
17. Krux, R., W. Homberg, M. Kalveram, M. Trompeter, M. Kleiner, and K. Weinert. 2005. "Die Surface Structures and Hydrostatic Pressure System for the Material Flow Control in High-Pressure Sheet Metal Forming." *Advanced Materials Research* 6–8: 385–92. https://doi.org/10.4028/www.scientific.net/AMR.6-8.385.
18. Pawlus, P., R. Reizer, M. Wieczorowski, and G. Krolczyk. 2020. "Material Ratio Curve as Information on the State of Surface Topography—A Review." *Precision Engineering* 65 (September): 240–58. https://doi.org/10.1016/j.precisioneng.2020.05.008.
19. Steitz, M., P. Stein, and P. Groche. 2015. "Influence of Hammer-Peened Surface Textures on Friction Behavior." *Tribology Letters* 58 (2): 1–8. https://doi.org/10.1007/s11249-015-0502-9.
20. Franzen, V., J. Witulski, A. Brosius, M. Trompeter, and A. E. Tekkaya. 2010. "Textured Surfaces for Deep Drawing Tools by Rolling." *International Journal of Machine Tools and Manufacture* 50 (11): 969–76. https://doi.org/10.1016/j.ijmachtools.2010.08.001.
21. Groche, P., J. Stahlmann, J. Hartel, and M. Köhler. 2009. "Hydrodynamic Effects of Macroscopic Deterministic Surface Structures in Cold Forging Processes." *Tribology International* 42 (8): 1173–79. https://doi.org/10.1016/j.triboint.2009.03.019.
22. Bay, N., D. D. Olsson, and J. L. Andreasen. 2008. "Lubricant Test Methods for Sheet Metal Forming." *Tribology International* 41 (9–10): 844–53. https://doi.org/10.1016/j.triboint.2007.11.017.
23. Olsson, D.D., N. Bay, and J.L. Andreasen. 2004. "Prediction of Limits of Lubrication in Strip Reduction Testing." *CIRP Annals – Manufacturing Technology* 53 (1): 231–34. https://doi.org/10.1016/S0007-8506(07)60686-6.
24. Trzepiecinski, T., Hirpa G. Lemu. 2020. "Recent developments and trends in the friction testing for conventional sheet metal forming and incremental sheet forming." *Metals* 10 (1): 47.
25. Bay, N. 2011. "Trends and Vision in Metal Forming Tribology." *Steel Research International* 82, Planetary (Special Edition): 15–26.
26. Nielsen, P. S., K.S. Friis, and N. Bay. 2011. "Testing and Modelling of New Tribo-Systems for Industrial Sheet Forming of Stainless Steels." *Proceedings of the Institution of Mechanical Engineers, Part J: Journal of Engineering Tribology* 225 (10): 1036–47. https://doi.org/10.1177/1350650111415331.
27. Sulaiman, M.H., R.N. Farahana, M.J.M. Ridzuan, H. Jaafar, L. Tajul, and J.A. Wahab. 2019. "Manufacturing Strategies and Tribology of Tailored Tool Surface Topographies for Enhanced Lubrication in Metal Forming." In *Proceedings of 3rd Mytribos Symposium*, 47–51.

28. Sulaiman, M.H., P. Christiansen, and N. Bay. 2018. "Development and Testing of Tailored Tool Surfaces for Sheet Metal Forming." In *8th JSTP International Seminar on Precision Forging (ISPF)*, 69–72.
29. Wanheim, T., and T. Abildgaard. 1980. "A Mechanism for Metallic Friction." In *Proceedings of the 4th International Conference on Production Engineering Tokyo*, 122–27.
30. Siebel, E. 1923. "Untersuchungen Über Bildsame Formänderung Unter Besonderer Berück- Sichtigung Des Schmiedens." *Maschinenbau/Betrieb* 9: 307–312.
31. Steitz, M., P. Stein, and P. Groche. 2015. "Influence of Hammer-Peened Surface Textures on Friction Behavior." *Tribology Letters* 58 (2): 1–8. https://doi.org/10.1007/s11249-015-0502-9.

10 Green Machining Techniques

A Review

Sangeeta Das
Girijananda Chowdhury Institute of Management and Technology
Guwahati, India

Shubhajit Das
National Institute of Technology
Arunachal Pradesh, India

CONTENTS

10.1 INTRODUCTION

During metal-cutting operations, heat is generated due to plastic deformation and friction between the tool-work piece interface that leads to high cutting temperature. The undesirable effects of high cutting temperature must be controlled or reduced through various methods like proper material and tool geometry selection, optimized combination of cutting speed and feed without affecting material removal rate (MRR), application of cutting fluid, application of a feasible technique whenever required. Cutting fluids have been inevitably used in metal-cutting operations for

DOI: 10.1201/9781003139386-10

many decades. During their initial phase of application, cutting fluids consisted of simple oils with occasional addition of animal fats to improve lubricating ability. However, the cutting fluids in present days consist of a formulated mixture of chemical additives, lubricants and water in order to facilitate complex cutting operations in metal-working industries. The cutting fluids have adverse effects on health and the environment and comprise 16–20% of the manufacturing costs [1]. The maintenance cost and disposing of cutting fluids in combination with their health and safety hazards eventually led to either eliminating or limiting the use of cutting fluids. The process of elimination of cutting fluids in machining operation is known as dry machining and the process involving limited use of cutting fluid is known as near-dry machining or minimum quality lubrication.

Though dry machining is one of the cleanest manufacturing techniques, the higher cutting variables limit its application. High tool wear, bad surface finish, work hardening and plastic deformation of chips are the disadvantages of dry machining. In dry machining, almost 20–30% of the total power is lost due to heat generation between tool surface and workpiece by plastic deformation and by friction between tool/chip on the rake and flank surfaces. Hence, it can be inferred that the use of metal-working fluid is indispensable in machining [2]. The vital functions of metal-working fluid are shown in Figure 10.1.

The conventional lubrication techniques require a huge quantity of cutting fluids, approximately 10–100 L/min, that needs immediate attention for achieving sustainability. In recent years, researchers have been trying to minimize the use of cutting fluids by adopting alternative cooling/lubrication techniques and focusing on the use of biodegradable, non-toxic and environment-friendly lubricants. Hence, sustainable manufacturing/green machining techniques were adopted almost 20 years ago in order to protect the environment, lessen health threats, meet the energy demand, maintain environmental balance and reduce the cost of manufacturing. Sustainable manufacturing transforms the existing resources into eco-friendly products. It provides operational and functional safety along with enhanced personal health. Sustainable manufacturing is also linked with preserving product and process quality [3]. Thus, green machining is the future of manufacturing companies to protect environmental laws and health regulations. The various sustainable machining methods are shown in Figure 10.2.

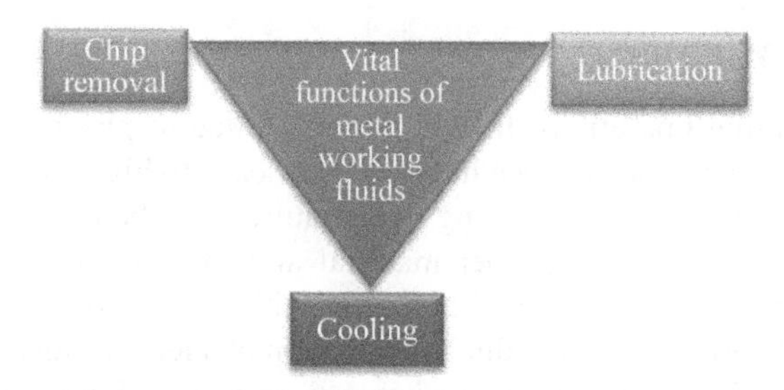

FIGURE 10.1 Functions of metal-working fluids.

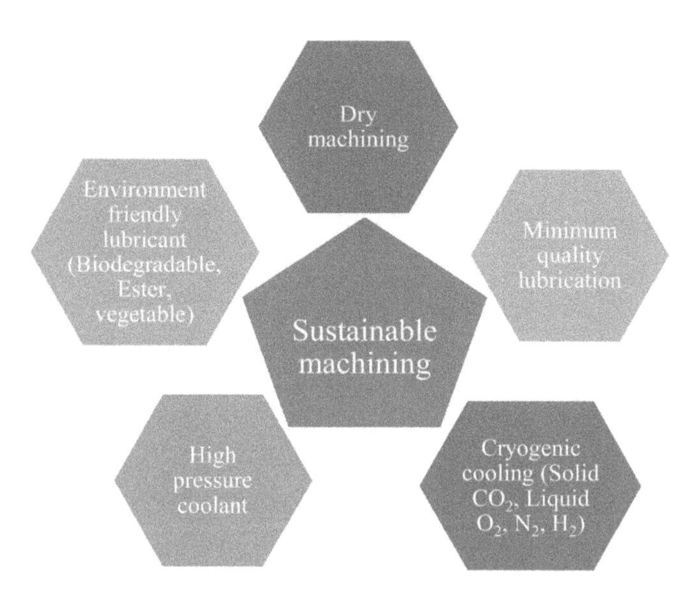

FIGURE 10.2 Sustainable machining techniques.

10.2 DRY MACHINING

Dry machining is one of the cleanest machining techniques, with advantages like elimination of the disposal problem of cutting fluid that leads to cleaner air and water along with improved health and environmental conditions. In spite of its various advantages, it is necessary to adopt suitable measures to replace cutting fluids in dry machining. In dry machining, it is challenging to achieve the basic functions of cutting fluids, such as heat removal, lubrication and cleaning of chips. The cleaning of chips from drilled holes is a major problem in dry drilling [4]. Hence, in order to overcome the thermal damage due to heat generation in dry machining, researchers have found various cooling and cutting tool technology with very less or no use of cutting fluids, as shown in Figure 10.3.

10.2.1 Cutting Tool Technology

The rapid evolution of cutting tool technology is very much necessary in order to meet the changing demands and hard-to-machine workpiece materials. In today's world, the requirements of machined workpieces are not only dimensional accuracy, delivery and quality, but also, they are made of hard materials that are difficult to machine. In other words, the evolution of workpieces is faster in comparison to the machine tools used to cut them. Hence, in the present manufacturing scenario, it is important to achieve advancements in cutting tool technology. Some of the factors that influence the development of cutting tools include challenging materials, complex geometries, large component sizes and specialized quality and performance requirements [5].

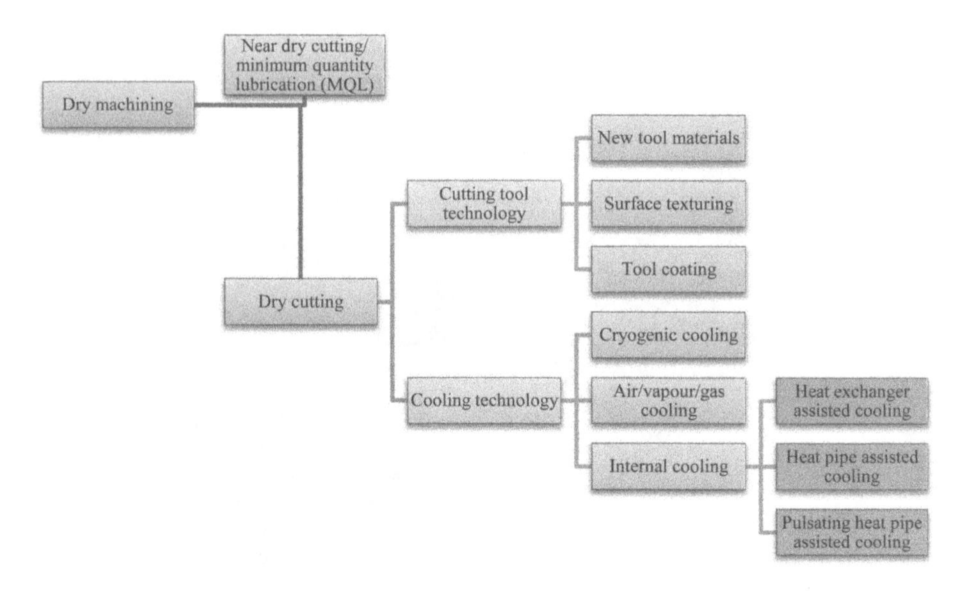

FIGURE 10.3 Dry machining techniques.

10.2.1.1 Surface Texturing

In the field of dry machining, surface texturing potentially improves the tribological conditions. It helps in improving the tribological properties of the cutting tools through the entrapment of debris, reduction in contact length between the tool and the workpiece and improved lubrication capacity, thereby reducing cutting force, tool wear and surface roughness of the workpiece [6]. The textures on the material surfaces can be created through different techniques as shown in Figure 10.4.

The various techniques to create surface textures have their own merits and demerits and are used for different applications. The techniques should be selected based on the requirement of dimensional precision, good surface quality, optimum production rate, compatibility with the workpiece, etc. The cost-effectiveness of each of the techniques is one of the main concerns of the product managers. Thomas and Kalaichelvan [7] studied the machining performance of micro-textured high speed steel (HSS) single point cutting tool on mild steel (EN3B) and aluminum (AA 6351) using a lathe machine. The textures were created on the surface using a Rockwell hardness tester, Vickers hardness tester and by scratching with a diamond dresser. They observed a decrease in cutting force and temperature with textured tools in comparison to non-textured tools. Sugihara et al. [8] created micro-dimples on the rake face of a WC-Co cemented carbide cutting tool using laser surface texturing. They found that the micro-dimples suppressed the chip adhesion by 10% in comparison to the non-textured tool in dry machining of aluminum alloys.

In spite of having numerous benefits of micro-textures on cutting tools, the performance of textured cutting tools greatly depends on micro-texture parameters. Durairaj et al. [9] carried out an experimental study by creating circular dimples on the rake face of tungsten carbide cutting tools by considering several parameters

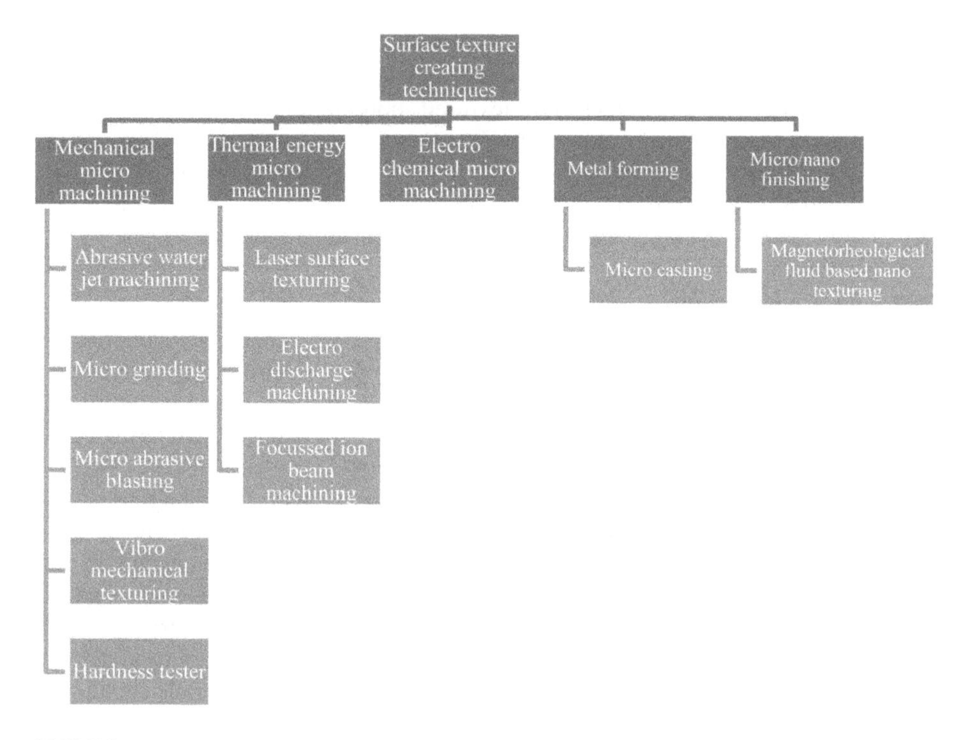

FIGURE 10.4 Techniques used to create surface textures [6].

like diameter and depth of dimples, pitch between two dimples and the distance of the dimples from the cutting edge. They experimented and studied the effect of these dimples during orthogonal cutting of aluminum using the Taguchi method and analysis of variance (ANOVA). They observed that the diameter of the circular dimples is the most influencing parameter, and there is 15% less smearing or adhesion of aluminum in the textured tool in comparison to non-textured tool. Micro-textures on the flute and margin side of a drill tool can reduce the sliding friction and eliminate the problem of cutting fluid reachability at the machining zone due to upward motion of chips sliding along the flute surface. Niketh and Samuel [10] found that the margin-textured tool is more efficient that the flute-textured and non-textured tool, thereby achieving a net thrust force reduction of 10–12% in dry, 15–20% in wet and 15–19% in minimum quantity lubrication (MQL) conditions.

Arslan et al. [11] reviewed techniques available for making surface textures along with their tribological effect on cutting tool performance. The improved cutting performance of textured tools is due to reduced friction and cutting force, crater wear, flank wear and increased shear angle. The surface textures can be favorable in both wet and dry cutting conditions. The texture geometries like shape and size of textures, spacing between them, its orientation and placement with respect to cutting edge in conjunction with cutting condition, tool geometry, cutting process, workpiece material and tool material influence the benefits of textured cutting tools. The textures placed parallel to the cutting edge show reduced friction and adhesive wear in comparison to the perpendicularly placed textures. The textures should be placed near the edge of cutting tools as the textures created at the edge may reduce the mechanical strength of the cutting tools. The cutting fluids can be completely removed by using solid lubricant-filled textured tools in dry machining [11]. Orra and Choudhury [12] carried out a machining investigation and found a decrease in frictional and cutting force and an increase in shear angle with horizontal, vertical and elliptical micro-textures filled with a solid lubricant MoS_2 in comparison to the results of machining without micro-textures or without solid lubricant.

10.2.1.2 Tool Coating

The hard coated cutting tools have been successfully used in manufacturing industries for more than four decades. The market distribution for cutting tool materials in the year 2013 is shown in Figure 10.5, and it is seen that 53% is covered by cemented carbide and 20% is covered by high-speed steel (HSS). The rest is covered by ceramics, cermets, carbon boron nitride, diamond PCD, etc. [13].

The development of the tool market is mainly to cope with the changing workpiece materials together with modified cutting parameters in order to increase productivity and sustainability of the machining processes. Coatings are applied on the cutting tools to protect them against abrasion, adhesion, oxidation etc. and to act as a diffusion barrier between the tool and the workpiece. The coating materials should be able to withstand high temperature to protect cutting tools against abrasion during a high temperature cutting operation. Moreover, the coated cutting tools should be capable of withstanding mechanical and thermal loads that may arise due to cyclic loading along with interrupted cuts. In present days, coating is provided in almost half of the cutting tools. Some of the commonly used coating tool materials are TiC, TiCN, TiN, Al_2O_3, TiAlN, diamond, TiB_2, AlCrN, etc. [13].

The tool cost is only around 3% of machining cost as is shown in Figure 10.6. The increase in tool life by applying a coating in order to reduce the number of tools

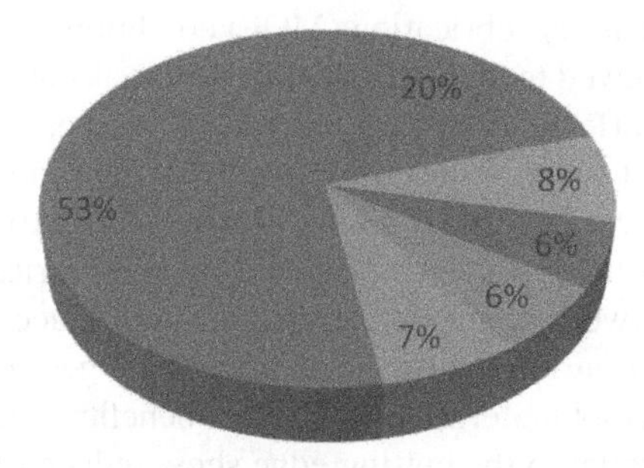

FIGURE 10.5 Market distribution of cutting tool materials in the year 2013.

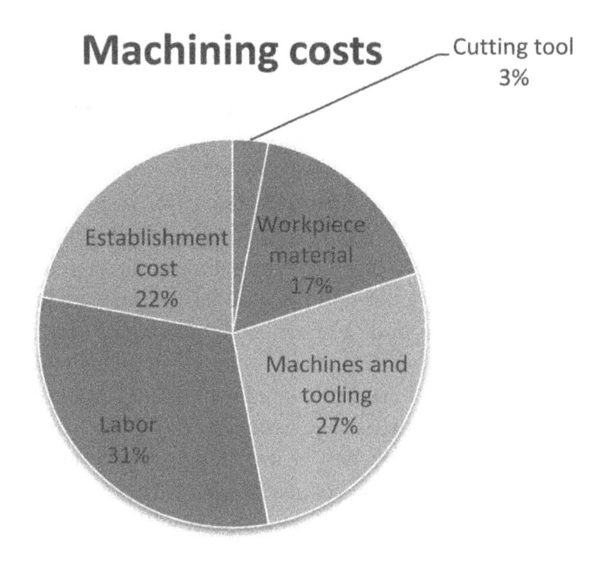

FIGURE 10.6 Distribution of machining cost during production.

required alone cannot have a great impact on the total production cost. Hence, it is necessary to develop more efficient tools that can reduce machining cost including tooling and labor cost during production. Moreover, with the use of more efficient tools, downtime required for tool changing and the amount of cutting fluids can be substantially reduced and the effective machining of hard-to-machine materials can be achieved [13].

The coating materials are applied on cutting tools by vapor depositing a layer of material having high hardness and wear-resistant properties on the surface of the conventional cutting tool. In the vapor deposition process, the elements of the coating material evaporate into atoms, molecules or ions in a vacuum that condenses on the substrate surface to form the desired coating. Vapor deposition techniques are broadly classified into chemical vapor deposition (CVD) and physical vapor deposition (PVD). The characteristics of CVD and PVD are shown in Table 10.1.

TABLE 10.1
Characteristics of CVD and PVD Techniques [14]

Characteristics	CVD	PVD
Form of coating material	Gaseous	Solid
Deposition temperature (°C)	450–1100	250–600
Gas pressure (Pa)	100–100000	0.01–10
Coating thickness	Thick	Thin
Coating residual stress	Tensile residual stress	Comprehensive residual stress
Applications	Solid carbide tools	High speed steel tools, carbide tools, PCBN tools, etc.

The high deposition temperature and gas pressure of the CVD technique together with its toxic effect on the environment restricts its wide application. Hence, the PVD technique is largely used due to its low deposition temperature, environment friendly nature and controlled coating composition and structure [14].

Capasso et al. [15] investigated the wear characteristics of two different coating systems, namely (a) a new nano-composite multi-layer $Ti_{25}Al_{65}Cr_{10}N/Ti_{20}Al_{52}Cr_{22}Si_{8}N$ PVD coating, and (b) an AlTiN benchmark coating deposited on cemented carbide tool and used to perform finish turning operation on Inconel DA718 aerospace alloy. They observed that the nano-composite coating performed better than the AlTiN coating at high cutting speed, whereas both performed at their best at speeds under 80 m/min. The adhesive and abrasive wear mechanism was confirmed during machining. But, the nano-composite-coated tool showed superior wear-resistant quality with reduced chipping intensity. Chekalova and Zhuravlev [16] proposed a method to improve the wear properties of a complex profile cutting tool through diffusion discrete coating, i.e., uneven coating, where the starting material is used as soft interlayer. The non-uniformity of the coating will contribute to an increase in tool deformation resistance and increases rigidity by retaining its strength which prevents spreading of cracks in the coating surface. Santecchia et al. [17] reviewed the wear-resistant property of titanium-nitride-based coating produced using CVD, PVD and thermal spraying techniques. The machine tools can be essentially coated with titanium nitride in order to reduce the application of cutting fluid in a dry high speed machining operation. The performance of coatings and their wear properties are influenced by substrate, deposition technique, temperature and humidity, counter-body properties, applied load and testing speed. The PVD or reactive plasma spraying technique should be used for multi-layer coatings, whereas CVD should be used for substrate with complex geometries.

10.2.1.3 New Tool Material

The cutting tools remain in direct contact with the workpiece during mechanical machining and are subjected to high stress, temperature and corrosion. This may lead to high strain rate limiting the useful life of cutting tools, especially single-point cutting tools. One of the most commonly used cutting tool materials is tungsten carbide. However, its main ingredients W and Co are listed among critical raw materials (CRMs) for EU, having a high risk of supply and they should be used cautiously for sustainability. Hence, several scientific actions need to be performed simultaneously like improving the production processes of CRMs, finding suitable candidates to partially or totally substitute the CRMs and increasing their recycling [18].

In this scenario, new tool materials with superior qualities or better functioning than existing ones need to be developed to increase the tool life and reduce the cost of product. A substantial amount of the total manufacturing cost in machining sophisticated turbine blades, automobile and aerospace parts, bio-medical implants, etc. constitute the cost of cutting tools. The tooling costs include the cost of relapping of tools and the cost of increased cycle time due to unloading and reloading of new tools. Hence, the present research in the field of manufacturing mainly focuses on developing new cutting tool materials that can bear cutting loads higher than the present limit. Also, in machining, the "smart tooling" concept has been used, in which sensor actuators are inserted within the cutting tools in order to monitor the tool beforehand. Another alternative method is to use tool coating materials to protect cutting tools by providing high hardness, toughness, and thermal and chemical stability during high speed machining. Most of the coating materials do not contain CRMs, whereas multi-elemental and high entropy coatings may include them. However, the amount of material needed for coating is comparatively less than the analogous bulk CRMs used in cutting tools [18].

Liu et al. [19] synthesized a novel TiB_2-based ceramic cutting tool material using TiC whiskers as the toughening phase. An addition of 30 vol% TiC whiskers increased the flexural strength and fracture toughness of the TiB_2-based ceramic composite tools. They obtained a very high value of flexural strength, fracture toughness and Vickers hardness of the cutting tool as 860 MPa, 7.9 $MPa{\cdot}m^{1/2}$ and 22.6 GPa, respectively at a combination of 30 vol% of whisker, sintering temperature of 1700°C and a holding time of 30 minutes. Traxel and Bandyopadhyay [20] designed and fabricated a diamond-reinforced carbide structure using directed energy deposition-based additive manufacturing. The synthesized composites are free from large-scale cracking and possess improved micro-structure and machining performance of the WC-Co-based material in comparison to a commercial carbide cutting tool. There occurs a substantial amount of reduction in build-up edge during machining of an aluminum workpiece with 2.5wt% and 5wt% diamond dust added to the WC-Co cutting tool. Tu et al. [21] fabricated a novel cubic boron nitride (CBN)-coated silicon nitride (Si_3N_4) cutting tool by electron cyclotron resonance (ECR) microwave plasma chemical vapor deposition (MPCVD) technique. During the dry-turning of hardened ductile iron, the synthesized cutting tool showed lower cutting force, cutting temperature and tool wear in comparison to commercially available titanium aluminum nitride (TiAlN)-coated tools.

10.2.2 Cooling Technology

In order to eliminate or reduce the use of cutting fluids, researchers are finding alternative cooling techniques to adapt to the concept of green machining. Green machining can become a trend in the future manufacturing field with focus on cooling technologies like cryogenic cooling, air/vapor gas cooling and internal cooling, as all of them perform promisingly well.

10.2.2.1 Cryogenic Cooling

The process of cryogenic cooling involves cooling a material far below 0°C, which can improve mechanical properties like wear resistance, strength and hardness. It is a commonly used cheap process to obtain good quality steel and tungsten carbide cutting tools. Cryogenic cooling works successfully at low cutting speed, whereas it is unable to increase tool life at higher speeds. Hence, cryogenic cooling coupled with minimum quality lubrication (MQL) can be effectively used to reduce tool wear [22]. Some of the cryogenic cooling materials include liquid hydrogen (boiling point −252.882°C), liquid nitrogen LN2 (boiling point −195.80°C), liquid oxygen (boiling point −182.97°C) and so-called dry ice (sublimation point −78.5°C). In practical conditions, the liquid nitrogen and dry ice are commonly used during machining due to their market availability and safety [23].

Based on the applications, cryogenic cooling approaches can be divided into four groups [24]:

1. *Cryogenic pre-cooling the workpiece and cryogenic chip cooling:* In this method, the workpiece and chips are cooled to change the property of chip material from ductile to brittle when exposed to low temperature to increase chip breakability.
2. *Indirect cryogenic cooling/cryogenic tool back-cooling/conductive remote cooling:* In this method, the cooling of a tool cutting point is done by heat transfer from a cryogenic liquid chamber at the tool face or the tool holder.
3. *Cryogenic spraying or jet cooling*: The main concerned area in this method is the tool-chip interface where cryogenic liquid is sprayed using nozzles. In this case, the consumption of cryogenic fluid can be high leading to increased production cost due to the flooding or spraying of coolant to the entire cutting area. However, in the cryogenic jet cooling method, micro-nozzles are used to direct the cryogenic liquid to the rake or flank surface of the tool that is subjected to maximum temperature.
4. *Cryogenic treatment of cutting tools:* Like heat treatment, cryogenic treatments are used to cool down the cutting tools to cryogenic temperature, where it is maintained for a long time followed by heating the cutting tools to room temperature to achieve superior wear-resistant property and dimensional stability.

Hong and Ding [25] investigated to find the best cryogenic cooling approach out of a few approaches like dry machining, cryogenic tool back cooling, emulsion cooling, precooling the workpiece, cryogenic flank cooling, cryogenic rake cooling and simultaneous rake and flank cooling for economical machining. They found that simultaneous flank and rake cooling is the most effective one for machining Ti-6Al-4V. Stampfer et al. [26] used liquid nitrogen in the rake face of the tool during orthogonal turning of Ti-6Al-4V. On comparison of cryogenic cooling with dry cutting, they found that the liquid nitrogen reduced tool-chip contact length, thermal stress in the cutting zone and width of flank wear in orthogonal cutting. Dhar et al. [27] investigated the effect of liquid nitrogen on tool wear, dimensional stability and surface quality in turning of C60 steel using uncoated carbide inserts of different geometric configurations. They observed a reduction in tool wear, thus increasing tool life, dimensional accuracy and surface finish, which may be due to reduced cutting zone temperature and better tool-chip interaction. Yong et al. [28] analyzed the performances of both cryogenically treated and untreated tungsten carbide tool inserts during the high-speed milling operation of medium carbon steel. They performed the machining operation under dry cutting as well as machining using coolants. They observed that cryogenically treated tools possessed better wear properties than the untreated one. Vicentin et al. [29] developed a new cooling system in which the refrigerant fluid, R-123, a hydro-chloro-fluoro-carbon (HCFC), flows inside the body of the cutting tool in turning a Cr–Ni–Nb–Mn–N austenitic steel, that is capable of phase change due to heat generation in machining. Yıldırım et al. [30] studied the influence of MQL, cryogenic cooling with liquid nitrogen and cryo-MQL on tool wear behavior, cutting temperature, surface roughness and chip morphology in a turning operation of alloy 625. They performed the experiments by varying cutting speed, depth of cut and feed rate and observed an improvement of 24.82% in surface roughness using cryo-MQL in comparison to cryogenic cooling. When compared with cryogenic cooling, the tool wears decreased by 50.67% and 79.60% using MQL and cryo-MQL, respectively.

10.2.2.2 Air/Vapor/Gas Cooling

In order to eliminate the use of cutting fluid completely for sustainable machining, compressed gas or air coolant can be successfully used. Cold air has been considered as a viable alternative to cutting fluid by various manufacturing industries as it can reduce the disposal cost of cutting fluid [31]. Numerous studies have been carried out regarding the advantages of using cold for machining. Ginting et al. [32] studied to determine the most effective cold air generation method out of the three methods, viz. vortex tube (VT), thermoelectric cooling (TEC) and cryogenic cooling using compressed air (CCA), for machining operations. They found that the TEC method is the most energy efficient one and is convenient to use. Air cooling is a special case of gas cooling that uses dry compressed air to replace the cutting fluid. Some of the commonly used coolants by different researchers in their studies include oxygen, CO_2, argon, water vapors, air and nitrogen [22]. A water vapor generator can be used to produce vapor that can be blown out through a nozzle to reach the tool-workpiece interface to improve machining performance [33].

Kumar and Gandotra [34] carried out experimentations to study the effect of cooling air produced by vortex tube during hard turning of AISI 4340 steel. They observed a substantial reduction in tool flank wear, cutting force and surface roughness when compared with dry cutting, wet cutting and minimum quality lubrication. Fan et al. [35] performed a machining operation of Inconel 718 under a combined water vapor and air cooling lubrication condition and studied the mechanisms related to tool-chip interface and tool wear mechanisms. They observed that a chemical film, consisting of new oxides Cr_2O_3, Fe_2O_3, Fe_3O_4, is deposited at the tool-chip interface that reduces adhesive and abrasive wear. The cutting temperature gets decreased and the tool wear occurs mainly due to mild oxidation wear of W and Co elements and the tool life is enhanced.

Li et al. [36] compared low frequency vibration-assisted drilling (LFVAD) assisted by forced air cooling with conventional drilling of CFRP/Ti6Al4V and observed slower flank wear rate and low cutting temperature for the air-cooled LFVAD technique. In both the methods, adhesion was the main wear mechanism, though chemical wear is more dominant in the LFVAD method. Singh et al. [37] performed a turning operation on Ti-3Al-2.5 V with coated carbide tools, and different cooling techniques were used to study sustainability of the techniques. They selected five techniques, viz. dry, pressurized compressed air-assisted wet cooling, cooling with Ranque-Hilsch-vortex-tube (RHVT), conventional MQL and wet oil cooling. From the obtained results based on working condition, energy efficiency, carbon emissions, tool wear, surface finish and chip analysis, they have concluded that the RHVT technique can be suitably used for environmental, economic and technological benefits.

Jozić et al. [38] studied the output parameters of end milling such as surface roughness, cutting force, flank wear and material removal rate under dry machining, machining in presence of cutting fluid and air-cooled machining using a Taguchi coupled grey relational analysis. They confirmed that air-cooled machining is the most suitable method for sustainable manufacturing. Nadolny and Kieraś [39] developed a new cooling and lubrication technique for the dry machining condition of internal cylindrical grinding of bearing rings. They simultaneously applied compressed cold air through a cold air gun (CAG) nozzle and a solid anti-adhesive and lubricating agent supplied to the grinding area as an impregnate in the grinding wheel. They used hexagonal boron nitride as the impregnate, and this technique simultaneously performs cooling and lubrication functions of cutting fluid.

10.2.2.3 Internal Cooling

In internal cooling, also known as an indirect cooling system, the heat generated at the tool-workpiece interface is not in direct contact with a heat sink. There are many problems associated with internal cooling. Firstly, a large amount of heat is generated in a small area on the tool surface which is difficult to remove with low flow rate single phase fluid. Also, it is next to impossible to install a heat sink in the cutting tool. The internal cooling method should also be economically viable, and the installation of the system on the machine tools should be easy for large-scale application. There are broadly two types of internal cooling, namely heat exchanger-assisted cooling and heat pipe-assisted cooling [33].

The internal cooling system has been developed by many researchers in which cooling channels are cut on the tool and fluid flowing through the channels carries away the heat from the cutting zone [40, 41]. All the studies reported a reduced cutting temperature and increased tool life. Another possible means of heat dissipation from the cutting zone is the application of a heat pipe that has low thermal resistance and can be fitted in a small space [42]. The grinding process involves high temperature due to heat generated in the grinding contact zone that may damage both the grinding wheel and the workpiece. The process of using coolants to reduce the temperature has adverse effects on health and environment and is also not efficient due to film boiling. Hence, it has been thought by different scientists to use a new cooling method in which a two-phase revolving heat pipe is incorporated within the grinding wheel disk. In a revolving heat pipe grinding wheel (RHPGW), the cylindrical grinding surface comprises the evaporator section, and cooled rotating sides are the condenser section [43]. Chen et al. [44] performed a simulation study on the mechanism of heat transfer of RHPGW rotating at a velocity of 45 m/s. They observed that natural convection and film condensation occurred under this condition. Further, they performed experiments on grinding of Inconel 718 using RHPGW and a normal grinding wheel without coolant. They observed that a grinding temperature below 100°C can be maintained using RHPGW with good quality workpiece and grinding wheel in comparison to a normal grinding wheel.

A vertical rotating heat pipe can be used in an abrasive milling tool to adopt green machining of hard-to-machine surfaces. Chen et al. [45] compared dry machining without rotating heat pipe with the rotating heat pipe abrasive milling tool and observed a lower evaporator section temperature of approximately 67% in the case of rotating heat pipe. A tool holder containing pulsating heat pipe was designed and fabricated by Wu et al. [46] to perform dry turning of Ti-6Al-4V alloy and was compared with the conventional one. Pulsating heat pipe is one of the most efficient heat conductors as heat is transported by latent heat of vaporization caused by a very small temperature difference. They observed a decrease in the wear of the cutting tool with the tool life increased by 20–30%. The highest recorded cutting temperature decreased by 10% in the case of the pulsating heat pipe-assisted tool holder in comparison to the conventional one. Researchers also used self-lubricating and self-cooling tools with the combined benefits of pulsating heat pipe and micro-texturing techniques. Wu et al. [47] used laser beam machining and electric discharge machining to fabricate the new self-lubricating and self-cooling tool. They assessed the machining performance of the synthesized tool in dry cutting of Ti-6Al-4V alloy and observed a low cutting temperature, cutting force and tool wear in comparison to the conventional one.

Rozzi et al. [48] developed an internal cooling system for a lathe turning operation where the cutting tool is cooled using a very small quantity of inert cryogenic cutting fluid that flows through a micro-channel heat exchanger placed under the cutting tool insert. They observed that the cutting tool life improved significantly at all cutting speeds. The cutting fluid should have large latent heat capacity in order to achieve maximum heat transfer in the tool. They concluded that the internal cooling method can be successfully used to increase tool life in comparison to the external cryogenic cooling method. Shu et al. [49] developed an internally cooled cutting tool and investigated it numerically and experimentally to determine tool-chip interface temperature. The analysis showed that the innovative cutting tool can be used to measure the cutting tool tip temperature and also aids in reducing and controlling the critical cutting temperature.

10.3 MINIMUM QUANTITY LUBRICATION (MQL)

MQL is an environment-friendly machining process that can serve as an alternative to wet machining for various tool and work piece combinations. It can reduce damages caused to the health of working personnel on the shop floor without deteriorating the surface quality of the machined surfaces. The effectiveness of using MQL should be studied for machining hard-to-machine materials like titanium- and nickel-based alloys in order to improve the existing technique. One of the most prevailing problems in MQL is that only a single supply of cutting fluid is provided to both the heat-affected flank and rake surfaces of a single point cutting tool, instead of supplying the required proportion of cutting fluid to these surfaces separately [50].

The machining cost can be reduced substantially by replacing the flood cooling technique by an MQL system. The use of the MQL system for machining with different cutting fluids was successfully implemented since its emergence. The cutting fluids in MQL are used in the form of undiluted oil or in combination of oil with water in varying ratios to cool and lubricate the tool-workpiece area. In order to achieve sustainable and green machining, the cutting fluid should be biodegradable, possessing high stability and lubricating effect. Based on superior biodegradation property, vegetable oils and synthetic esters are extensively used as cutting fluids in an MQL system. In comparison to the traditional metal-working fluids, the vegetable oils possess some advantageous qualities; for example, they act as better pressure absorbents,

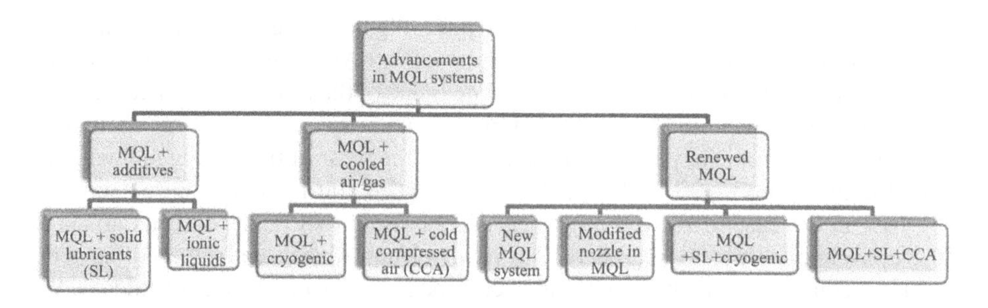

FIGURE 10.7 Methods to improve the MQL system.

increase material removal rate and provide less loss from vaporization and misting. Moreover, synthetic esters possess high boiling and flash point and low viscosity and behave in a similar manner to vegetable oils. Sometimes it has been seen by many researchers that the performance of synthetic oil is better than vegetable oils [51].

In general, the MQL technique involves atomization of a mixture of oil and pressurized air and spraying it directly into the cutting zone at a flow rate of 1000 ml/h and less [50]. Many researchers have experimented with MQL conditions using a variety of cutting fluid combinations including biodegradable nano-fluids for different machining processes [52, 53]. They observed a better tribological and machining performance for the MQL system in comparison with other conditions like dry machining and machining with flooded cutting fluids [54, 55]. In spite of their various advantages, the MQL systems have some major limitations that forced the researchers to work towards the advancements of the system. The improvements in MQL systems is broadly classified into three categories, viz. MQL with additives, MQL with cooled air/gas and renewed MQL system as shown in Figure 10.7. In recent years, research has been carried out using a hybrid MQL system in conjunction with cryogenic liquid, solid lubricants and compressed cold air. Solid lubricants like micro-fluids and nano-fluids and cryogenic liquids like carbon dioxide and liquid nitrogen are used in the advancements of MQL machining technique [51].

10.4 CONCLUSION

Green machining techniques for achieving sustainability in manufacturing are under investigation and more developments are expected in industries in the near future. Some of the green techniques include dry machining, minimum quantity lubrication (MQL), cryogenic cooling, replacement of traditional lubricants by biodegradable liquid and solid lubricants, usage of coated and textured tools, etc. These techniques can be used to substantially reduce or eliminate the use of cutting fluids that may cause environmental and health hazards and create disposal difficulties Dry machining can be successfully implemented by industries keeping in mind that the production rate, tool life, cost and surface finish achieved by the dry-cutting technique are comparable to that of wet machining. This chapter includes a detailed discussion of the limitations of the conventional flood cooling method in machining and the techniques adopted to overcome these limitations.

REFERENCES

1. Sivarajan, S. and Padmanabhan, R. (2014). Green machining and forming by the use of surface coated tools. *Procedia Engineering*, 97, 15–21.https://doi.org/10.1016/j.proeng.2014.12.219
2. Sultana, N., Dhar, N. R., and Zaman, P. B. (2019). A review on different cooling/lubrication techniques in metal cutting. *American Journal of Mechanics and Applications*, 7, 4, 71–87. https://doi.org/10.11648/j.ajma.20190704.11
3. Singh, G., Aggarwal, V., and Singh, S. (2020). Critical review on ecological, economical and technological aspects of minimum quantity lubrication towards sustainable machining. *Journal of Cleaner Production*. https://doi.org/10.1016/j.jclepro.2020.122185.
4. Dixit, U. S., Sarma, D. K., and Davim, J. P. (2011). Dry machining. *Springer Briefs in Applied Sciences and Technology*, 19–28. https://doi.org/10.1007/978-1-4614-2308-9_3
5. https://www.mmsonline.com/articles/the-new-rules-of-cutting-tools----introduction, accessed on 07.12.2020.
6. Ranjan, P. and Hiremath, S. S. (2019). Role of textured tool in improving machining performance: A review. *Journal of Manufacturing Processes*, 43, 47–73.
7. Thomas, S. J., and Kalaichelvan, K. (2017). Comparative study of effect of surface texturing on cutting tool in dry cutting. *Materials and Manufacturing Processes*. https://doi.org/10.1080/10426914.2017.1376070
8. Sugihara, T., Singh, P., and Enomoto, T. (2017). Development of novel cutting tools with dimple textured surfaces for dry machining of aluminum alloys. *Procedia Manufacturing*, 14, 111–117. https://doi.org/10.1016/j.promfg.2017.11.013
9. Durairaj, S., Guo, J., Aramcharoen, A., and Castagne, S. (2018). An experimental study into the effect of micro-textures on the performance of cutting tool. *The International Journal of Advanced Manufacturing Technology*, 98(1–4), 1011–1030. https://doi.org/10.1007/s00170-018-2309-y
10. Niketh, S. and Samuel, G. L. (2018). Drilling performance of micro textured tools under dry, wet and MQL condition. *Journal of Manufacturing Processes*, 32, 254–268. https://doi.org/10.1016/j.jmapro.2018.02.012
11. Arslan, A., Masjuki, H. H., Kalam, M. A., Varman, M., Mufti, R. A., Mosarof, M. H., Khuong, L. S., and Quazi, M. M. (2016). Surface texture manufacturing techniques and tribological effect of surface texturing on cutting tool performance: A review. *Critical Reviews in Solid State and Materials Sciences*, 41(6), 447–481. https://doi.org/10.1080/10408436.2016.1186597
12. Orra, K. and Choudhury, S. K. (2018). Tribological aspects of various geometrically shaped micro-textures on cutting insert to improve tool life in hard turning process. *Journal of Manufacturing Processes*, 31, 502–513. https://doi.org/10.1016/j.jmapro.2017.12.005
13. Bobzin, K. (2017). High-performance coatings for cutting tools. *CIRP Journal of Manufacturing Science and Technology*, 18, 1–9. https://doi.org/10.1016/j.cirpj.2016.11.004
14. Deng, Y., Chen, W., Li, B., Wang, C., Kuang, T., and Li, Y. (2020). Physical vapor deposition technology for coated cutting tools: A review. *Ceramics International*. https://doi.org/10.1016/j.ceramint.2020.04.168
15. Capasso, S., Paiva, J. M., Junior, E. L., Settineri, L., Yamamoto, K., Amorim, F. L., Torres, R. D., Covelli, D., Fox-Rabinovich, G., and Veldhuis, S. (2019). A novel method of assessing and predicting coated cutting tool wear during Inconel DA 718 turning. *Wear*, 202949. https://doi.org/10.1016/j.wear.2019.202949
16. Chekalova, E., and Zhuravlev, A. (2019). Increasing the wear resistance of a complex profile cutting tool by applying a diffusion discrete coating. *Materials Today: Proceedings*. https://doi.org/10.1016/j.matpr.2019.08.053

17. Santecchia, E., Hamouda, A. M. S., Musharavati, F., Zalnezhad, E., Cabibbo, M., and Spigarelli, S. (2015). Wear resistance investigation of titanium nitride-based coatings. *Ceramics International*, 41(9), 10349–10379. https://doi.org/10.1016/j.ceramint.2015.04.152
18. Rizzo, A., Goel, S., Grilli, M. L., Iglesias, R., Jaworska, L., Lapkovskis, V., Novak, P., Postolnyi, B.O., and Valerini, D. (2020). The critical raw materials in cutting tools for machining applications: A review. *Materials*, 13(6), 1377. https://doi.org/10.3390/ma13061377
19. Liu, B., Wei, W., Gan, Y., Duan, C., and Cui, H. (2020). Preparation, mechanical properties and microstructure of TiB2 based ceramic cutting tool material toughened by TiC whisker. *International Journal of Refractory Metals and Hard Materials*, 93, 105372. https://doi.org/10.1016/j.ijrmhm.2020.105372
20. Traxel, K. D. and Bandyopadhyay, A. (2020). Diamond-reinforced cutting tools using laser-based additive manufacturing. *Additive Manufacturing*, 101602. https://doi.org/10.1016/j.addma.2020.101602
21. Tu, L., Tian, S., Xu, F., Wang, X., Xu, C., He, B., Zuo, D., and Zhang, W. (2020). Cutting performance of cubic boron nitride-coated tools in dry turning of hardened ductile iron. *Journal of Manufacturing Processes*, 56, 158–168. https://doi.org/10.1016/j.jmapro.2020.04.081
22. Chetan, Ghosh, S., and Venkateswara Rao, P. (2015). Application of sustainable techniques in metal cutting for enhanced machinability: A review. *Journal of Cleaner Production*, 100, 17–34. https://doi.org/10.1016/j.jclepro.2015.03.039
23. Krolczyk, G. M., Maruda, R. W., Krolczyk, J. B., Wojciechowski, S., Mia, M., Nieslony, P., and Budzik, G. (2019). Ecological trends in machining as a key factor in sustainable production – A review. *Journal of Cleaner Production*. https://doi.org/10.1016/j.jclepro.2019.02.017
24. Yildiz, Y. and Nalbant, M. (2008). A review of cryogenic cooling in machining processes. *International Journal of Machine Tools and Manufacture*, 48(9), 947–964. https://doi.org/10.1016/j.ijmachtools.2008.01.008
25. Hong, S. Y. and Ding, Y. (2001). Cooling approaches and cutting temperatures in cryogenic machining of Ti-6Al-4V. *International Journal of Machine Tools and Manufacture*, 41(10), 1417–1437. https://doi.org/10.1016/s0890-6955(01)00026-8
26. Stampfer, B., Golda, P., Schiebl, R., Maas, U., and Schulze, V. (2020). Cryogenic orthogonal turning of Ti-6Al-4V. *The International Journal of Advanced Manufacturing Technology*, 111, 359–369. https://doi.org/10.1007/s00170-020-06105-z
27. Dhar, N. R., Islam, S., Kamruzzaman, M., and Paul, S. (2006). Wear behavior of uncoated carbide inserts under dry, wet and cryogenic cooling conditions in turning C-60 steel. *Journal of the Brazilian Society of Mechanical Sciences and Engineering*, 28(2), 146–152. https://doi.org/10.1590/s1678-58782006000200003
28. Yong, A. Y. L., Seah, K. H. W., and Rahman, M. (2006). Performance of cryogenically treated tungsten carbide tools in milling operations. *The International Journal of Advanced Manufacturing Technology*, 32(7–8), 638–643. https://doi.org/10.1007/s00170-005-0379-0
29. Vicentin, G. C., Sanchez, L. E. A., Scalon, V. L., and Abreu, G. G. C. (2011). A sustainable alternative for cooling the machining processes using a refrigerant fluid in recirculation inside the toolholder. *Clean Technologies and Environmental Policy*, 13(6), 831–840. https://doi.org/10.1007/s10098-011-0359-z
30. Yıldırım, Ç. V., Kıvak, T., Sarıkaya, M., and Şirin, Ş. (2020). Evaluation of tool wear, surface roughness/topography and chip morphology when machining of Ni-based alloy 625 under MQL, cryogenic cooling and CryoMQL. *Journal of Materials Research and Technology*. https://doi.org/10.1016/j.jmrt.2019.12.069
31. Rubio, E. M., Agustina, B., Marín, M., and Bericua, A. (2015). Cooling systems based on cold compressed air: A review of the applications in machining processes. *Procedia Engineering*, 132, 413–418. https://doi.org/10.1016/j.proeng.2015.12.513

32. Ginting, Y. R., Boswell, B., Biswas, W. K., and Islam, M. N. (2016). Environmental generation of cold air for machining. *Procedia CIRP*, 40, 648–652. https://doi.org/10.1016/j.procir.2016.01.149
33. Wu, Z., Yang, Y., Su, C., Cai, X., and Luo, C. (2016). Development and prospect of cooling technology for dry cutting tools. *The International Journal of Advanced Manufacturing Technology*, 88(5–8), 1567–1577. https://doi.org/10.1007/s00170-016-8842-7
34. Kumar, S. and Gandotra, S. (2020). Effect of cooling air on machining performance during hard turning. *Materials Today: Proceedings*. https://doi.org/10.1016/j.matpr.2020.06.263
35. Fan, Y., Hao, Z., Lin, J., and Yu, Z. (2015). New observations on tool wear mechanism in machining Inconel 718 under water vapor + air cooling lubrication cutting conditions. *Journal of Cleaner Production*, 90, 381–387. https://doi.org/10.1016/j.jclepro.2014.11.049
36. Li, C., Xu, J., Chen, M., An, Q., El Mansori, M., and Ren, F. (2019). Tool wear processes in low frequency vibration assisted drilling of CFRP/Ti6Al4V stacks with forced air-cooling. *Wear*, 426–427, 1616–1623. https://doi.org/10.1016/j.wear.2019.01.005
37. Singh, R., Dureja, J. S., Dogra, M., Kumar Gupta, M., Jamil, M., and Mia, M. (2020). Evaluating the sustainability pillars of energy and environment considering carbon emissions under machining ofTi-3Al-2.5 V. *Sustainable Energy Technologies and Assessments*, 42, 100806. https://doi.org/10.1016/j.seta.2020.100806
38. Jozić, S., Bajić, D., and Celent, L. (2015). Application of compressed cold air cooling: achieving multiple performance characteristics in end milling process. *Journal of Cleaner Production*, 100, 325–332. https://doi.org/10.1016/j.jclepro.2015.03.095
39. Nadolny, K., and Kieraś, S. (2020). New approach for cooling and lubrication in dry machining on the example of internal cylindrical grinding of bearing rings. *Sustainable Materials and Technologies*, 24, e00166. https://doi.org/10.1016/j.susmat.2020.e00166
40. Peng, R., Jiang, H., Tang, X., Huang, X., Xu, Y., and Hu, Y. (2019). Design and performance of an internal-cooling turning tool with micro-channel structures. *Journal of Manufacturing Processes*, 45, 690–701. https://doi.org/10.1016/j.jmapro.2019.08.011
41. Bleicher, F., and Reiter, M. (2018). Wear reduction on cutting inserts by additional internal cooling of the cutting edge. *Procedia Manufacturing*, 21, 518–524. https://doi.org/10.1016/j.promfg.2018.02.152
42. Uhlmann, E., Riemer, H., Schröter, D., Sammler, F., and Richarz, S. (2017). Substitution of coolant by sing a closed internally cooled milling tool. *Procedia CIRP*, 61, 553–557. https://doi.org/10.1016/j.procir.2016.11.267
43. Chen, J., Fu, Y., Qian, N., Ching, C. Y., Ewing, D., and He, Q. (2020). A study on thermal performance of revolving heat pipe grinding wheel. *Applied Thermal Engineering*, 116065. https://doi.org/10.1016/j.applthermaleng.2020.116065
44. Chen, J., Fu, Y., He, Q., Shen, H., Ching, C. Y., and Ewing, D. (2016). Environmentally friendly machining with a revolving heat pipe grinding wheel. *Applied Thermal Engineering*, 107, 719–727. https://doi.org/10.1016/j.applthermaleng.2016.07.030
45. Chen, J., Fu, Y., Gu, Z., Shen, H., and He, Q. (2017). Study on heat transfer of a rotating heat pipe cooling system in dry abrasive-milling. *Applied Thermal Engineering*, 115, 736–743. https://doi.org/10.1016/j.applthermaleng.2016.12.138
46. Wu, Z., Xing, Y., Liu, L., Huang, P., and Zhao, G. (2020). Design, fabrication and performance evaluation of pulsating heat pipe assisted tool holder. *Journal of Manufacturing Processes*, 50, 224–233. https://doi.org/10.1016/j.jmapro.2019.12.054
47. Wu, Z., Deng, J., Su, C., Luo, C., and Xia, D. (2014). Performance of the micro-texture self-lubricating and pulsating heat pipe self-cooling tools in dry cutting process. *International Journal of Refractory Metals and Hard Materials*, 45, 238–248. https://doi.org/10.1016/j.ijrmhm.2014.02.004

48. Rozzi, J. C., Sanders, J. K., and Chen, W. (2011). The experimental and theoretical evaluation of an indirect cooling system for machining. *Journal of Heat Transfer*, 133(3), 031006. https://doi.org/10.1115/1.4002446
49. Shu, S., Cheng, K., Ding, H., and Chen, S. (2013). An innovative method to measure the cutting temperature in process by using an internally cooled smart cutting tool. *Journal of Manufacturing Science and Engineering*, 135(6), 061018. https://doi.org/10.1115/1.4025742
50. Banerjee, N. and Sharma, A. (2018). A comprehensive assessment of minimum quantity lubrication machining from quality, production, and sustainability perspectives. *Sustainable Materials and Technologies*, 17, e00070. https://doi.org/10.1016/j.susmat.2018.e00070
51. Hamran, N. N. N., Ghani, J. A., Ramli, R., and Haron, C. H. C. (2020). A review on recent development of minimum quantity lubrication for sustainable machining. *Journal of Cleaner Production*, 122165. https://doi.org/10.1016/j.jclepro.2020.122165
52. Singh, J. and Chatha, S. S. (2020). Tribological behaviour of nanofluids under minimum quantity lubrication in turning of AISI 1055 steel. *Materials Today: Proceedings*. https://doi.org/10.1016/j.matpr.2020.09.156
53. Khunt, C. P., Makhesana, M. A., Patel, K. M., and Mawandiya, B. K. (2021). Performance assessment of vegetable oil-based minimum quantity lubrication (MQL) in drilling. *Materials Today: Proceedings*, 44(Part 1), 341–345. https://doi.org/10.1016/j.matpr.2020.09.741
54. Virdi, R. L., Chatha, S. S., and Singh, H. (2020). Experimental investigations on the tribological and lubrication behaviour of minimum quantity lubrication technique in grinding of Inconel 718 alloy. *Tribology International*, 106581. https://doi.org/10.1016/j.triboint.2020.106581
55. Virdi, R. L., Chatha, S. S., and Singh, H. (2020). Machining performance of Inconel-718 alloy under the influence of nanoparticles based minimum quantity lubrication grinding. *Journal of Manufacturing Processes*, 59, 355–365. https://doi.org/10.1016/j.jmapro.2020.09.056

11 Future Outlooks in Green Tribology

T.V.V.L.N. Rao
Madanapalle Institute of Technology & Science
Madanapalle, India

Salmiah Binti Kasolang
Universiti Teknologi MARA
Shah Alam, Malaysia

Guoxin Xie
Tsinghua University
Beijing, China

Jitendra Kumar Katiyar
SRM Institute of Science and Technology
Kattankulathur, India

Ahmad Majdi Abdul Rani
Universiti Teknologi PETRONAS
Seri Iskandar, Malaysia

CONTENTS

11.1 INTRODUCTION

The future outlooks of green tribology should examine economic, environmental and social impact. Green tribology has the potential to improve the economy through energy conservation, environmental sustainability and improved social quality of life [1]. A comprehensive perspective on how green tribological design increases economy through energy conservation and environmental sustainability is presented. A recent review on the emerging area of green tribology for its future

DOI: 10.1201/9781003139386-11

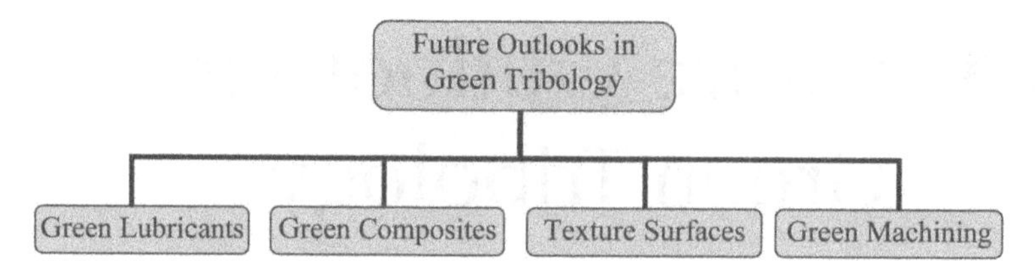

FIGURE 11.1 An overview of future outlooks in green tribology.

development discusses the following aspects: energy, lubrication, sustainability and life cycle assessment [2]. The principles of tribological design can appreciably increase energy conservation. Significant energy savings and lifetime expansion of machineries come from application of tribological design principles in transportation and power generation. A large proportion (one-fifth) of world total energy is dissipated in friction, while one-third of all energy in transportation is dissipated in friction [3]. A sizeable portion of the energy can be conserved by application of new and emerging technologies in green tribology. Leonardo da Vinci's ideas, principles and philosophies about friction reduction are applicable in modern day tribology [4]. The tribological design principles, theories, practices and formulations of Leonardo da Vinci will help advance progress in the field of green tribology. Environmental and sustainability welfare can be attained through green tribological design principles, providing more feasible long-term development of our societies [5].

The latest progress in green tribology may lead to a robust foundation of future developments to go green in tribology. The principles of green tribological design can appreciably increase its impact on future developments for sustainability. An overview of future outlooks in green tribology is presented in Figure 11.1. The future outlooks in green lubricants, green composites, texture surfaces and green machining will establish progress in green tribology. Following the brief overview based on the recent developments on green tribology with emphasis on economic, environmental and social impact, the following sections provide further explicit directions for future outlooks in green tribology.

11.2 GREEN LUBRICANTS

An overview of future outlooks in green tribology on lubricants is presented in Figure 11.2. Following the recent developments in green lubricants, the explicit directions for future outlooks are: (i) green lubricants synthesis from bio-based esters and additives, (ii) green additives synthesis from bio-based thickener agents and waxes, (iii) green lubricants mixtures with nano-particle additives and with bio-based nano-particle additives, and (iv) green ionic liquid lubricants with external electric fields and nano-particle additives.

11.3 GREEN COMPOSITES

An overview of future outlooks in green tribology on composites is presented in Figure 11.3. Following the recent developments in green composites, the explicit directions for future outlooks are: (i) green composites synthesis from bio-based

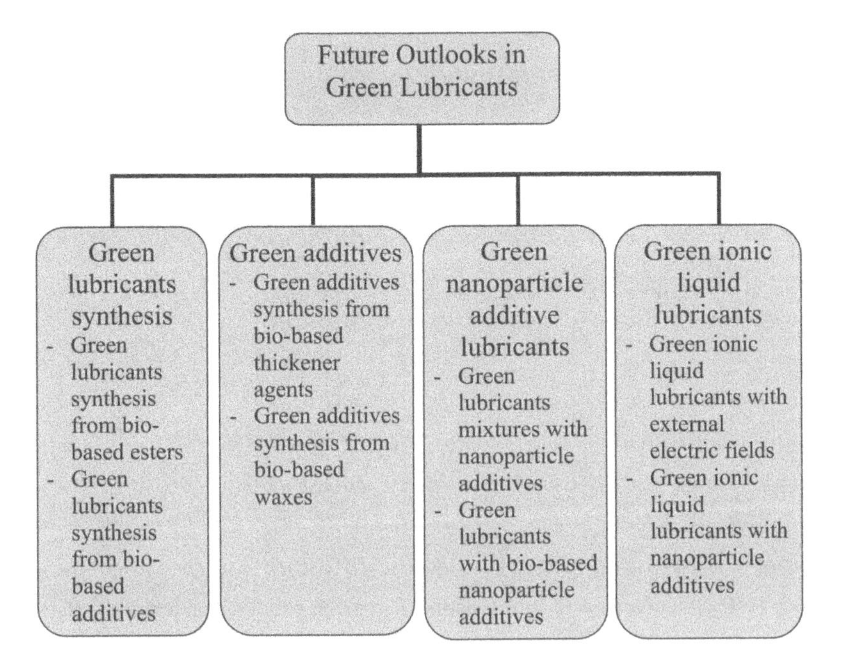

FIGURE 11.2 An overview of future outlooks in green tribology on lubricants.

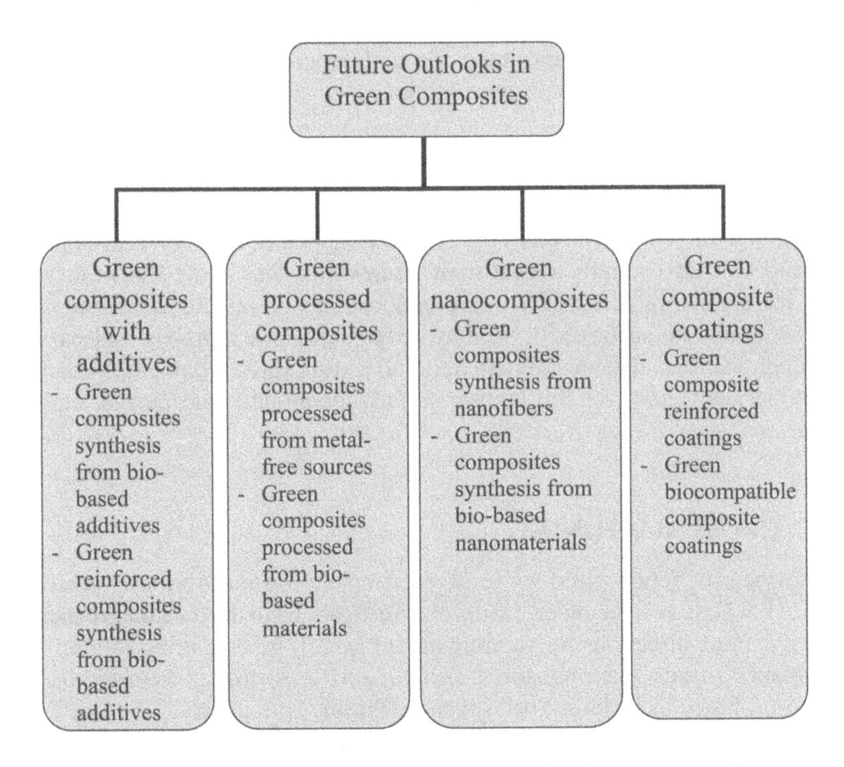

FIGURE 11.3 An overview of future outlooks in green tribology on composites.

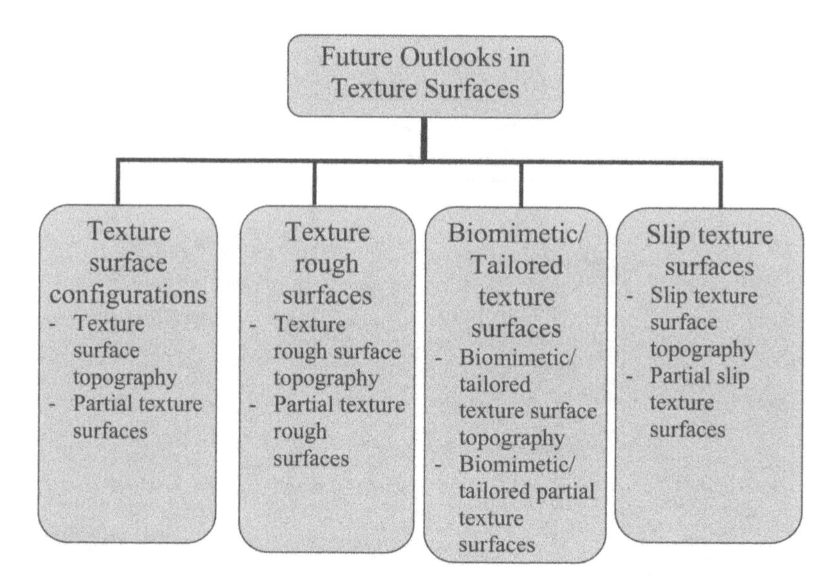

FIGURE 11.4 An overview of future outlooks in green tribology on texture surfaces.

additives and reinforced composites synthesis, (ii) green composites processed from metal-free sources and bio-based materials, (iii) green composites synthesis from nano-fibers and bio-based nano-materials, and (iv) green composite reinforced coatings and bio-compatible composite coatings.

11.4 TEXTURE SURFACES

An overview of future outlooks in green tribology on texture surfaces is presented in Figure 11.4. Based on the recent status of investigations in the area of texture surfaces, the explicit directions for future outlooks are: (i) texture surface topography and partial texture surfaces, (ii) texture rough surface topography and partial texture rough surfaces, (iii) bio-mimetic/tailored texture surface topography and partial texture surfaces, and (iv) slip texture surface topography and partial slip texture surfaces.

11.5 GREEN MACHINING

An overview of future outlooks in green tribology on machining is presented in Figure 11.5. Based on the recent status of investigations in the area on green machining, the explicit directions for future outlooks are: (i) green cutting fluids synthesis from bio-based oils and emulsifiers, (ii) green MQL synthesis from bio-based oils and bio-oil-based nano-fluids, (iii) green machining with micro-/nano- textured and coated tools, and (iv) green machining using bio-oil-based nano-fluids and green additives for nano-lubricated machining.

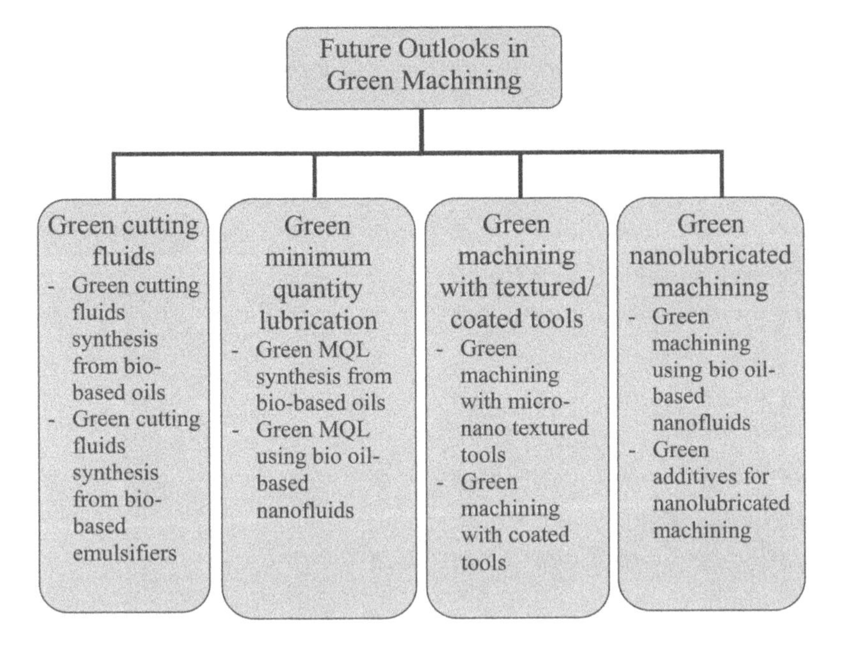

FIGURE 11.5 An overview of future outlooks in green tribology on machining.

11.6 CONCLUSION

The future outlooks and prospects of green tribology for sustainability are highlighted in green lubricants, green composites, texture surfaces and green machining. The detailed directions in the future outlooks are established based on recent developments in green tribology. The emphasis on green tribology in the areas of lubricants, composites, surfaces and machining will greatly affect the economic, environmental and social impact.

REFERENCES

1. R. Shah, M. Woydt, N. Huq and A. Rosenkranz. Tribology meets sustainability, *Industrial Lubrication and Tribology*, 2020, https://doi.org/10.1108/ILT-09-2020-0356.
2. A. Anand, M. Irfan Ul Haq, K. Vohra, A. Raina and M.F. Wani. Role of green tribology in sustainability of mechanical systems: A state of the art survey, *Materials Today: Proceedings*, 4, 2, Part A, 2017, 3659–3665.
3. K. Holmberg and A. Erdemir. The impact of tribology on energy use and CO_2 emission globally and in combustion engine and electric cars, *Tribology International*, 135, 2019, 389–396.
4. S. Betancourt-Parra. Leonardo da Vinci's tribological intuitions, *Tribology International*, 153, 2021, 106664.
5. I. Tzanakis, M. Hadfield, B. Thomas, S. M. Noya, I. Henshaw, and S. Austen. Future perspectives on sustainable tribology, *Renewable and Sustainable Energy Reviews*, 16, 6, 2012, 4126–4140.

Index